A survey of techniques

INTRODUCTION TO

DIFFERENTIAL EQUATIONS

This is a companion volume to

DIFFERENTIAL EQUATIONS

THEORY AND APPLICATIONS

hereafter referred to as DETA

INTRODUCTION TO
DIFFERENTIAL EQUATIONS

by

Ray Redheffer
University of California at Los Angeles

with

Dan Port

JONES AND BARTLETT PUBLISHERS

BOSTON   LONDON

Permissions

We are indebted to Charles and Ray Eames for the X ray photo of a nautilus shell that is correlated with a logarithmic spiral in Chapter 3 and forms part of our logo. The satellite orbit of Chapter 1, Figure 1 was computed by M.C. Davidson. With permission from the McGraw-Hill Book Company, we have used a few problems and excerpts from the Sokolnikoff-Redheffer text *Mathematics of Physics and Modern Engineering.* The quotation on page 14 from Libby's article on radiocarbon dating is reprinted with permission from *The Encyclopædia Britannica,* 14th edition, copyright 1965 by Encyclopædia Britannica, Inc. Vol. 18, pages 904–905.

Editorial, Sales, and Customer Service Offices

Jones and Bartlett Publishers
20 Park Plaza
Boston, MA 02116

Jones and Bartlett Publishers International
P.O. Box 1498
London W6 7RS
England

Library of Congress Cataloging-in-Publication Data

Redheffer, Raymond M.
    Introduction to differential equations / by Ray Redheffer with
    Dan Port
        p.   cm.
    Includes index.
    ISBN 0-86720-289-0
    1. Differential equations.   I. Port, Dan.   II. Title.
    QA371.R342 1992
    515' .35--dc20   91-40004
                                                              CIP

Cover Design: Hannus Design Associates
Cover Illustration: X ray photo of a nautilus shell, courtesy of Charles and Ray
                Eames.

Printed in the United States of America
96 95 94 93 92   10 9 8 7 6 5 4 3 2 1

To Heddy

TABLE OF CONTENTS

E= elementary   I= intermediate   A= advanced

The most basic sections are printed in color

Preface

When David Hilbert presided over lectures at Göttingen, he used to tell the speaker: "All we want is the raisins from the cake." This book is based on a similar idea. It is an introductory version of the Jones and Bartlett text "Differential Equations: Theory and Applications" hereafter referred to as DETA. It is not an afterthought, but was planned by both the author and publisher before the first book went into production.

The guiding principle is to omit all topics that are not obviously suitable for a basic introductory course; if any subject does not meet that criterion, out it goes. In an honors or terminal course under a single instructor, supplementary topics add to the learning environment. They are a source of enrichment. But in a large class consisting of many sections, or in a service course attended by students from other departments, such topics complicate the preparation of a syllabus and invite unauthorized departure from it.

For want of a better term, a course that interlocks with others is referred to as a *standardized course*. The essential feature is that the coverage is clearly prescribed and must be carefully followed.

The dilemma of standardization versus enrichment is familiar to all who have been involved in the planning of basic courses. A possible answer is to have two books, closely correlated, but with different goals. One is a shared copy for consultation by the instructor and by those few students who may want to look beyond the standard curriculum. The other, shorter, version is the prescribed text.

During the four years of its composition, DETA was used by me and others in essentially page-proof form. Preparation of the introductory version was guided not only by professional opinion (see Acknowledgement) but also by classroom experience. Fourier series and boundary-value problems are now presented without the complications entailed by a general development of Sturm-Liouville theory. The discussion of nonlinear problems is based mainly on local rather than global methods and emphasizes problems of ecology. A concise introduction to vector spaces has been added to one of the chapters on linear algebra. All chapters on applications have been shortened to the minimum that seems to me necessary for insight and motivation. This list is far from complete, but it shows that the changes are not merely cosmetic.

There are conflicting and often prejudiced currents of thought as to what should be proved in a book of this kind. The decision in DETA was to prove nearly everything (and my belief that no elementary text does this was one of the reasons for writing it.) "Everything" includes the theorem that a convergent power-series can be differentiated term by term, proof of convergence for power-series and generalized power-series solutions, a convergence theorem for Fourier series, uniqueness of the generating function for the Laplace transform, development of the Picard procedure in the vector case, proof that a Hermitian matrix can be diagonalized even when it has repeated eigenvalues, the Dahlquist theory of linear multistep formulas, and much more.

These topics are unsuited for a standardized course in most institutions. Hence they are omitted. But reference to the correct pages of DETA has much the same effect as if they were in an appendix. DETA has the same style and notation as the introductory version and adjoins it in an almost seamless way. Providing an easily accessible reference for difficult proofs was, to me, a major attraction of the dual-book approach when it was set as a goal about five years ago. Within that framework, this book retains the same commitment to logical development as its predecessor.

A few remarks on goals and philosophy may not be amiss. This book is written for average students who have had one year of calculus. The guiding principle is to accept students as they are and bring them to the point where we want them to be. Weak backgrounds at the start of a course need not remain weak as the course progresses. In recognition of this fact, each major topic shows a gradual progression in the level at which it is treated. The chapters are not of uniform difficulty because students are not static. Growth is the goal.

Prepublication reviewers of DETA were divided on the question whether linear algebra and infinite series should be developed or assumed. My answer is to provide a self-contained minicourse on each subject, but to keep it separate from the main body of the text. Their inclusion harms no one, but their omission would make the book inaccessible to some. In view of this decision, the book can be used in a course entitled "Differential Equations and Linear Algebra."

It is not always obvious whether new ideas should be presented first in their general form or introduced in special cases and then generalized. Both approaches are followed here. Among topics that are presented first in their general form are the notion of linearity and the theory of matrix differential equations. But the Wronskian, linear dependence, and eigenvectors are taken up first in the second-order case.

The following comments pertain to layout and format. Colored print is used only to improve articulation and readability, not to highlight key equations or statements. In my experience such highlighting encourages students to rush to the homework without reading the text and it makes a negative rather than a positive contribution to competence and scholarship.

While nearly everyone agrees that applications are an important source of motivation, not everyone wants to allow time for their inclusion. The solution here is to present applications with the same care and attention to detail as the rest of the text, but to put them in separate chapters. Any applied chapter can be taken up fully or in part, concurrently with or after the chapter that precedes it. No applied chapter is a prerequisite for anything. These points are worth emphasizing, because the treatment of applications caused more confusion among reviewers of DETA than any other aspect of the book. Many thought that, since applied chapters have the same appearance and format as others, they must be accorded the same emphasis. This is opposite to my intent.

A survey of applications can be found in the front endpapers. This survey is at the same time an overview of a course entitled "Mathematical Models" that was taught several times using DETA. It dovetails nicely with a prior or concurrent course on technique in which applications are not emphasized.

Applications remind us that engineers, physicists and other mathematical scientists are among those whose undergraduate education includes a course in differential equations. What such students need especially is an efficient mathematical technique. For example, the systematic use of complex numbers allows easy solution of resonance problems that are much harder to solve by undetermined coefficients in the real domain. Conversation with colleagues in physics and engineering suggests that failure to teach the confident and skillful use of complex numbers is a shortcoming of many mathematics courses. In the thought that this criticism may contain a measure of truth, complex numbers are used here whenever their use leads to simplification.

Additional topics related to mathematical technique include a simplified procedure for computing power-series solutions, the evaluation of Fourier coefficients and Laplace transforms by differentiation, and a development of the matrix exponential without infinite series. It reduces the 3-by-3 case to the computation of three matrix products even when there are repeated eigenvalues. No eigenvectors and no matrix inversions are needed.

Because of their special role in the educational process, some problems are accompanied by the letter P for "propædeutic." This is a word of Greek origin suggesting a combination of *pedagogical, preview* and *preparatory.* The response to P problems in my classes has been uniformly favorable. They are designed to ensure that students' first contact with a new topic will be positive and reinforcing. They also promote the use of efficient techniques from the very beginning.

In any section that students are expected to know, my experience indicates that it is advisable to assign all the P problems and a substantial proportion of the remainder. Additional problems are done in class, in part as a means of encouragement and support, in part to make sure students understand what is wanted. If the basic sections highlighted in the Table of Contents of this book are fully mastered, much of the rest of the text becomes immediately accessible. Mastery is the key.

Acknowledgement

The contributions of Basil Gordon and Dan Port, as acknowledged in DETA, carry over to this introductory version. Dan Port also took responsibility for checking new material and for preparation of the index.

In DETA I referred to Jones and Bartlett as "the kind of publisher that authors look for but seldom find." This favorable opinion has been confirmed and strengthened by more recent experience. An added plus is that J & B have provided advice and support from an individual of extraordinary helpfulness and competence: Dave Mallis, the founder and president of Publishing Experts. Also appreciated is the opportunity for unlimited consultation with Vice President Carl Hesler and others of the J & B staff. It is characteristic of their philosophy as publishers that this expert assistance was offered, not to thwart or alter, but to implement, my wishes as author.

Now that the project is drawing to a close, it is time to thank those who were involved at the beginning, as well as those who later shared their views with me. Chief among the former is Richard Jones, who was then an editor of West Publishing Company. He persuaded me to undertake a book on differential equations and provided valuable feedback in the early stages. Some persons in the following list responded to questionnaires sent out by Richard Jones starting about five years ago, others to a questionnaire that Carl Hesler sent out last year for suggestions as to how DETA could best be shortened. Some gave prepublication reviews solicited by J & B and other publishers; several helpful but anonymous reviews unfortunately cannot be acknowledged by name. A few individuals are included merely because, on their own initiative, they offered comments or encouragement at a time when these were especially welcome:

Richard Askey, Roger Chalkley, Steven Dunbar, Allan Edelson, Gary Etgen, Graeme Fairweather, William Fitzgibbon, David Fonkin, Juan Gatica, Ian Graham, John Gregory, Leon Hall, Henry Helson, Terry Herdman, Edmund Hlawka, Elgin Johnston, Victor Klee, Harold Levine, Warren Loud, Jack Maki, Stephen Merrill, Robert Mitchell, Jeff Morgan, William Olmstead, Kenneth Ostrom, Jimmie Poxon, Murray Protter, Moshen Razzaghi, Robert Rogers, James Serrin, Ralph Showalter, Donald Taranto, Robert Turner, Carroll Wilde.

The voice of the intellect is a soft one, but it does not rest until it has gained a hearing.

—Sigmund Freud

Thoughts that come with doves' footsteps guide the world.

—Friedrich Nietzsche

Chapter 1
INTRODUCTION

THE DERIVATIVE of a function represents a rate of change, and an equation involving derivatives is called a differential equation. Many natural laws give a relation between the rates of change of various quantities, rather than between the quantities themselves. It should be no surprise, then, that differential equations have had a major impact on the history of science and represent one of the main avenues by which mathematical methods are applied to the real world. This chapter describes some of the uses of differential equations, first in general terms, then in the context of a specific equation.

1. A survey of applications

The subject of differential equations has its primary historical origin in Newtonian mechanics, and we begin by saying something about Newton and his achievements. Isaac Newton was born on Christmas Day, 1642. It was a premature birth and, according to tradition, he was "small enough to fit into a quart mug." From this modest beginning, he became one of the chief architects of modern science. Newton's importance for this book is that much of his scientific work was based on differential equations, which he applied with astonishing skill to the study of Nature. For example, Newton's second law of motion

$$(\text{force}) = (\text{mass})(\text{acceleration})$$

is a differential equation. When combined with his law of gravitation, Newton's laws of motion enabled him to compute the orbits of the planets, the moon, and comets. As part of this same development he went on to estimate the weight and density of the sun and moon, to calculate the tides, and to explain the precession of the equinoxes.

The orbit of the moon is perturbed by the pull of the sun. Taking account of this perturbation, Newton explained two ancient observations that had gone unexplained since the time of Ptolemy. He also began the theory of planetary perturbations. An interesting sidelight is that the planet Uranus was discovered about fifty years after Newton's death, and about seventy years after that, perturbations in the orbit of Uranus led to the discovery of Neptune. Today the theory of perturbations is a major subject in differential equations and is applied to such diverse topics as weather prediction, electronic orbits, and ecological systems.

When only the sun and a single planet are considered, the differential equations obtained by Newton can be solved in a way that would now be considered elementary, though at that time the solution was a spectacular achievement. As a result, Newton could give a rather brief derivation of the three laws of Kepler, which Kepler had deduced by a lifetime spent in the study of astronomical observations. The problem of determining the motion of two gravitating masses, such as the sun and a planet, is called the *two-body problem.* The *three-body problem* is to determine the motion of three gravitating masses, such as earth-moon-sun. This problem is not in any sense elementary. It has occupied the attention of many outstanding mathematicians since Newton and still has its puzzles today. It was the only problem, Newton said, that made his head ache.

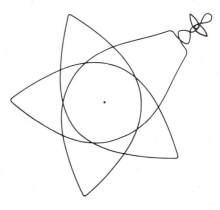

Figure 1

As you would surmise, the corresponding many-body problem for the solar system as a whole is even harder. For example, nobody has been able to prove that the solar system is stable, though the differential equations can be formulated with great precision. (Stability in this context means that things will stay much as they are now, without anything catastrophic such as a collision of two planets to upset the picture.) The differential equations that Newton formulated for the study of astronomy are the same as those that are now used to predict the orbits of man-made satellites like the Moon Lander. Here too we are dealing with the many-body problem, and one reason why space travel is technically feasible is that today's computers have enough power to solve the relevant differential equations numerically.

Figure 1 shows a periodic earth-moon orbit of a satellite. This orbit lies in a plane. In general, orbits in the three-body problem do not lie in a plane, a fact that leads to even further complication.

Since Newton hated controversy, he hesitated to publish his results, and leadership in the field of differential equations quickly passed to the European continent, where it remained for over one hundred years. Some of the leading names are Leibniz, Johann and Daniel Bernoulli, Euler, Laplace, Lagrange, Fourier, Poincaré, Picard, Liapunov, and Volterra. It is chiefly through their work that the subject of elementary differential equations attained the form it has today.

These later researchers showed that differential equations apply not only to Newtonian mechanics, but to a truly astonishing variety of scientific fields. The main requirement is that the phenomena being described should vary continuously, rather than in discrete jumps. For example, differential equations apply to fluid flow, propagation of sound waves, and the flow of electricity in a cable. Fourier's book of 1822 entitled *The Analytic Theory of Heat* has been called the Bible of mathematical physics. It is devoted, in large part, to a study of various differential equations. By means of differential equations, Maxwell predicted the existence of electromagnetic waves (radio waves) before they were observed experimentally by Hertz. The theory of geometric optics depends on differential equations, and so does the theory of deformation of elastic structures (beams and the like). Chemical reactions ... radioactive decay ... propagation of neutrons in a reactor ... the list goes on and on.

The examples mentioned above are confined to such fields as physics, chemistry, and engineering, which are sometimes referred to as the "exact sciences." What is meant by this phrase, for purposes of the present text, is that one can formulate the relevant differential equations with a high degree of certainty. Though the equation may be hard to solve, as is the case for the many-body problem, there is no serious doubt as to what the equation should be. To be sure, in setting up the equations of motion for a satellite the effects of relativity are usually neglected. But these effects are so minute that only an experiment of extreme sensitivity can detect the difference between the mechanics of Newton and that of Einstein. Similar remarks apply to other topics, such as the theory of heat flow. Although the molecules of a substance do not form a continuum, and hence do not correspond precisely to the mathematical model, the error is negligible in all ordinary cases. The equations for heat flow are not subject to controversy.

There are other fields, no less important, in which differential equations are still useful, but are not known with the same certainty as in the fields mentioned above. Among these are topics connected with economics, ecology, biology, and medicine. Often in such fields it is by no means obvious what the differential equation should be, and in not a few cases the formulation of the equation is more difficult than its solution.

For example, the successful program of eradicating smallpox received an early boost from Daniel Bernoulli, who set up a differential equation for the progress of an infectious disease. This equation enabled him to predict, in 1760, that a program of inoculation would increase life expectancy by about three years. (He also made some contributions to the theory of probability in the course of his calculation.) Even now, nobody would claim to know an exact differential equation for the problem he considered, in the sense that the equations of mathematical physics are exact. Nevertheless his work was of great importance, since it gave a scientific justification for the risk of vaccination.

Another example of this kind is the application of differential equations to the use of pesticides. It may happen that the target population—insects, for example—is actually increased by the pesticide, instead of being de- creased. The trouble is that other insects that prey on the target population are killed off too. The reduction of predators may help the target population more than the pesticide harms it. Differential equations have given much insight into problems such as this, though their correct formulation is not as clear-cut as in problems of physical science.

Similar remarks apply when differential equations are used to study the fluctuations of animal populations, such as the Arctic fox, or the harvesting of renewable resources such as fish in the Adriatic. (Both these examples have played a role in the development of mathematical ecology.) Further examples are given by the mathematics of heart physiology, the transmission equations for nerve impulses, the growth laws for tumors—again the list could go on and on.

This section was written, in part, to alert you to the fact that when differential equations are applied to scientific problems, they may have varying degrees of reliability. Though strongly committed to a logical development of the subject of differential equations, we feel no obligation to examine the logical structure of the different fields to which these equations are applied. With that in mind, the derivations are presented as briefly as clarity allows, and it is left for you to decide whether you think the underlying assumptions are justified.

To conclude this general discussion, we remark that it would be a mistake to think that differential equations are important only in connection with other fields of science. The *Mathematical Reviews* publishes abstracts of about 75 new papers on differential equations per month, and the number has been increasing from year to year. Differential equations occupy a central place in modern mathematics, and they are interesting in their own right apart from applications.

2. Exponential growth

To bring the foregoing remarks into sharper focus, we consider a specific differential equation in detail. It is an equation with which you are probably familiar and it has been chosen, in part, for that reason. The equation is

$$\frac{dy}{dx} = ky, \qquad k = \text{const.}$$

This is called the equation of exponential growth (or decay) because its solutions are exponential functions.

On any interval on which $y \neq 0$ the equation can be written

$$\frac{dy}{y} = k \, dx.$$

If $y > 0$ on the interval, integration gives

$$\ln y = kx + C$$

where C is constant and where, as elsewhere in this book, ln denotes the natural logarithm; that is, the logarithm to base e. Taking exponentials we get

$$y = e^{kx+C} = e^{kx}e^{C} = e^{kx}c.$$

Here $c = e^C$ is another constant, necessarily positive because the exponential function takes only positive values. However, when we check the solution, we find that the sign of c does not matter. Namely,

$$y = ce^{kx} \quad \text{gives} \quad \frac{dy}{dx} = cke^{kx} = k(ce^{kx}) = ky.$$

In writing the equation as $dy/y = k\,dx$ we put everything involving y on one side of the equation and everything involving x on the other. It is said, then, that the equation is separated, and the procedure illustrates a general method known as the method of separation of variables.

In checking the equation we compute dy/dx from the given formula and verify that the equation holds at all values of x. This illustrates what is meant by a solution of a differential equation. A formal definition is given in Chapter 2.

We now give another method. When $y > 0$ it is possible to write $y = e^u$ where u, like y, is a differentiable function. By the chain rule

$$\frac{d}{dx}e^u = e^u \frac{du}{dx}.$$

Substituting $y = e^u$ into the differential equation $dy/dx = ky$ we get

$$e^u \frac{du}{dx} = ke^u.$$

Since an exponential never vanishes, the factor e^u can be divided out to give

$$\frac{du}{dx} = k.$$

Integration gives $u = kx + C$, where C is constant, and $y = e^u$ leads again to the solution found above.

The above calculation illustrates the method of substitution. As in this case, quite often a differential equation can be simplified by an appropriate change of variable. Although at first glance the correct choice of variable may seem mysterious, it can be suggested by the form of the equation or by the form of the integrals encountered in a more direct approach. When applied problems are formulated as differential equations, the question whether the solution is simple or complicated depends strongly on the choice of variables.

The foregoing methods can be modified so as to work when $y < 0$ throughout the interval, instead of $y > 0$; but they both break down if y changes sign. We now present another method that is free of this defect. Write the differential equation in the form

$$\frac{dy}{dx} - ky = 0$$

and multiply by e^{-kx}. The left side turns out to be the derivative of $e^{-kx}y$, so that we get

$$0 = e^{-kx}\left(\frac{dy}{dx} - ky\right) = e^{-kx}\frac{dy}{dx} + y(-ke^{-kx}) = \frac{d}{dx}(e^{-kx}y).$$

It is a familiar fact of calculus that if the derivative of a function is 0 on an interval then the function is constant on that interval. Hence

$$e^{-kx}y = c, \qquad c = \text{const.},$$

and multiplying by e^{kx}, we get the same solution as before. This procedure does not assume $c > 0$, it does not assume $y \neq 0$, and it shows that every solution of the differential equation is given by the formula for some value of c. We summarize by writing

(1) $$\frac{dy}{dx} = ky \iff y = ce^{kx}$$

where k and c are constant. The symbol $\iff$ means "if and only if." It gives a brief way of saying that two statements are logically equivalent in the sense that either can be deduced from the other. The solutions for $k = 1$ and $k = -1$ are shown graphically in Figure 2.

From the standpoint of general procedure, the function e^{-kx} is called an integrating factor, because multiplying by this factor gives a function that is readily integrated. The method of integrating factors applies to broad classes of differential equations.

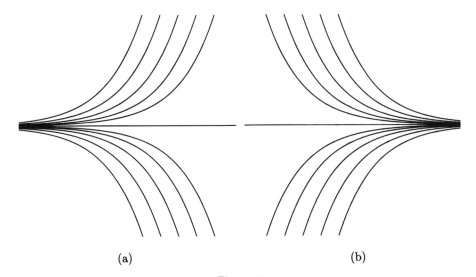

(a)   (b)

Figure 2

Since the solution (1) contains the arbitrary constant c, the function y is not uniquely determined by the differential equation and some other condition is needed. The most common condition is to require that the solution curve pass through some specified point (x_0, y_0) of the (x, y) plane, so that $y = y_0$ when $x = x_0$. This gives, in succession,

$$ce^{kx_0} = y_0, \quad c = y_0 e^{-kx_0}.$$

Substitution into $y = ce^{kx}$ gives

$$y = y_0 e^{-kx_0} e^{kx} = y_0 e^{k(x-x_0)}.$$

Anticipating terminology that will be discussed again later, we remark that the family ce^{kx} is called the *complete solution* of $dy/dx = ky$ because every member of the family satisfies the differential equation, and every solution of the differential equation is a member of the family. The family ce^{kx} is called a *general solution* of the differential equation because the constant c in it can be chosen so as to satisfy the arbitrarily specified condition $y = y_0$ at the arbitrarily specified point $x = x_0$. (Hence, in this case the complete solution and the general solution coincide.) The statement that $y = y_0$ when $x = x_0$ is called an initial condition. It is often written in the form

$$y(x_0) = y_0.$$

Generally speaking, an initial condition holds at a single place or time, while the differential equation holds at all places or times. Both a differential equation and an initial condition are needed, as a rule, to determine a particular solution, that is, a solution of the form $y = \phi(x)$ not containing any arbitrary constants. Depending on context, the word "solution" might mean a general solution such as $y = ce^x$ or a particular solution such as $y = 2e^x$.

PROBLEMS

Problems marked P, for "preview", are intended to make later problems, or later parts of the text, easier for you to understand. Please read these problems whether you do them or not. Knowledge of them is assumed from that point on.

1P. In this book, derivatives are often denoted by primes, thus:

$$y' = \frac{dy}{dx}, \qquad y'' = \frac{d^2 y}{dx^2}, \qquad y''' = \frac{d^3 y}{dx^3}, \qquad y^{iv} = \frac{d^4 y}{dx^4}$$

When the independent variable is the time t, derivatives are often indicated by dots placed above the letters, thus:

$$\dot{u} = \frac{du}{dt}, \qquad \ddot{u} = \frac{d^2 u}{dt^2}$$

This notation goes back to Newton, who called u the *fluent* and $\dot{u}$ the *fluxion*. It is seldom used in this book, but may be used by your instructor and you are likely to meet it in other courses. For higher derivatives $y^{(j)}$ is used, with the convention that $y^{(0)} = y$. Thus,

$$y^{(j)} = \frac{d^j y}{dx^j}, \qquad j = 0, 1, 2, 3, 4, \cdots$$

Check that the functions e^{x^2}, $\sin x$, x^3, x^5, te^t satisfy the following differential equations, respectively:

$$y' = 2xy, \qquad y'' + y = 0 \qquad y^{(4)} = 0 \qquad y^{(5)} = 120 \qquad \ddot{u} + u = 2\dot{u}$$

2. If $y' = ky$ on an interval, where k is constant, show that y cannot change sign on the interval. That is, $y > 0$ throughout the interval, or $y = 0$ throughout the interval, or $y < 0$ throughout the interval.

Hint: Use (1). It will be found that the three possible behaviors correspond to three possibilities for the constant c.

3. A family of negative solutions for $y' = 25y$ is obtained by setting $y = -e^u$, where u is a differentiable function of x. Show that a graph of u versus x gives a family of parallel straight lines. What is their common slope?

4. Find a family of positive solutions of $y' = 6y$

 (a) by setting $y = e^u$

 (b) by use of a suitable integrating factor

 (c) by separation of variables.

5. Find a family of negative solutions of $y' + 7y = 0$

 (a) by setting $y = -e^u$

 (b) by use of a suitable integrating factor

 (c) by separation of variables.

  Hint for (c): The integral of dy/y is $\ln|y| + c$.

6. Solve the problem $y' + 4y = 0$, $y(0) = 3$ by setting $y = 3e^u$ and also by separation of variables.

7. Solve the problem $y' - 4y = 0, y(0) = -3$ by setting $y = -3e^u$ and also by separation of variables.

8P. According to the text, e^{2x} is an integrating factor for $y' + 2y = 0$. Using this integrating factor, solve $y' + 2y = 6$.

Hint: Multiplication by e^{2x} gives $(e^{2x}y)' = 6e^{2x}$.

9P. Find a constant a such that $y = a$ satisfies $y' + 2y = 6$. Add to this the complete solution of $y' + 2y = 0$ and thus obtain the family of solutions found in Problem 8. The procedure illustrated here applies to a broad class of equations known as linear differential equations with constant coefficients.

10. By the procedure suggested in Problem 8 solve $y' + 3y = 8e^x$.

11. By trying $y = ae^x$ where a is constant, obtain a solution of $y' + 3y = 8e^x$. Add to this the complete solution of $y' + 3y = 0$ and thus obtain the result of Problem 10.

12. Solve $y' + 1 = y + e^{2x}$, $y(0) = 3$, and check.

13P. The left-hand curves in Figure 2 have the equations

$$y = e^{x-c}, \qquad y = 0, \qquad y = -e^{x-c}$$

so that changing c merely translates the curve $y = e^x$ or $y = -e^x$ horizontally. Similarly interpret the right-hand curves. What choice of c gives $y = 0$?

14P. Variation of parameters. According to the text, the family $y = ce^{kx}$ represents the complete solution of $y' = ky$. We are going to solve $y' = ky + f \left(x \right)$? by the substitution $y = ue^{kx}$. Since the constant parameter c is now replaced by a variable u, this is called the method of *variation of parameters*.

Show that the suggested substitution leads to $u' = e^{-kx}f(x)$. If f is continuous, an integration yields u and then $y = ue^{kx}$ is the solution sought.

15. (abc) Solve Problems 8, 10, and 12 by variation of parameters.

Hint for 8: Check that the substitution $y = ue^{-2x}$ yields $u' = 6e^{2x}$.

———————————————————ANSWERS———————————————————

2. The three cases correspond to $c > 0$, $c = 0$, $c < 0$ respectively.
3. Since $u' = 25$, the lines are parallel and the common slope is 25.
4. $y = e^{6x+C}$, $y = ce^{6x}$ with $c > 0$, $y = e^{6x+C}$
5. $y = -e^{-7x+C}$, $y = ce^{-7x}$ with $c < 0$, $y = -e^{-7x+C}$
6. $y = 3e^{-4x}$ 7. $y = -3e^{4x}$ 8. $y = 3 + ce^{-2x}$ 9. $a = 3$
10. $y = 2e^x + ce^{-3x}$ 11. $a = 2$ 12. $y = 1 + e^x + e^{2x}$
13. Partial answer: None.

3. Four examples

In numerous applications the independent variable is the time t rather than a space variable x. When this is the case we write

$$y' = \frac{dy}{dt}, \quad y'' = \frac{d^2y}{dt^2}$$

and so on. The main result (1) of the last section gives

(1) $$y' = ky \iff y = y_0 e^{kt},$$

where k is constant and where $y_0 = y(0)$ is the initial value. In the following examples we assume that the initial value is nonzero.

(a) Growth

Many organisms grow at a rate proportional to their size. If S denotes the size at time t, this means $dS/dt = kS$, where k is constant. It is a remarkable fact that this law of growth is wholly determined if we know the size at two different times. To see why, let the size be S_0 at time $t = 0$ and S_1 at time $t = t_1$. By change of notation in (1),

(2) $$S = S_0 e^{kt}.$$

Since the size is S_1 when $t = t_1$ we have

(3) $$S_1 = S_0 e^{kt_1}.$$

We could find k by taking logarithms in (3), but it is more efficient to find e^k instead. This is done as follows. By (3),

$$e^{kt_1} = \frac{S_1}{S_0}, \quad \text{hence} \quad e^k = (e^{kt_1})^{1/t_1} = \left(\frac{S_1}{S_0}\right)^{1/t_1}.$$

Using this value for e^k in (2) we get the desired result,

$$S = S_0(e^k)^t = S_0\left(\frac{S_1}{S_0}\right)^{t/t_1}.$$

As a check and aid to memory, note that $S(0) = S_0$, $S(t_1) = S_1$ by inspection, and the variable t occurs as an exponent.

(b) Bacteria

In a colony of bacteria each bacterium divides into two in a time interval of average length T. To find T it would be hopelessly inefficient, and also inaccurate, to watch a bacterium through a microscope until you see it divide. What you really want is an average over many bacteria, not just a single observation.

This problem is solved as follows. Using a microscope slide ruled in squares, one can count the bacteria present at two different times and use a formula to compute T. The basis for the computation is a differential equation, but there is a logical difficulty because the function N, being integer-valued, is not differentiable unless it is constant. To deal with this difficulty we replace N by a differentiable approximation, still called N, and we suppose that the hypothesis about doubling means

$$\frac{dN}{dt} = kN, \qquad k = \text{const.}$$

After these preliminaries let there be N_0 bacteria at time $t = 0$ and N_1 at time $t = t_1$. Since the initial population N_0 doubles in time T

$$0 \le t \le T \quad \text{corresponds to} \quad N_0 \le N \le 2N_0$$

and, by hypothesis,

$$0 \le t \le t_1 \quad \text{corresponds to} \quad N_0 \le N \le N_1.$$

Let us write the differential equation in the separated form

$$k\,dt = \frac{dN}{N}$$

and integrate between corresponding limits. The result is

$$\int_0^T k\,dt = \int_{N_0}^{2N_0} \frac{dN}{N}, \qquad \int_0^{t_1} k\,dt = \int_{N_0}^{N_1} \frac{dN}{N}$$

or, equivalently,

$$kT = \ln 2N_0 - \ln N_0, \qquad kt_1 = \ln N_1 - \ln N_0.$$

Since $\ln 2N_0 - \ln N_0 = \ln 2$, dividing the first equation by the second gives

$$\frac{T}{t_1} = \frac{\ln 2}{\ln N_1 - \ln N_0}.$$

Clearly, T can be found from this. The calculation illustrates the important method of *integrating between limits*, which is justified in Problem 12.

Many investigations involve integer-valued variables like the number N of bacteria in this example. In the following discussion of radioactivity, the primary variable is the number of atoms in a sample of material. Ecology is concerned with the number of individuals in a population, and further examples are easily given. In such cases the differential equation applies not to the original variable, but to a differentiable approximation thereto. Since this matter has now been brought to your attention, it will not be mentioned again.

(c) Radiocarbon dating

In a radioactive material each atom has a certain probability of disintegrating into a lower state. For example, radium goes through a number of states the last of which is lead. If you have twice as much material, twice as many atoms will disintegrate, on the average, in a given time interval. If you have three times as much material, three times as many will disintegrate, and so on. These considerations lead to the differential equation

$$\frac{dA}{dt} = kA, \qquad k = \text{const.}$$

where A is the amount of material at time t. The constant k is negative because the amount A decreases with time—the equation describes decay not growth. A measure of the rate of decay is given by the *half-life, H.* This is the time needed for A to decay to $A/2$, in other words, for half the material to change into the new form. As an illustration, the half-lives of some materials are:

radium	carbon 14	uranium
1620 years	5568 years	4,510,000,000 years

Obviously one cannot wait until half the material is gone to get the half-life. Instead, one measures the amount present at two different times. If the amount A_0 is present at time $t = 0$ and A_1 at time $t = t_1$, then

$$0 \leq t \leq H \quad \text{corresponds to} \quad A_0 \geq A \geq \frac{1}{2}A_0,$$

$$0 \leq t \leq t_1 \quad \text{corresponds to} \quad A_0 \geq A \geq A_1.$$

Since $\ln(1/2) = -\ln 2$, a development almost identical to that given for bacteria above leads to the formula

$$\frac{H}{t_1} = \frac{\ln 2}{\ln A_0 - \ln A_1}.$$

This is the way the half-life H is determined in practice.

If a radioactive material is combined with another material that is not radioactive, the ratio of radioactive material to the other satisfies the same differential equation as does the amount of radioactive material. More important, the rate of decomposition of the mixture is the same as that for the radioactive material alone. Suppose, for example, that a sample of carbon consists of A grams of carbon 14, which is radioactive, combined with C grams of ordinary carbon, which is not radioactive. The above remarks mean

$$\frac{dA}{dt} = kA \iff \frac{d}{dt}\left(\frac{A}{C}\right) = k\frac{A}{C}, \quad \frac{dA}{dt} = kA \iff \frac{d}{dt}(A + C) = kA.$$

The equations follow from the fact that C is constant, independent of time. The easy verification is left for you.

These remarks give the mathematical basis for a method of dating historical events, which was discovered by the Nobel prize winner Willard Libby and is known as the method of *radiocarbon dating*. The best way to explain the procedure is to present it in the words of the discoverer himself. The following is quoted from an article that Libby wrote for the Encyclopædia Britannica:

"Radiocarbon age dating, developed in the late 1940s at The University of Chicago, is an example of the application of one of the newest sciences (atomic energy) to one of the oldest (archeology). The technique involves measuring the relative activities of radioactive carbon (C^{14}) in (1) present-day living organic matter and (2) the sample under investigation, and multiplying the logarithm of this ratio by the rate at which the activity of C^{14} decays with time. Careful measurements have shown that the activity of any given preparation of carbon-14 is reduced by exactly one-half during each interval of $5,568 \pm 30$ years. This value is called the half-life of C^{14}.

"Radiocarbon is produced in nature by an indirect process involving the interaction of cosmic rays from outer space with the nitrogen in the earth's atmosphere. The competing processes of formation and of decay of C^{14} have been going on for so long that the equilibrium has been established, and the world inventory of C^{14} is estimated at about 70 metric tons. Radiocarbon therefore has been introduced into the biosphere, and all living matter contains a small quantity of radiocarbon which averages 15.3 ± 0.1 disintegrations per minute per gram of contained carbon. This activity remains constant throughout the life of the organic matter because of the above-mentioned equilibrium processes.

"However, at death the introduction of radiocarbon into the specimen ceases, while the normal decay of the contained radiocarbon continues according to the half-life mentioned above. Therefore an archeological specimen (for example, a mummy or a tree) which yields 7.65 disintegrations per minute per gram of carbon instead of 15.3 is judged to be $5,568 \pm 30$ years old. If the material shows only one-fourth the radiocarbon content of living matter, the age of the specimen is $11,136 \pm 60$ years, etc.

"Age dating has revealed that there lived in North America over 9,000 years ago people who were capable of performing the finest basketry work; that the continents of North America and Europe were both glaciated in their northern latitudes 11,000 years ago; and that the calendars of the ancient Babylonians and the Mayas can be correlated with that of the Christians."

(d) Drifting of a canoe

This problem pertains to a canoe in still water on a windless day. For ease of exposition, you are nominated as the paddler. At time $t = 0$ you stop paddling just as you reach a line of flags stretched over the water to mark the finish line of a race. By time $t = t_0$ the canoe has drifted a distance s_0 beyond the line, and by time $t = 2t_0$ it has drifted a distance s_1 beyond the line. Both s_0 and s_1 are known to you. The problem is to show that the maximum distance the canoe can drift after you stop paddling is

$$s_\infty = \frac{s_0^2}{2s_0 - s_1}.$$

That is, if you wait long enough, the distance will be arbitrarily close to that value, but you will never quite reach $s = s_\infty$. The surprise of the result is that it does not require any knowledge of t_0, or of the velocity v_0 that the canoe had when you stopped paddling, or of the magnitude of the resistance that the water offers to the motion, or of the mass of the loaded canoe!

To see where the formula comes from, let us accept the fact that for low-speed motion the force of resistance due to the water is proportional to the velocity v. Since the acceleration is dv/dt, Newton's law gives

$$m\frac{dv}{dt} = -k_1 v, \qquad k_1 = \text{const.},$$

where m is the mass of the canoe. The constant $-k_1$ is negative because the force tends to retard the motion, hence is in a direction opposite to that associated with the velocity v. Resistance of this type is called *viscous damping*. Setting $k_1 = km$ and dividing by m we get

(4) $$\frac{dv}{dt} = -kv \quad \text{or} \quad v = v_0 e^{-kt}.$$

It is important that k is a *positive constant*. Aside from this, the value of k will not enter into the final result, and that is what we meant when we said the result is independent of the resistance and of the mass.

Since the displacement s satisfies $ds/dt = v$ by the definition of velocity, the second formula (4) gives

(5)
$$s = \int_0^t v\, dt = v_0 \frac{e^{-kt}}{-k}\bigg|_0^t = \frac{v_0}{k}(1 - e^{-kt}).$$

The formula shows that $s < v_0/k$ and that

$$s_\infty = \lim_{t\to\infty} s(t) = \frac{v_0}{k}.$$

Thus, our problem is to determine v_0/k given

(6)
$$s_0 = \frac{v_0}{k}(1 - e^{-kt_0}), \quad s_1 = \frac{v_0}{k}(1 - e^{-2kt_0}).$$

These formulas follow from (5) together with the hypothesis involving s_0 and s_1. If the second equation (6) is divided by the first the result is

$$\frac{s_1}{s_0} = \frac{1 - e^{-2kt_0}}{1 - e^{-kt_0}} = 1 + e^{-kt_0}$$

which gives

$$1 - e^{-kt_0} = 2 - (1 + e^{-kt_0}) = 2 - \frac{s_1}{s_0}.$$

Hence, by the first equation (6),

$$\frac{v_0}{k} = \frac{s_0}{1 - e^{-kt_0}} = \frac{s_0}{2 - (s_1/s_0)} = \frac{s_0^2}{2s_0 - s_1}.$$

Since $s_\infty = v_0/k$, this is the desired result.

PROBLEMS

1. The size of a growing organism, such as a clam, can be measured by its length L, its area A, or its volume V. If the shape remains unchanged, so that the clam remains similar to itself as it grows,

$$A = aL^2, \qquad V = bL^3 \qquad a, b \text{ const.}$$

As an illustration, for a cube of side L we have $a = 6$ and $b = 1$. For a sphere of diameter L the values are $a = \pi$, $b = \pi/6$. Show that if the first of the following laws of growth holds, where $k = k(t)$ is any function of t, then the second and third hold automatically:

$$\frac{dL}{dt} = kL \qquad \frac{dA}{dt} = 2kA \qquad \frac{dV}{dt} = 3kV$$

$$2aL\, dL/dt$$

The point is that the result is independent of the shape.

Hint: $dA/dt = (d/dt)(aL^2) = 2aL\,dL/dt$.

2. If the equation for radioactive decay is written $dA/dt = -KA$ and the equation for bacterial growth is written $dN/dt = KN$, where K is a positive constant, show that the half-life H and the doubling time T satisfy

$$KH = KT = \ln 2$$

3. The rate of decomposition of a substance is proportional to the amount y at time t, so that $dy/dt = -ky$ where k is a positive constant. Find k if y changes from 1000 grams to 500 grams in half an hour, t being measured in hours.

4. Lambert's law of absorption. The percentage of incident light absorbed in passing through a layer of material is proportional to the thickness of the material. If one inch of material reduces the light to $1/3$ its original intensity, how much additional material is needed to reduce the intensity to $1/27$ of its initial value? Obtain the answer by inspection, and check by use of an appropriate differential equation.

5. If 16 grams of a radioactive substance were present at time $t = 1$ year, and 2 grams at time $t = 4$ years, how much was present initially, and what is the half-life? Solve by inspection, and check by use of a differential equation.

6. A geometric sequence is a sequence $a,\ ab,\ ab^2, \cdots, ab^n, \cdots$ such that the ratio of two consecutive terms is constant, and an arithmetic sequence is a sequence $A,\ A + B,\ A + 2B, ..., A + nB, ...$ such that the difference of two consecutive terms is constant. If y is a solution of the growth equation $dy/dt = ky$ with k constant, show that the function $y(t)$ transforms any arithmetic sequence into a geometric sequence. That is,

$$y(A + Bn) = ab^n, \qquad n = 0, 1, 2, \cdots$$

If $y(0) = 1$ express a and b in terms of k, A, B. This problem sheds light on Problems 4 and 5 above.

7. In this problem we assume no wind, still water, and viscous damping. The time t is in seconds, the displacement s in feet. A motorboat is going 30 miles per hour and at time $t = 0$ the engine is turned off. At $t = 10$ the speed has dropped to 15 miles per hour. Using a hand calculator if you wish, find the velocity v at time t and thus get the displacement s when $t = 10, 20$, and ∞. Verify that your three values for s agree with the result obtained in the canoe problem of the text.

Note: 30 miles per hour equals 44 feet per second.

8. Newton's law of cooling. When the temperatures are not large enough to involve radiation, the rate at which a body cools is proportional to the difference between the temperature T of the body and the temperature T^* of the surrounding medium, which is assumed constant. This is Newton's law of cooling. An easy way to solve problems involving Newton's law is to set $y = T - T^*$. Then $dy/dt = ky$, $k = $ const., which can be solved with ease. Knowing T at various times t_0, t_1 we can find the corresponding values y_0, y_1 from $y = T - T^*$, and once y is found, we get T from $T = y + T^*$. If $T = T_0$ when $t = 0$, get a formula for T at any later time.

9. The temperature of a metal ingot fell from $120°$ to $70°$ in one hour when it was allowed to cool in air at a temperature of $20°$. How long will it take to cool to $40°$? $30°$? $20°$? Does it matter whether the given temperatures are stated in degrees C or degrees F? Use Problem 8.

10. Time of death. When a dead body is discovered in an environment of constant temperature, it is sometimes possible to estimate the time of death by Newton's law of cooling. Namely, measure the temperature T_0 at the time $t = 0$ when the body is discovered, and the temperature T_1 at a later time t_1. Extrapolate back to get the time of death t_d when the temperature was presumably $T_d = 98.6°F$. The positive quantity $|t_d|$ gives the time elapsed from the time of death to the time the body was discovered. Setting $y_i = T_i - T^*$, $i = 0, 1, d$, where T^* is the ambient temperature, obtain the formula

$$\frac{|t_d|}{t_1} = \frac{\ln(y_d/y_0)}{\ln(y_0/y_1)}$$

Hint: By Problem 8 show $(y_d/y_0) = (y_0/y_1)^{t_d/t_1}$ and take logs.

11. The temperature of a body upon discovery is $85°F$ and after one hour the temperature is $80°F$. If the ambient temperature is $70°F$, about how long after death was the body found? Use Problem 10, taking $T_d = 99°F$. How much does the answer change if you take $T_d = 98.6°F$?

12. Assuming $y_0 > 0$ and $y > 0$, evaluate the integrals on the left and thus get the other two equations:

$$\int_{y_0}^{y} \frac{ds}{s} = \int_{x_0}^{x} k \, ds \qquad \frac{dy}{dx} = ky \qquad y(x_0) = y_0$$

13. If a and b are constant, the equation $dy/dt = ay + b$ is encountered in connection with the spread of rumors, Newton's law of cooling, certain types of chemical reactions, problems of population growth with immigration, and the theory of investment. If $a \neq 0$ show that the substitution $u = ay + b$ gives the differential equation on the left, hence gives y as on the right:

$$\frac{du}{dt} = au \qquad ay + b = (ay_0 + b)e^{at}$$

14. A curve $y = f(x)$ in Cartesian coordinates is such that the projection of part of the tangent on the x axis has a constant length $b > 0$, as shown in Figure 3. Geometrically, this means that an increased distance from the x axis is just compensated by an increased steepness of the curve, so that b remains constant. Obtain the differential equation $dy/dx = y/b$ for the curves of this family and deduce that $y = ce^{x/b}$.

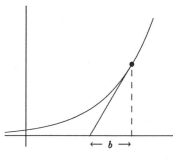

Figure 3

————————ANSWERS————————

3. $k = \ln 4$ 4. two inches additional   5. 32 grams, 1 year

6. $a = e^{kA}$, $b = e^{kB}$ 7. $s \ln 2 = 440(1 - 2^{-t/10})$

8. $T - T^* = (T_0 - T^*)e^{kt}$

9. About 2.32, 3.32, ∞ hours. The choice of units does not matter.

11. $t_d = \ln(29/15)/\ln(15/10) \approx 1.63$ hours; $t_d \approx 1.59$ hours. In both cases the sensible answer is *about an hour and a half.*

> The heart and soul and all the poetry of Natural Philosophy are embodied in the concept of mathematical beauty.
>
> —d'Arcy Wentworth Thompson

> To myself I seem to have been only like a boy playing on the seashore, and diverting myself in now and then finding a smoother pebble or a prettier shell than ordinary, whilst the great ocean of truth lay all undiscovered before me.
>
> —Isaac Newton

Chapter 2
METHODS OF SOLUTION

THE WORDS *linear* and *nonlinear* describe an important classification of mathematical and scientific problems. Linear problems lend themselves to comprehensive and satisfying theories, while nonlinear problems often require special devices. Both types are encountered in applications. Differential equations provide one of the few sources of significant nonlinear problems for which simple solutions are available. After setting forth the basic terminology of the subject, we give a number of elementary techniques by which differential equations can be solved.

1. Terminology

Differential equations are divided into two classes, ordinary and partial. An *ordinary differential equation* involves one independent variable and derivatives with respect to that variable. A *partial differential equation* involves more than one independent variable and corresponding partial derivatives. The order of the highest derivative in a differential equation is the *order* of the equation. For example, the first of the equations

$$(1) \qquad x^2 \frac{d^2 y}{dx^2} + 10y = 6x \frac{dy}{dx}, \quad \frac{\partial^4 u}{\partial x^4} + \frac{\partial^4 u}{\partial y^4} = 2$$

is an ordinary differential equation of order 2 and the second is a partial differential equation of order 4. Unless the context makes it obvious that a partial differential equation is intended, the phrase "differential equation" in this book means "ordinary differential equation."

Using notation introduced in Chapter 1 we write

$$y' = \frac{dy}{dx}, \quad y'' = \frac{d^2 y}{dx^2}, \quad \cdots, \quad y^{(n)} = \frac{d^n y}{dx^n}.$$

For example, $(x^5)'' = (5x^4)' = 20x^3$. In this notation the first equation (1) takes the form

$$(2) \qquad x^2 y'' + 10y = 6xy'.$$

Differential equations are usually considered on an open interval, that is, on an interval $a < x < b$ that does not contain its endpoints. Such an interval is denoted by the letter I and the statement that x is on the interval

I is written $x \in I$. A function $\phi(x)$ is said to be a *solution* of a differential equation on I if upon the substitution $y = \phi(x)$, $y^{(j)} = \phi^{(j)}(x)$ the equation is identically satisfied; that is, it holds for all $x \in I$. As an illustration, $y = x^2$ and $y = x^5$ are solutions of (2) because

$$x^2(2) + 10x^2 = 6x(2x), \quad x^2(20x^3) + 10x^5 = (6x)(5x^4).$$

These equations hold for all values of x, hence the solutions are valid on the interval $(-\infty, \infty)$.

It was stated above that differential equations are divided into two classes, ordinary and partial. They are also divided into two other classes, linear and nonlinear. A linear equation is an equation of the form

$$p_n y^{(n)} + p_{n-1} y^{(n-1)} + \cdots + p_1 y' + p_0 y = f$$

where the coefficients p_j and f are functions of x alone, not involving y. To emphasize this point one often writes $p_j(x)$ and $f(x)$ instead of p_j and f. For example, the linear equations of order 1 and 2 are respectively

$$p_1(x)y' + p_0(x)y = f(x), \quad p_2(x)y'' + p_1(x)y' + p_0(x)y = f(x).$$

The functions p_j and f are assumed to be defined on one and the same interval, finite or infinite. If $f = 0$, the linear equation is said to be *homogeneous*.

An equation is considered to be linear if it can be put into the above form by reversible algebraic operations. For example, the equations

$$y'' = y' + 1, \quad y = y' \sin x, \quad x^2(y'' + x) = xy' - y$$

are linear because they can be put into the respective forms

$$y'' - y' = 1, \quad (\sin x)y' - y = 0, \quad x^2 y'' - xy' + y = -x^3.$$

The second of these equations is homogeneous, the others are inhomogeneous. Examples of nonlinear equations are

$$(y')^2 = x + y, \quad yy' = x, \quad y' = \sin y, \quad y' = x + e^y.$$

The most general differential equations of orders 1 and 2 are

$$F(x, y, y') = 0, \quad F(x, y, y', y'') = 0,$$

respectively. For theoretical analysis these general equations are often assumed to be solvable for the highest derivative, thus:

$$y' = f(x, y), \quad y'' = f(x, y, y').$$

But it is not advisable to impose this condition on the linear equation, because in important cases the coefficient of the highest-derivative term may vanish at one or more points. For example, (2) has solutions valid on $(-\infty, \infty)$, although it cannot be solved for y'' on any interval containing the origin.

Differential equations of the kind commonly encountered in applications have not just one or two solutions, but infinitely many. More specifically, an equation of order 1 usually has a family of solutions

$$y = \phi(x, c)$$

depending on a single parameter c. A second-order equation usually has a family of solutions

$$y = \phi(x, c_1, c_2)$$

depending on two parameters c_1, c_2, and so on. These parameters are like constants of integration. As an illustration, the equation $y'' = 0$ has a family of solutions

$$y = c_1 x + c_2$$

depending on two constants of integration, c_1 and c_2.

One of the simplest ways of determining the constants is to specify the value of y and some of its derivatives at a single point x_0. For an equation of order 1 there is one constant c, in general, and we need a single condition

$$y(x_0) = y_0$$

to determine c. For an equation of order 2 there are usually two constants and we need two conditions

$$y(x_0) = y_0, \quad y'(x_0) = y_1,$$

to determine them, and so on.

The values y_j are called *initial values*, the equations $y^{(j)}(x_0) = y_j$ are called *initial conditions*, and the initial conditions together with the differential equation constitute an *initial-value problem*. For example, the initial-value problems associated with equations of orders 1 and 2 are

$$\begin{cases} y' = f(x, y) \\ y(x_0) = y_0 \end{cases} \qquad \begin{cases} y'' = f(x, y, y') \\ y(x_0) = y_0, \ y'(x_0) = y_1 \end{cases}$$

respectively. It is assumed that x_0 is on the interval I where the equation is to hold, or sometimes that x_0 is an endpoint of this interval. The reason for the term "initial value" is that in many problems x denotes the time t, and $x_0 = t_0$ is the time at which the process starts.

So far we have considered only single differential equations containing the single unknown function y. Many problems lead to two or more differential equations in two or more unknowns; in other words, they lead to a *system* of differential equations. For example,

$$\begin{cases} u' = f(x, u, v) \\ v' = g(x, u, v) \end{cases}$$

is a system of two first-order differential equations in the two unknowns u and v.

The theory of first-order systems actually embraces the entire theory of differential equations, because an equation of higher order can always be reduced to a first-order system by taking y and some of its derivatives as new unknowns. For example, the second-order equation

$$y'' = f(x, y, y')$$

is equivalent to the first-order system

$$\begin{cases} u' = v \\ v' = f(x, u, v) \end{cases}$$

with $u = y$ and $v = y'$. From this point of view the most efficient development of the subject of differential equations is to start with first-order systems. However, such an approach would seem intolerably abstract to many readers and would rob the subject of much of its charm.

EXAMPLE 1. Obtain a two-parameter family of solutions of the second-order linear equation $y'' = 6$. Integration gives $y' = 6x + c_1$, where c_1 is constant, and integrating again gives the two-parameter family

(3) $$y = 3x^2 + c_1 x + c_2, \qquad c_1, c_2 \text{ const.}$$

It is obvious that every function (3) satisfies $y'' = 6$ and almost obvious that every solution of $y'' = 6$ is in the family (3). A proof is given in Problem 11.

In general, a family of functions is the *complete solution* of a differential equation if:

    1. every member of the family satisfies the equation, and

    2. every solution of the equation is a member of the family.

The family of functions given by (3) is the complete solution of $y'' = 6$.

As illustrated in this example, the solution of a differential equation often involves one or more integrations, and for this reason the graph of a solution is called an *integral curve*. By (3) the integral curves of the equation $y'' = 6$ are all parabolas.

EXAMPLE 2. Find the solution of $y'' = 6$ satisfying general initial conditions

$$y(x_0) = y_0, \quad y'(x_0) = y_1.$$

Using (3) one could determine c_1 and c_2 from the two equations $y(x_0) = y_0$, $y'(x_0) = y_1$. But a more efficient method is to note that the gist of (3) is that $y = 3x^2$ plus an arbitrary linear function. Thus one can equally well write

$$y = 3(x - x_0)^2 + a(x - x_0) + b$$

since this too is $3x^2$ plus an arbitrary linear function. The latter equation yields, by inspection, $b = y(x_0) = y_0$, $a = y'(x_0) = y_1$. Hence

$$y = 3(x - x_0)^2 + y_1(x - x_0) + y_0.$$

As explained more fully in Chapter 6, a family of solutions that can be used to satisfy arbitrarily specified initial conditions at the arbitrary point x_0 is called a *general solution*. The family (3) is a general solution of $y'' = 6$.

EXAMPLE 3. The linear homogeneous equation $y'' + y = 0$ has the obvious solutions $\cos x$ and $\sin x$. Check that

$$y = c_1 \cos x + c_2 \sin x$$

is also a solution for every choice of the constants c_1 and c_2. This follows from the calculation

$$y'' = (c_1 \cos x + c_2 \sin x)'' = (-c_1 \sin x + c_2 \cos x)' = -c_1 \cos x - c_2 \sin x = -y.$$

EXAMPLE 4. Find all constant solutions $y = c$ of

$$y'' = (y^2 - 1)\sin(y + y').$$

Since $y' = y'' = 0$ when $y = c$, we must have

$$0 = (c^2 - 1)\sin c.$$

This is satisfied by $c = 1$, $c = -1$, and $c = m\pi$ where m is an integer, and by no other values.

PROBLEMS

1. Here are twelve differential equations:

(a) $y'' + 4y = 8$ (e) $y^{iv} = 0$ (i) $x^2 y'' + 35y = 11xy'$

(b) $(y')^2 + y^2 = 1$ (f) $(y'')^2 = 4$ (j) $(y' - y)(y' - 2y) = 0$

(c) $y'' + 3 = 4y$ (g) $y^{iv} = y$ (k) $(y')^2 - 3y^2 = 2yy'$

(d) $y' = y(1 - y)$ (h) $x^2 y^{iv} = 2y''$ (l) $y' = \sin y$

Verify that the sum of the orders of these equations is 25. (The problem is worded so you can check your work without being told the answers.)

2. Exactly half of the equations in Problem 1 are linear, and two-thirds of these are both linear and homogeneous. Which are linear, and which are both linear and homogeneous?

3. Some of the equations in Problem 1 have a solution of the form $y = c$, where c is constant. Which ones are they, and what are the values of c? A solution $y =$ const. is called a *stationary solution*.

4. Verify that the given function is a solution of the differential equation immediately below it:

(a) $\begin{cases} y = xe^x \\ y' = e^x + y \end{cases}$ (b) $\begin{cases} y = \tan x \\ y' = 1 + y^2 \end{cases}$ (c) $\begin{cases} y = e^x - e^{-x} \\ (y')^2 = 4 + y^2 \end{cases}$

5. Verify that the given function is a solution of the differential equation immediately below it:

(a) $\begin{cases} y = e^{7x} \\ y'' = 7y + 6y' \end{cases}$ (b) $\begin{cases} y = x^7 \\ x^2 y'' = 6xy' \end{cases}$ (c) $\begin{cases} y = xe^{-x} \\ y'' + 2y' + y = 0 \end{cases}$

6P. Verify that the equation $y'' + y' = 2y$ is satisfied by $y = e^{ax}$ if the constant a satisfies the quadratic equation $a^2 + a - 2 = 0$. Thus get the two solutions $y = e^x$, $y = e^{-2x}$. The behavior shown here is typical of all linear homogeneous differential equations with constant coefficients. That is, the substitution $y = e^{ax}$ always leads to an algebraic equation for a.

7. Equations (gjk) of Problem 1 have solutions of the form $y = e^{ax}$ where a is constant. Find the values of a.

8. Equations (bg) of Problem 1 have solutions of the form $y = \sin \omega x$ where ω is a positive constant. What are the possible values of ω?

9P. Check that $y = x^a$ satisfies $x^2 y'' = 2y$ if the constant a satisfies the equation $a^2 - a - 2 = 0$. Thus get the two solutions x^2, x^{-1}. Note that the first is valid on $(-\infty, \infty)$ but the second on $(-\infty, 0)$ or $(0, \infty)$ only. The behavior shown here is typical for a broad class of linear homogeneous equations known as equations of *Euler type*. For this class the substitution $y = x^a$ always leads to an algebraic equation for a.

10. Equations (efhi) of Problem 1 admit a solution $y = x^a$ where a is a nonzero constant. What are the possible values of a?

11. If $y'' = 6$ check that $(y' - 6x)' = 0$, and hence $y' - 6x = c_1$ where c_1 is constant. Using this, check that $(y - 3x^2 - c_1 x)' = 0$, hence $y - 3x^2 - c_1 x = c_2$ where c_2 is also constant. This shows that the solution obtained in Example 1 is the complete solution. We used the fact that if the derivative of a function is 0 on an interval then the function is constant on the interval. In the present case the interval is $(-\infty, \infty)$.

12. By repeated integration as in Example 1 show that

$$y''' = 24x \iff y = x^4 + c_1 x^2 + c_2 x + c_3, \qquad c_i \text{ const.}$$

A justification can be given as in Problem 11 above, but if the process of justification has been seen once, it need not be repeated (and you are not asked to repeat it).

13. Solve the initial-value problem $y''' = 24x$, $y(1) = y'(1) = y''(1) = 1$

(a) by determining the constants in the result of Problem 12.

(b) by writing the result of Problem 12 in the equivalent form

$$y = x^4 + a(x - 1)^2 + b(x - 1) + c, \qquad a, b, c \text{ const.}$$

and determining the constants a, b, c.

(c) Show that your two answers agree; no answer is given in the text.

14. Check that the equation $y'' + y = 0$ becomes $u' = v$, $v' = -u$ upon the substitution $u = y$, $v = y'$. Using these latter equations and the chain rule, verify that $u^2 + v^2$ is constant.

15. The functions $y = e^x$ and $y = \cos x$ cannot be solutions of one and the same first-order equation $y' = f(x, y)$ on any interval containing the origin. Why not?

————————————————ANSWERS————————————————
---

2. aceghi are linear, eghi are homogeneous
3. (a) $c = 2$ (b) $c = 1$ or -1 (c) $c = 3/4$ (d) $c = 0$ or 1 (e) any c will do
   (g) $c = 0$ (h) any c will do (ijk) $c = 0$ (l) $c = n\pi$, n any integer
7. (g) $a = 1$ or -1 (j) $a = 1$ or 2 (k) $a = -1$ or 3
8. (b) $\omega = 1$ (g) $\omega = 1$
10. (e) $a = 1$ or 2 or 3 (f) $a = 2$ (h) $a = 1$ or 4 (i) $a = 5$ or 7
15. Because $f(0, 1)$ would have to be both 1 and 0.

---

2. Implicit solutions

In the first-order equation

$$\frac{dy}{dx} = f(x, y),$$

the left-hand member dy/dx denotes the slope of the solution curve and the right-hand member $f(x, y)$ gives the value of the slope at the point (x, y) on the curve. It is as if we had a traffic policeman stationed at each point (x, y) directing traffic to flow along the direction specified by $f(x, y)$. From this point of view, a *solution* is a curve that obeys the law at each of its points.

The formulation by use of dy/dx breaks down when the tangent is vertical, since a vertical tangent corresponds to $dy/dx = \infty$. But the geometric interpretation remains meaningful. To deal with this matter, note that a curve in the (x, y) plane can be described, not only by $y = \phi(x)$, but also by $x = \psi(y)$. The first of the following conditions is meaningless, but the second is permissible:

$$\frac{dy}{dx} = \phi'(x) = \infty, \qquad \frac{dx}{dy} = \psi'(y) = 0.$$

Thus, there is merit in treating x and y symmetrically, so that the independent variable can be either x or y.

The same conclusion is suggested when differential equations are applied to practical problems. Nature has no cognizance of coordinate systems, which merely provide a framework for the mathematical description of an underlying reality. If a problem seems intractable when we insist on a solution $y = \phi(x)$, but easy when we allow $x = \psi(y)$, it could mean that we have made an inappropriate choice of independent and dependent variables in the initial formulation. These remarks motivate the following discussion.

A first-order equation is said to be in *differential form* if it is written

(1) $$P(x, y)dx + Q(x, y)dy = 0.$$

A differentiable function $y = \phi(x)$ is a *solution* of (1) on an interval I if the substitution

(2) $$y = \phi(x), \quad dy = \phi'(x)\,dx$$

makes the left-hand member of (1) vanish for $x \in I$. A differentiable function $x = \psi(y)$ is a *solution* of (1) on an interval J if the substitution

(3) $$x = \psi(y), \quad dx = \psi'(y)\,dy$$

makes the left-hand member of (1) vanish for $y \in J$. The advantage of writing the equation in differential form is that it allows solutions of either type; indeed, we may have $y = \phi(x)$ at some points on the solution-curve and $x = \psi(y)$ at others.

If an equation

(4) $$F(x,y) = c, \qquad c = \text{const.}$$

can be solved for $y = \phi(x)$ or for $x = \psi(y)$ in the neighborhood of each point (x,y) satisfying (4), and if the corresponding function ϕ or ψ satisfies (1) in the sense explained above, (4) is said to give an *implicit solution* of (1). The term "implicit" distinguishes solutions (4) from the explicit solutions $y = \phi(x)$ or $x = \psi(y)$.

As a matter of terminology, we mention that the set of points satisfying an equation (4) is called the *locus* of the equation. The word "locus" is Latin for "place" and refers to the places in the (x,y) plane where the equation is true. For example, the locus defined by $x^2 + y^2 = 1$ is a circle centered at the origin.

Implicit solutions of the form $F(x,y) = c$ are commonly obtained from an equation $dF = 0$, as explained next. If $F(x,y)$ has continuous partial derivatives, an *exact differential* is an expression of the form

$$dF = \frac{\partial F}{\partial x}\,dx + \frac{\partial F}{\partial y}\,dy.$$

For example, each of the following is an exact differential:

$$d(x^2 + y^3) = 2x\,dx + 3y^2\,dy, \quad d(x^3y^5) = 3x^2y^5\,dx + 5x^3y^4\,dy$$

$$d\left(\frac{y}{x}\right) = \frac{x\,dy - y\,dx}{x^2}, \qquad x \neq 0.$$

It is an important fact that, under suitable conditions, the equations $F = \text{const.}$ and $dF = 0$ are equivalent, that is,

$$dF = 0 \iff F(x,y) = c,$$

where c is constant. A precise statement and proof of conditions under which this holds requires a deeper knowledge of two-variable calculus than is assumed at this point, and we merely give a few examples. The general theory is outlined in Section 14.1.

EXAMPLE 1. The equation $y\,dx + x\,dy = 0$ can be written $d(xy) = 0$, giving the implicit solution $xy = c$ where c is constant. The integral curves form a set of rectangular hyperbolas as shown in Figure 1. There is one and only one solution through each point (x_0, y_0) other than the origin, and the solution can be written in the explicit form $y = c/x$ or $x = c/y$ as the case may be. There are two solutions through the origin, namely $y = 0$ and $x = 0$.

EXAMPLE 2. Consider the equation $x\,dy - y\,dx = 0$. If $x \neq 0$ the equation can be multiplied by $1/x^2$ and if $y \neq 0$ it can be multiplied by $1/y^2$ to give, respectively,

$$(5) \qquad \frac{x\,dy - y\,dx}{x^2} = 0 \quad \text{or} \quad \frac{x\,dy - y\,dx}{y^2} = 0.$$

These in turn are equivalent to

$$(6) \qquad d\left(\frac{y}{x}\right) = 0, \quad -d\left(\frac{x}{y}\right) = 0,$$

respectively. The solutions are $y/x = c$ in the first case and $x/y = c$ in the second, where c is constant. The integral curves are shown in Figure 2. There is one and only one integral curve through each point (x_0, y_0) other than the origin, but there are infinitely many through the origin.

The factors $1/x^2$ and $1/y^2$ in (5) are called *integrating factors* because they reduce the left side to an exact differential (6) which is readily integrated. Under mild hypotheses it is seen that an equation (1) always admits an integrating factor, but construction of the factor in the general case requires prior knowledge of the solution. The usefulness of the method does

Figure 1

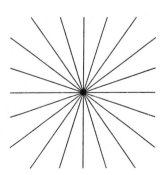

Figure 2

not depend so much on general theory as on the fact that in many problems an integrating factor can be found more or less by inspection.

EXAMPLE 3.   The equation $x\,dx + y\,dy = 0$ can be written

$$\frac{1}{2}d(x^2 + y^2) = 0$$

to give the implicit solution

(7) $x^2 + y^2 = c,$ c const.

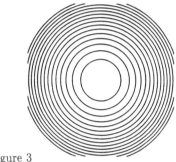

Figure 3

If $c < 0$ the locus is empty, that is, has no points. If $c = 0$ the locus consists of a single point $x = y = 0$. However, this does not give a solution, since it is not a differentiable curve of the form $y = \phi(x)$ or $x = \psi(y)$. If $c > 0$, which is the only interesting case, the integral curves are circles centered at the origin as shown in Figure 3. There is no integral curve through the origin and exactly one through every point (x_0, y_0) other than the origin.

As discussed more fully in the next section, an equation of the form

$$f(x)\,dx = g(y)\,dy$$

is said to be *separated* and a solution is given, in general, by

$$\int f(x)\,dx = \int g(y)\,dy + c,\qquad c \text{ const.}$$

The above equation $x\,dx = -y\,dy$ falls into this pattern.

EXAMPLE 4.   Discuss the relationship of the three equations

(8) $\dfrac{dy}{dx} = -\dfrac{x}{y},\quad \dfrac{dx}{dy} = -\dfrac{y}{x},\quad x\,dx + y\,dy = 0.$

The first equation implies that $y = \phi(x)$ where ϕ is differentiable, and no integral curve of the equation can contain a point (x, y) where $y = 0$. The second equation requires $x = \psi(y)$ where ψ is differentiable and no integral curve of the equation can contain a point (x, y) where $x = 0$. The third equation allows solutions of either type and leads to the family of circles obtained in Example 3.

In the neighborhood of any point (x, y) of the circle we can express y as a differentiable function of x, or x as a differentiable function of y, using one of the relations

$$y = (c - x^2)^{1/2},\quad x = (c - y^2)^{1/2},\quad x = -(c - y^2)^{1/2},\quad y = -(c - x^2)^{1/2}.$$

To be sure, one can get additional functions $y = \phi(x)$ by taking $y > 0$ on part of the circle and $y < 0$ on the remainder. However, such a function is not differentiable and does not satisfy any equation (8). See Figure 4.

EXAMPLE 5. The position of a moving point at time t is given by

$$x = \sin t, \quad y = \cos t.$$

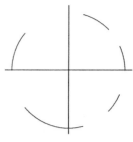

Figure 4

Check that $x\,dx + y\,dy = 0$ where t, rather than x or y, is taken as the independent variable. The proof is simple. Since $d(\sin t) = \cos t\,dt$ and $d(\cos t) = -\sin t\,dt$, we get

$$x\,dx + y\,dy = (\sin t)(\cos t\,dt) + (\cos t)(-\sin t\,dt) = 0$$

as required. The corresponding locus is $x^2 + y^2 = 1$, which agrees with the results of the preceding examples. Since a locus $F(x, y) = c$ can generally be described as the path of a moving point, a locus is often called an *orbit*.

In applications it is seldom necessary to distinguish the forms

$$y = \phi(x, c), \quad F(x, y) = c, \quad x = \psi(y, c)$$

as carefully as in the examples above; the main purpose of these examples is to explain what an implicit solution is. Since the matter of interpretation has now been sufficiently emphasized, the following problems are concerned, not with the interpretation of implicit solutions, but with methods by which they are discovered.

PROBLEMS

1P. A sum such as $du + dv + dw$ equals $d(u + v + w)$, hence $du + dv + dw = 0$ has the implicit solution $u + v + w = c$. Using this idea, obtain an implicit solution of

$$(2x + y - y^{-1})dx + (x + 2y + xy^{-2})dy = 0, \quad y \neq 0$$

Hint: By rearrangement (and this is by far the hardest step) the equation can be written

$$2(x\,dx + y\,dy) + (x\,dy + y\,dx) + \left(\frac{x\,dy - y\,dx}{y^2}\right) = 0$$

Results in the text now give the solution with ease.

2P. Obtain an implicit solution of $(xy)^3(x\,dy + y\,dx) = dx + dy$.

Hint: If $u = xy$ the left side is $u^3\,du$. The method illustrated here is general.

3. The following problems are to be solved as in Problems 1 or 2, but they do not require any rearrangement. Obtain implicit solutions by inspection:

(a) $d(x^2y) = x\,dy + y\,dx$ (d) $2xy\,dy + y^2\,dx = dx + dy$

(b) $d(xy^2) = x\,dx + y\,dy$ (e) $2x\,dx + 6y^2\,dy = x\,dy + y\,dx$

(c) $x^2dy + 2xy\,dx = x\,dx$ (f) $(xy)^2(x\,dy + y\,dx) = 2x\,dx$

Hint: For (c), compute the left side of (a).

4. Rearrange suitably and thus get an implicit solution:

(a) $(y - 2x + 1)dx = (2y + 1 - x)dy$ (c) $(y - x^3)dx = (y^3 - x)dy$

(b) $(y + y^{-1})dx = x(y^{-2} - 1)dy$ (d) $(e^x + x^{-2}y)dx = (x^{-1} - e^{-y})dy$

5. One of the following has an integrating factor $1/x^2$ and the other has an integrating factor $1/y^2$. Decide which is which, and solve:

(a) $x\,dy = (y + x^4)dx + x^4y^2d(xy)$ (b) $(x^4y^7 + y)dx = (x - x^5y^6)dy$

6. The orbit of an implicit solution of $(3x^2 - y)dx + (3y^2 - x)dy = 0$ contains the point $(1,1)$. Show that it also contains $(-1,1), (1,-1), (0,1), (1,0)$.

7. Show that the orbits of $(x + 1)dx + (y + 1)dy = 0$ are circles. What is the radius of the orbit containing the point $(1,1)$?

8. If a is constant, $x > 0$ as needed, and $y = \phi(x)$ is differentiable, then

$$x^{a-1}(x\,dy + ay\,dx) = d(x^a y)$$

Hence the equation $x\,dy + ay\,dx = f(x)dx$ can be solved if both sides are multiplied by x^{a-1}. Using this method, solve the following explicitly, in the form $y = \phi(x, c)$:

(a) $x\,dy + 2y\,dx = 5x^3dx$ (b) $x\,dy - y\,dx = x^2dx$ (c) $x\,dy - 2y\,dx = dx$

Note that division by dx gives linear equations. The procedure illustrates a general method that will be discussed later.

9. Solve $y\,dx + 3x\,dy = 14y^4dy$ in the form $x = \psi(y, c)$ as follows: Interchange x and y to get an equation of the form considered in Problem 8, solve that equation in the form $y = \phi(x, c)$, and interchange x and y in the latter solution.

10. Solve Problem 9 directly, without interchanging x and y, by a procedure similar to that in Problem 8.

11. In comparing solutions of form $y = \phi(x, c)$ and $x = \psi(y, c)$ for Problems 9 and 10, which statement is nearest the truth?

(a) The first is much harder to get than the second.

(b) The second is much harder to get than the first.

(c) The difficulty is about the same.

The situation illustrated here is typical, and it is one of the reasons for introducing implicit solutions and for interchanging the roles of x and y.

————————————————————ANSWERS————————————————————

1. $x^2 + y^2 + xy - x/y = c$, $\quad y \neq 0$ $\qquad$ 2. $(xy)^4 - 4(x + y) = c$

3. (a) $\quad x^2 y - xy = c$ $\qquad$ (b) $\quad 2xy^2 - x^2 - y^2 = c$ $\quad$ (c) $\quad 2x^2 y - x^2 = c$

$\quad$ (d) $\quad xy^2 - x - y = c$ $\quad$ (e) $\quad x^2 + 2y^3 - xy = c$ $\quad$ (f) $\quad (xy)^3 - 3x^2 = c$

To avoid fractions, some of the equations have been multiplied by a constant and c has been renamed; for example, $2c$ is an arbitrary constant if c is.

4. (a) $\quad xy - x^2 - y^2 + x - y = c$ $\quad$ (b) $\quad (x/y) + xy = c$, $\quad y \neq 0$

$\quad$ (c) $\quad 4xy - x^4 - y^4 = c$ $\qquad\qquad$ (d) $\quad (y/x) - e^x + e^{-y} = c$, $\quad x \neq 0$

5. (a) $3(y/x) - x^3 - (xy)^3 = c$, $\quad x \neq 0$ $\qquad$ (b) $5(x/y) + (xy)^5 = c$, $\quad y \neq 0$

7. $2\sqrt{2}$ $\qquad$ 8. (a) $x^3 + cx^{-2}$, $\quad x \neq 0$ $\quad$ (b) $x^2 + cx$ $\quad$ (c) $cx^2 - 1/2$

9. $x = 2y^4 + cy^{-3}$, $\quad y \neq 0$

11. (a) The equation for y as function of x is of degree 7 and cannot be solved by radicals.

3. A survey of basic procedures

This section illustrates the use of integration to find the solution of differential equations. With that object in mind, we begin by giving a brief review of the concepts of *integration* and *solution*.

The word "integral" has two meanings. The *indefinite* integral

$$\int f(x)\, dx$$

denotes any function F such that $F' = f$. The set of all such functions is

$$\int f(x)\,dx + c, \qquad c = \text{const.}$$

On the other hand the *definite* integral

$$\int_a^b f(x)\,dx$$

is defined by a limiting process that has nothing to do with differential equations. It is true that

$$\frac{d}{dx}\int_a^x f(t)\,dt = f(x)$$

if f is continuous, but this is the fundamental theorem of calculus, not a definition.

For precision of notation the variable of integration should be distinguished from the upper limit, and that is why we used t instead of x in the integrand above. However it would be inconvenient to insist on this throughout our study of differential equations, and in simple cases in which there is no danger of confusion we agree that

$$\int_a^x f(x)\,dx = \int_a^x f(t)\,dt.$$

The symbol on the left is defined by the symbol on the right. Naturally, a similar convention holds with other letters, such as y, in place of x.

A *solution* of a differential equation

$$\frac{dy}{dx} = f(x, y)$$

is a function $y = \phi(x)$ that satisfies the equation on an interval I. To know the solution one must know both ϕ and I. In nonlinear problems of interest in applications, however, the interval of existence usually depends on the initial value $y(x_0)$ and is difficult to determine. Aside from the problem of establishing an interval of validity, it may be virtually impossible to express the solution in the explicit form $y = \phi(x)$.

Much of the difficulty disappears if we allow solutions $x = \psi(y)$ or $F(x, y) = c$. For this reason first-order equations are often restated in differential form

$$dy = f(x, y)\,dx.$$

Methods based on such a restatement do not give incorrect information, but they often give information in a more accessible form than would be the case if we insisted on the original formulation $y' = f(x, y)$. The latter requires

$y = \phi(x)$ by definition of the term "solution." The use of alternatives such as $x = \psi(y)$ or $F(x, y) = c$ is such a powerful aid in the study of nonlinear problems that no attempt to avoid it will be made in this text. One should be aware, however, that a reformulation of the differential equation is involved.

In this section and at many subsequent places in the text we make the convention that the letters

$$c,\; C,\; c_i,\; C_i, \qquad i = 0, 1, 2, 3, \cdots$$

if introduced without explanation, always denote constants.

EXAMPLE 1. Using indefinite integrals, solve the initial-value problem

$$\frac{dy}{dx} = \frac{1}{1 + x^2}, \qquad y(x_0) = y_0.$$

Integration and the initial condition give the two equations

$$y = \tan^{-1} x + c, \quad y_0 = \tan^{-1} x_0 + c$$

respectively. The second equation gives c and the first then gives the solution,

$$(1) \qquad\qquad y = \tan^{-1} x - \tan^{-1} x_0 + y_0.$$

The integral curves form a family of parallel curves defined on $(-\infty, \infty)$. To get the graph of the curve through (x_0, y_0) all we have to do is translate the graph of

$$y = \tan^{-1} x$$

upward by the amount $y_0 - \tan^{-1} x_0$. (If this quantity is negative the "upward" translation is really downward.) See Figure 5(a).

EXAMPLE 2. Solve the preceding problem by definite integrals. Writing the equation in differential form

$$dy = \frac{1}{1 + x^2} dx,$$

we note that y_0 corresponds to x_0 and y to x. Integrating between corresponding limits gives

$$(2) \qquad \int_{y_0}^{y} dy = \int_{x_0}^{x} \frac{dx}{1 + x^2} = \tan^{-1} x \Big|_{x_0}^{x} = \tan^{-1} x - \tan^{-1} x_0.$$

Since the left-hand side is $y - y_0$, this agrees with (1).

Both the above methods apply to the general equation

$$\frac{dy}{dx} = f(x), \quad y(x_0) = y_0$$

(a)

Figure 5

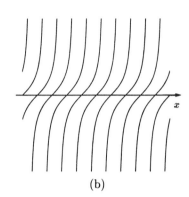

(b)

if f is continuous, but there is a subtle difference that we now explain. If no elementary function F such that $F' = f$ is at hand, the method of indefinite integration does not lead anywhere; it simply restates the question. But the method of definite integrals gives the explicit solution

$$y = y_0 + \int_{x_0}^{x} f(t)\, dt.$$

That $y(x_0) = y_0$ is obvious, and $y' = f(x)$ follows from the fundamental theorem of calculus. The same result would be obtained by writing $dy = f(x)\, dx$ and integrating between corresponding limits.

As an illustration, the solution of the initial-value problem on the left below is given by $y = S(x)$, where $S(x)$ is the integral on the right:

(3) $$\frac{dy}{dx} = \sin x^2, \quad y(0) = 0; \qquad S(x) = \int_{0}^{x} \sin t^2\, dt.$$

There is no elementary function F such that $F'(x) = \sin x^2$, but the function $S(x)$ defined by (3) gives a perfectly good solution. It is called the *Fresnel sine integral* and is important in geometric optics.

An equation of form $g(y)\, dy = f(x)\, dx$ is said to be *separated,* and an equation that can be put into this form by reversible algebraic transformations is called a *separable equation.* A separated equation can be solved, in general, by writing

$$\int g(y)\, dy = \int f(x)\, dx + c$$

and carrying out the indicated integrations.

EXAMPLE 3. By separation of variables solve the initial-value problem

$$\frac{dy}{dx} = 1 + y^2, \quad y(x_0) = y_0.$$

Separating variables and integrating, we get

$$\frac{dy}{1+y^2} = dx, \quad \tan^{-1} y = x + c, \quad \tan^{-1} y_0 = x_0 + c$$

where the last equality follows from the initial condition. This last equality gives c and the second equation then gives the solution

$$y = \tan(x + c).$$

See Figure 5(b). The same result would be obtained by integrating between corresponding limits, thus:

(4)
$$\int_{y_0}^{y} \frac{dy}{1+y^2} = \int_{x_0}^{x} dx.$$

Some separable equations arise so often that it is advisable to short-circuit the separation procedure by a suitable substitution. An equation of this kind is

$$y' = p(x)y$$

where p is continuous. Writing it in the form $y'/y = p$ we note that the left side is the derivative of $\ln y$. This suggests the substitution $u = \ln y$, or equivalently $y = e^u$. The result of substituting $y = e^u$ into the equation $y' = py$ is

$$e^u u' = pe^u, \quad u' = p, \quad u = \int p(x)\,dx + C.$$

Hence

$$y = e^u = e^{\int p(x)\,dx + C} = e^{\int p(x)\,dx} e^C = ce^{\int p(x)\,dx}$$

where $c = e^C$. Although this equation requires $c > 0$, actually the sign of c does not matter. For many purposes only a single solution is needed, as in the following example.

EXAMPLE 4. Setting $y = e^u$ obtain a positive solution of $y' = y \sin x$. The substitution gives, in succession,

$$u' e^u = e^u \sin x, \quad u' = \sin x, \quad u = -\cos x, \quad y = e^{-\cos x}.$$

No additive constant is used in the equation $u = -\cos x$ because only a single solution (a particular solution) is asked for.

Many second-order differential equations can be reduced to first-order equations by an appropriate substitution. For example, an equation of form

$$y'' = f(x, y')$$

can be reduced to a first-order equation by the substitution $v = y'$. Since $y'' = (y')' = v' = dv/dx$ the first-order equation is

$$\frac{dv}{dx} = f(x, v).$$

When $v = dy/dx$ is found from this, an integration gives y.

An equation of form

$$y'' = g(y, y')$$

can also be reduced to a first-order equation, provided $v = y'$ can be expressed as a differentiable function of y. The basis of the method is the chain rule

$$y'' = \frac{dv}{dx} = \frac{dv}{dy}\frac{dy}{dx} = \frac{dv}{dy}v.$$

Using this in $y'' = g(y, y')$ we get the first-order equation

$$v\frac{dv}{dy} = g(y, v).$$

Once v is known, an integration gives y as before.

We have used the notation $v = y'$ because in many problems the independent variable is the time t, and $y' = dy/dt$ denotes the velocity. The notation v for velocity is standard in engineering and in physics. In mathematics, however, the notation $dy/dx = p$ is often used. The above transformations then become

$$\frac{d^2y}{dx^2} = \frac{dp}{dx} = \frac{dp}{dy}\frac{dy}{dx} = p\frac{dp}{dy}.$$

The expression $p\, dp/dy$ is more subtle than dp/dx, because y is thought to be a function of x in the differential equation and is treated as an independent variable in the chain rule. A mathematical justification is given in DETA, page 652. Here we show how the method is used.

EXAMPLE 5. When a second-order equation is reduced to an equation of first order, a solution of the latter is often called a *first integral* of the original differential equation. Obtain a first integral for the equation

$$2y'' = 3y^2.$$

Setting $p = y'$ and $y'' = p\, dp/dy$ we get, in succession,

$$2p\frac{dp}{dy} = 3y^2, \quad 2p\, dp = 3y^2\, dy, \quad p^2 = y^3 + c.$$

The latter equation gives the desired first integral. Since $p = dy/dx$ we have reduced the original second-order equation to a separable first-order equation,

$$\left(\frac{dy}{dx}\right)^2 = y^3 + c.$$

PROBLEMS

1P. Suppose a function $y = \phi(x)$ satisfies $dy = 4y \sin 2x \, dx$ and the initial condition $y(\pi) = e$. It is required to find $y(\pi/6)$, the value of y when $x = \pi/6$.

(a) Separate variables and integrate to get $\ln y = c - 2 \cos 2x$, $y > 0$. (Note that if you add a constant c_1 on the left and c_2 on the right when evaluating the indefinite integrals, the result is the above formula with $c = c_2 - c_1$, which is no more general.) By use of the initial condition show that $c = 3$ and then get $y(\pi/6)$.

(b) The initial condition $y(\pi) = e$ means that $x = \pi$ corresponds to $y = e$. Likewise, $x = \pi/6$ corresponds to $y = a$, where $a = y(\pi/6)$ is the unknown value that is to be found. Integrating between corresponding limits gives

$$\int_e^a \frac{dy}{y} = \int_\pi^{\frac{\pi}{6}} 4 \sin 2x \, dx$$

Evaluate the definite integrals and solve the resulting equation for a. Does this produce the same answer as found above?

(c) If $x \, dy + 3y \, dx = 0$ and $y(-\pi) = e$ you can't find $y(\pi)$. Why not?

2. Interpret in differential form and obtain implicit solutions $F(x, y) = c$ by separation of variables:

(a) $\dfrac{dy}{dx} = \dfrac{6(x^5 + x)}{5(y^4 + 1)}$ (b) $\dfrac{dy}{dx} = \dfrac{1 + \cos x}{1 + e^y}$ (c) $\dfrac{dy}{dx} = \dfrac{1}{x(4y^3 + 1)}$

3. Obtain the solution satisfying $y(0) = 1$ in explicit form $y = \phi(x)$ and give the interval or intervals of existence:

(a) $y' = y^2$ (b) $y' = \dfrac{1 + y}{1 + x}$ (c) $y' = 3(ye^x)^3$

4. Using the method of integration between limits, express the solution satisfying $y(0) = y_0$:

(a) $\dfrac{dy}{y} = \dfrac{dx}{1 + x^2}$ (b) $e^y \, dy = \dfrac{dx}{1 + x}$ (c) $(\tan y) \, dy = 2x \, dx$

5. Obtain a single positive solution by the substitution $y = e^u$:

(a) $y' = 8x^7 y$ (b) $y' = \dfrac{2xy}{x^2 + 1}$ (c) $y' = y \sec x$ (d) $y' = y \tan x$

6. Obtain a first integral by the substitution $y' = v$, $y'' = v'$:

(a) $y'y'' = x$ (b) $y'' = e^{y'}$ (c) $y'' = \dfrac{y'}{x}$ (d) $y'' = e^x(y')^2$

7. Obtain a first integral by the substitution $y' = p$, $y'' = p\,dp/dy$:

(a) $y'' = 2yy'$ (b) $y'' + y = 1$ (c) $y'' = e^y y'$ (d) $yy'' = 1$

8. Solve $y' = 1 + y^2$ by setting $y = \tan u$.

9. If x and y are interchanged, dy/dx becomes dx/dy and hence the equation solved in Example 1 becomes, after the interchange, the equation solved in Example 3. If you replace (x, y) by (y, x) and (x_0, y_0) by (y_0, x_0) in the solution of Example 1, do you get the solution of Example 3?

—————————————ANSWERS—————————————

1. (b) Yes, both give $y(\pi/6) = e^2$. (c) Because dx/x is not integrable on any interval containing the point $x = 0$, the solution $y = -e(\pi/x)^3$ satisfying $y(-\pi) = e$ cannot be continued beyond $x = 0$. The point $x = 0$ is called a *singularity* for the function $1/x$. In solving differential equations you may not in general integrate across a singularity.

2. (a) $y^5 + 5y - x^6 - 3x^2 = c$ (b) $y + e^y - x - \sin x = c$
   (c) $y^4 + y - \ln|x| = c$

3. (a) $y = 1/(1 - x)$ on $(-\infty, 1)$ or $(1, \infty)$ (b) $y = 1 + 2x$ on $(-\infty, \infty)$
   (c) $y = (3 - 2e^{3x})^{-1/2}$ on $(-\infty, a)$ where $a = (1/3)\ln(3/2)$

4. (a) $\ln y/y_0 = \tan^{-1} x$ (b) $e^y - e^{y_0} = \ln(1 + x)$ (c) $\sec y = (\sec y_0)e^{x^2}$

5. (a) $y = e^{x^8}$ (b) $y = x^2 + 1$ (c) $y = |\sec x + \tan x|$ (d) $y = |\sec x|$

6. (a) $v^2 - x^2 = c$ (b) $e^{-v} + x = c$ (c) $v = cx$ (d) $e^x + (1/v) = c$

7. (a) $p = 0$ or $p - y^2 = c$ (c) $p = 0$ or $p - e^y = c$
   (b) $p^2 + y^2 - 2y = c$ (d) $p^2 - \ln(y^2) = c$

8. It will be found that $u' = 1$, hence $u = x + c$ and $y = \tan(x + c)$.

9. Yes, and the method is valid in general.

———————————————————————————————————

4. The first-order linear equation

We shall give a systematic method of solving the linear equation

(1) $$y' + py = f$$

on a given interval I, where f and p are continuous. The method depends on construction of a nonvanishing function r such that

(2) $$r(y' + py) = (ry)'$$

(and hence such that the left side of (2) can be easily integrated). Multiplying out the left side and carrying out the differentiation on the right we see that (2) is equivalent to

$$ry' + rpy = ry' + r'y.$$

This holds if $rpy = r'y$ and hence if $r' = rp$. But the equation $r' = rp$ was solved in the preceding section, by setting $r = e^u$. The result was

$$(3) \qquad\qquad r = e^u \quad \text{where} \quad u = \int p(x)\, dx.$$

We do not add a constant because only one function r is needed here.

If (1) is multiplied by r in (3) the result is

$$r(y' + py) = rf \quad \text{or} \quad (ry)' = rf$$

where the second equation follows from $r' = rp$. An integration gives

$$(4) \qquad\qquad ry = \int r(x)f(x)\, dx + c, \qquad c = \text{const.}$$

Since the function $r = e^u$ in (3) does not vanish we can divide by r to get y.

EXAMPLE 1.   Solve the equation $y' + 2xy = f(x)$ where f is continuous. The solution is in four steps.

First step: Find $u = \int 2x\, dx = x^2$ and $r = e^u = e^{x^2}$.

Second step: Multiply the equation by r to get

$$e^{x^2} y' + e^{x^2} 2xy = e^{x^2} f(x).$$

Third step: Verify that the above reduces, as it should, to

$$(e^{x^2} y)' = e^{x^2} f(x).$$

Fourth step: Integrate to get

$$e^{x^2} y = \int e^{x^2} f(x)\, dx + c.$$

Division by e^{x^2} now gives y.

EXAMPLE 2.   On an interval I where $x \neq 0$ solve

$$(5) \qquad\qquad xy' + 2y = 12x^4.$$

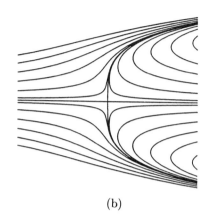

(a)   Figure 6   (b)

Since the coefficient of y' is 1 in the general theory, we divide by x to get

$$y' + \frac{2}{x}y = 12x^3.$$

The four steps of the solution are:

First step: $\int 2/x \, dx = 2\ln|x| = \ln x^2, \quad r = e^u = e^{\ln x^2} = x^2.$

Second step: $x^2 y' + 2xy = 12x^5.$

Third step: $(x^2 y)' = 12x^5.$

Fourth step: $x^2 y = 2x^6 + c, \quad y = 2x^4 + cx^{-2}.$

EXAMPLE 3. Solve the equation

(6) $$y \, dx + 2x \, dy = 12y^4 \, dy$$

by interchanging x and y. The result of the interchange is

$$x \, dy + 2y \, dx = 12x^4 \, dx.$$

Division by dx gives Equation (5), which was solved in the preceding example. Interchanging x and y in the solution found above gives a solution of (6),

(7) $$x = 2y^4 + cy^{-2}, \qquad y \neq 0.$$

The condition $y \neq 0$ comes from the condition $x \neq 0$ that was imposed in (5), and is clearly necessary for the formula (7) to be meaningful. Nevertheless, Equation (6) has the solutions $x = 2y^4$ and $y = 0$, both of which go through the origin. See Figure 6(a) for (5), Figure 6(b) for (6).

Since (6) is in differential form, a solution $x = \psi(y)$ is just as satisfactory as $y = \phi(x)$. An attempt to get the latter requires that we solve the sixth-degree equation

$$2y^6 - xy^2 + c = 0.$$

This is a cubic in y^2, hence it is solvable by radicals, but the resulting equation $y = \phi(x)$ is vastly more complicated than (7).

The preceding calculation illustrates a matter of personal taste with which not everyone will agree. If equation (6) is written in the form

$$y\frac{dx}{dy} + 2x = 12y^4$$

it is linear with x as the unknown function and y as the independent variable. Hence it can be solved, just as it is, by the usual method for linear equations.

For psychological reasons, the suggestion to interchange x and y is made from time to time in this text. It is only a suggestion, however, and you are not obliged to follow it.

If p and f are continuous functions of x and n is constant, the *Bernoulli equation*

$$y' + py = fy^n, \qquad y > 0$$

can be reduced to a linear equation by taking $y = u^m$ for a suitable constant m. This substitution gives $y' = mu^{m-1}u'$ by the chain rule and hence

$$mu^{m-1}u' + pu^m = fu^{mn}.$$

Dividing by u^{m-1} yields

$$mu' + pu = fu^{mn-m+1}.$$

If we choose m so that the exponent $mn - m + 1$ on the right is 0, we get a linear equation for u. The condition $m(n - 1) + 1 = 0$ gives $m = 1/(1 - n)$ and the final result is:

$$(8) \qquad mu' + pu = f, \quad y = u^m, \quad m = \frac{1}{1 - n}.$$

The method breaks down if $n = 1$, but then the original equation is linear:

$$y' + (p - f)y = 0, \qquad n = 1.$$

The condition $y > 0$ ensures that the function $u = y^{1/m}$ is meaningful.

When faced with a specific Bernoulli equation, it is better to set $y = u^m$ and repeat the procedure than to memorize the result (8). The general technique of dealing with differential equations is of more value than the formula.

EXAMPLE 4. Given that $y > 0$, reduce $y' + y = xy^3$ to a linear equation. The solution is in three steps:

First step: $y = u^m$: $mu^{m-1}u' + u^m = xu^{3m}$.

Second step: Divide by u^{m-1}: $mu' + u = xu^{2m+1}$.

Third step: $2m + 1 = 0$, $m = -\frac{1}{2}$: $-\frac{1}{2}u' + u = x$.

Multiplying the last equation by -2 we see that

$$y = u^{-1/2} \quad \text{where} \quad u' - 2u = -2x.$$

Although the method requires $y \neq 0$, the function $y = 0$ satisfies the differential equation. Note that $u > 0$ is needed to get a real y, but the basic equation $y^{-2} = u$ allows $y < 0$ as well as $y > 0$. Indeed, $-y$ satisfies the equation if y does. In general, the question whether $y > 0$ or $u > 0$ is required depends on m.

The following example illustrates a somewhat different procedure for the Bernoulli equation, which produces the correct value of m at once:

EXAMPLE 5. If the equation $y' + y = xy^3$ of Example 4 is divided by y^3 the result is
$$y^{-3}y' + y^{-2} = x.$$

This suggests the substitution $u = y^{-2}$, which is the same as that in Example 4. Since
$$u' = -2y^{-3}y'$$

the term $y^{-3}y'$ can be expressed in terms of u' and the equation for u so obtained agrees with that found above.

PROBLEMS

1P. This problem is concerned with the two equations

$$y' + ay = f(x), \qquad y' + \frac{b}{x}y = f(x)$$

where a and b are constant, f is continuous, and in the second case $x > 0$. Both these equations were considered, with special choices f, in the foregoing pages.

(a) Taking $r = e^u$ where $u' = a$ or b/x, obtain the integrating factors

$$r = e^{ax}, \qquad r = x^b$$

(b) Verify that, when the equations are multiplied by r, they become

$$(e^{ax}y)' = e^{ax}f(x), \qquad (x^by)' = x^bf(x)$$

(c) Integrate and solve for y to get

$$y = e^{-ax}\int e^{ax}f(x)\,dx + ce^{-ax}, \qquad y = x^{-b}\int x^bf(x)\,dx + cx^{-b}$$

2. In this and subsequent problems it is assumed that x or y is restricted to an interval on which the relevant functions are continuous. Solve:

(a) $y' + y\cos x = \cos x$ (b) $y' + y\tan x = 2\cos^2 x$ (c) $y' + y\sec x = \cos x$

3. Check that if the equation $(x^2 + 1)y' + 2xy = 3x^2$ is solved by the standard procedure for linear equations, you get the original equation back again. What is going on here? Solve the equation.

4. Find the solution satisfying $y(0) = -1$:

(a) $y' + xy = x$ (b) $y' + 2xy = e^{-x^2}$ (c) $(1 + x^3)y' + 3x^2y = \sin x$

5. Show that the following equations become linear in y if x and y are interchanged. Thus get solutions $x = \psi(y)$ to the original equations:

(a) $(8e^y - 3x)dy = dx$ (b) $y\,dx + 2x\,dy = 6y^4\,dy$

If solutions $y = \phi(x)$ are wanted, it will be found that the first leads to a quartic equation for e^y and the second to a cubic equation for y^2.

6. Reduce to a linear equation by the substitution $y = u^m$ as in Examples 4 or 5 and check that your result agrees with the formula (8):

(a) $y' + y\cos x = xy^{4/5}$ (b) $y' + e^xy = y^{-4}\sin x$ (c) $y' + x^2y = y^5$

7. Using (8) if you wish, reduce to a linear equation and solve completely:

(a) $x^2y' + y^2 = xy$, $y(e) = e$ (b) $xy' + y = 2x^3y^6$, $y(2) = (0.005)^{1/5}$

8. (a) Verify that the equation $dx + 3x\,dy = 9e^{2y}x^{2/3}\,dy$ becomes an equation of Bernoulli type if x and y are interchanged. Solve that equation and interchange x and y in the solution to get

$$x = e^{-3y}(e^{3y} + c)^3$$

(b) Find the solution satisfying $y(1) = 0$ in explicit form $y = \phi(x)$ for $x > 0$ and check by substitution into the original equation. Although we have not insisted upon it in this text, a solution obtained after a long sequence of calculations should be checked whenever the check is easy.

9P. Variation of parameters. If v is a positive solution of $v' + pv = 0$ we are going to solve $y' + py = f$ by the substitution $y = uv$. Show that this substitution yields $u' = f/v$. An integration gives v and then $y = uv$ is the solution sought.

10. (abc) Solve the equations of Problem 2 by variation of parameters.

REVIEW PROBLEMS————————————————————————————

11. In each equation $y' = dy/dx$, $y'' = d^2y/dx^2$:

(a) $y'' = (y')^2 + y$ (d) $yy'' = \sin y$ (g) $y'' + (y')^2 = 3$
(b) $y'' = (y')^2 + x$ (e) $xy'' = \sin y'$ (h) $y'' + (y')^2 = x^2$
(c) $y'' = (y' + 1)^2$ (f) $xy'' = \sin y$ (i) $y'' + y' \sin y' = 1$

Six of these can be reduced to first-order equations by the substitution $y' = v$, $y'' = dv/dx$. Five can be reduced to first-order equations by the substitution $y' = p$, $y'' = p\,dp/dy$. One can't be reduced by either substitution. Write the first-order equation if it can be found, or write "no go" if both substitutions fail.

12. Show that the solution of $2y'' = 3y^2$, $y(0) = 0$, $y'(0) = 1$ is given implicitly by

$$\int_0^y \frac{dt}{\sqrt{1 + t^3}} = x$$

The integral belongs to an important class of functions known as *elliptic integrals*. It is not expressible in terms of elementary functions.

13. Discuss the relative difficulty of solving $y'' = \sin x$, $y'' = \sin y$. No answer is provided.

14. Using the formula

$$\tan(A + B) = \frac{\tan A + \tan B}{1 - \tan A \tan B}$$

obtain the explicit solution on the right for the initial-value problem on the left:

$$y' = 1 + y^2, \quad y(x_0) = y_0, \qquad y = \frac{y_0 \cos(x - x_0) + \sin(x - x_0)}{\cos(x - x_0) - y_0 \sin(x - x_0)}$$

15. Show that the differential equation on the left has the family of solutions on the right:

$$(x^2 + 1)\frac{dy}{dx} + y^2 + 1 = 0, \qquad y = \frac{c - x}{1 + cx}$$

16. This problem pertains to the equation $y' + y = y^{a+1}$ where a is a nonzero constant. The equation has the stationary solution $y = 1$ and, if $a > -1$, the stationary solution $y = 0$. Setting these cases aside, consider the equation on an interval on which $y > 0$ and $y \neq 1$.

(a) By separating variables obtain the implicit solution

$$\int \frac{dy}{y(y^a - 1)} = x + c$$

(b) Using Bernoulli's procedure with $n = a + 1$ obtain the explicit solution

$$y = (1 + ce^{ax})^{-1/a}$$

(c) Check both results by substituting the latter function into the integral. The calculation is simplified if you use $dy/y = d(\log y)$.

───────────────── ANSWERS ─────────────────

2. (a) $y = 1 + ce^{-\sin x}$ (b) $y = \sin 2x + c \cos x$
   (c) $y = (c + x - \cos x)/(\sec x + \tan x)$

3. The left side is already in the form $(ry)'$. $y = (x^3 + c)/(x^2 + 1)$

4. (a) $y = 1 - 2e^{-x^2/2}$ (b) $y = (x - 1)e^{-x^2}$ (c) $y = (-\cos x)/(1 + x^3)$

5. (a) $x = 2e^y + ce^{-3y}$ (b) $x = y^4 + c/y^2$

6. (a) $y = u^5$ where $5u' + u \cos x = x$ (b) $y = u^{1/5}$ where $u' + 5e^x u = 5 \sin x$
   (c) $y = u^{-1/4}$ where $u' - 4x^2 u = -4$.

7. (a) $y = x/\ln x$ (b) $y = (5x^3 + 5x^5)^{-1/5}$ 8. (b) $y = (1/6)\ln x$

11. (a) $p\,dp/dy = p^2 + y$ (e) $x\,dv/dx = \sin v$
    (b) $dv/dx = v^2 + x$ (f) no go
    (c) $p\,dp/dy = (p + 1)^2$ (g) $p\,dp/dy + p^2 = 3$ or $dv/dx + v^2 = 3$
    or $dv/dx = (v + 1)^2$ (h) $dv/dx + v^2 = x^2$
    (d) $yp\,dp/dy = \sin y$ (i) $p\,dp/dy + p \sin p = 1$ or $dv/dx + v \sin v = 1$

And our own elders, may God have mercy upon them, used
to say, "The study of mathematics is for the mind like soap
for the clothes, which washes away from them dirt and cleans
the spots and stains."

—Ibn Khaldun

Chapter 3

APPLICATIONS

$\mathbf{A}$PPLICATIONS of differential equations can be organized by the type of differential equation or by the nature of the application. Both methods of organization are used here. We begin with some problems from biomathematics, military strategy, the theory of epidemics, and ecology, all of which lead to a first-order linear equation. Topics connected with geometry are discussed next. They include the spiral form of a snail shell, the relation between ray optics and wavefront optics in a parabolic mirror, and the equations for orthogonal trajectories.

1. Linear equations

In a sense, most of the equations considered up to now are equivalent to linear equations, because they were reduced either to an integration problem of the form

$$(1) \qquad \frac{dy}{dx} = f(x), \quad y = \int f(x)\,dx + c$$

or to a succession of such problems. The first-order linear equation itself was solved by two integrations and the Bernoulli equation was linearized by the substitution $y = u^m$. If x and y are interchanged, an equation

$$(2) \qquad \frac{dy}{dx} = g(y), \qquad g(y) \neq 0,$$

goes over into (1) with $f(x) = 1/g(x)$. The equations

$$(3) \qquad y'' = g(y'), \quad y'' = g(y)$$

take the form (2) upon the substitution $v = y'$, with

$$y'' = \frac{dv}{dx} \quad \text{or} \quad y'' = v\frac{dv}{dy}$$

as the case may be.

The foregoing remarks summarize the highlights of Chapter 2 and include every type of equation that will be needed in Chapter 3. In this section we consider five problems that lead directly to linear equations without the special devices mentioned above.

(a) A principle of similitude

Interaction between a solid object and its environment generally takes place at the surface. For example, the rate at which a mothball evaporates is proportional to the area and so is the rate at which a fuel pellet burns. Some microscopic organisms obtain nutrients through a permeable membrane forming the boundary of the organism, and here again the rate of growth is more or less proportional to the surface area. If V is the volume of the object at time t and A its surface area, the differential equation governing processes of this kind is

$$(4) \qquad \frac{dV}{dt} = kA, \qquad k = \text{const.}$$

Suppose an object obeys the differential equation (4) and stays similar to itself as its size changes. Then, with no further assumptions, one can show that a graph of its length L versus time will necessarily be a straight line. To see why this is so, recall that for similar figures the area is proportional to the square and the volume to the cube of the linear dimensions. Thus,

$$A = aL^2, \quad V = bL^3$$

where a and b are constant. Differentiating the second equation by the chain rule and substituting into (4) we get

$$3bL^2 \frac{dL}{dt} = kaL^2 \quad \text{hence} \quad \frac{dL}{dt} = \frac{ka}{3b}.$$

This shows that $L = mt + c$ where $m = ka/(3b)$. Thus

$$L = mt + c, \quad A = a(mt + c)^2, \quad V = b(mt + c)^3.$$

These represent (part of) a line, parabola, and cubic curve, respectively.

The only relevant feature of the shape of the object is given by the single constant a/b. For a sphere of diameter L or a cube of edge L the value of a/b is 6, so that for sphere-like or cube-like objects one has, approximately, $k = m/2$. The slope m is given by an experimental graph of L versus t and $m/2$ then gives the growth constant k.

(b) The square law of military strategy

Two opposing forces of x_0 soldiers and y_0 soldiers, respectively, have a military encounter in an open field without cover or concealment. This means that each soldier can see and fire upon the opposing soldiers without hindrance. The soldiers of a given force do not impede one another, so that if the size of the force is doubled, its effectiveness is also doubled. If the size is tripled, its effectiveness is tripled, and so on. The effectiveness of the x force is measured by the rate of decrease dy/dt of the y force and vice versa. If the soldiers of both forces have equal fighting skill, on the average, these considerations lead to the equations

$$\frac{dx}{dt} = -ky, \quad \frac{dy}{dt} = -kx$$

where k is a constant measuring the soldiers' effectiveness.

Multiply the first of the above equations by x, the second by y, and subtract. The result is

$$x\frac{dx}{dt} - y\frac{dy}{dt} = 0 \quad \text{or} \quad \frac{d}{dt}(x^2 - y^2) = 0.$$

This gives

$$x^2 - y^2 = c$$

where c is constant.

To interpret the result, suppose the battle continues until the smaller force, say the y force, is annihilated. If the larger x force ends up with x_1 survivors and the smaller y force with $y_1 = 0$ survivors the equation

$$x_0^2 - y_0^2 = x_1^2 - y_1^2 = x_1^2$$

shows that $\sqrt{x_0^2 - y_0^2} = x_1$. Hence $\sqrt{x_0^2 - y_0^2}$ measures the numerical advantage of the superior force. This is one form of the *square law*, which was proposed by Lanchester in 1916.

If a superior force z_0 meets a force x_0 and then a second force y_0, the square law indicates that there are $\sqrt{z_0^2 - x_0^2}$ z-survivors after the first battle and $\sqrt{(z_0^2 - x_0^2) - y_0^2}$ after the second. Thus, for the encounter to be on equal terms $z_0^2 = x_0^2 + y_0^2$. This is another version of the square law.

The assumptions underlying the square law are tolerably well fulfilled in old-fashioned naval battles, in which there is no concealment and every ship can act, more or less, without hindrance from its neighbors. Application of the second form of the square law to a naval battle of historical importance is discussed in the problems.

(c) The spread of disease

Differential equations have been applied with success to the study of epidemics and contagious disease. The first important application of this kind was made by Daniel Bernoulli in 1760. Because of its historical significance and because it is carried out at a high level of sophistication, we present Bernoulli's line of thought in full.

The disease in question is smallpox, which for most of recorded history was a serious medical problem. The last known case was in 1978, and the conquest of smallpox is one of the major medical achievements of this century. The disease is extremely contagious, but confers virtually complete immunity on anyone who has caught it and recovered. It is this last characteristic that makes vaccination so effective, and finally made it possible to eradicate the disease. The fact that smallpox confers immunity is an essential part of Bernoulli's analysis.

Bernoulli starts with a population of people all born at time $t = 0$. It is assumed that at time t

$x(t)$ people are alive,

$y(t)$ are alive and have not yet had smallpox.

The latter group, which we refer to as the y-population, are still susceptible to the disease. Suppose susceptible persons contract smallpox at the rate a, so that the y-population decreases from this cause alone at the rate

$$\frac{dy}{dt} = -ay.$$

These people leave the y-population either by getting smallpox and recovering, or by getting it and not recovering. Denote the fraction of the latter by b where $0 < b < 1$. For example if $b = 1/4$ this means that on the average $1/4$ of those getting smallpox die from it. The population as a whole thus declines, from this cause alone, at the rate

$$\frac{dx}{dt} = -aby.$$

Finally, suppose the death rate from all other causes is a time-dependent function $d(t)$ that is the same for both the x- and y- populations. If only this cause were acting, without smallpox, we would have

$$\frac{dx}{dt} = -d(t)\,x, \quad \frac{dy}{dt} = -d(t)\,y.$$

It is a well-accepted principle that rates add in cases such as this, so that

$$\frac{dx}{dt} = -aby - d(t)\,x, \quad \frac{dy}{dt} = -ay - d(t)\,y.$$

These are Bernoulli's equations.

To eliminate the unknown function $d(t)$, multiply the first equation by y and the second by x and subtract. The result is

$$y\,\frac{dx}{dt} - x\,\frac{dy}{dt} = -aby^2 + axy.$$

The form of the left side suggests that $1/y^2$ is an integrating factor. If we multiply by $1/y^2$ the result is

$$\frac{d}{dt}\left(\frac{x}{y}\right) = -ab + a\frac{x}{y}.$$

Thus the ratio $z = x/y$ satisfies the linear equation

$$\frac{dz}{dt} = -ab + az, \qquad z(0) = 1.$$

The initial condition results from an assumption that just after birth no survivors have smallpox (it is generally fatal to infants) so that $x(t) = y(t)$ for small t. The steady-state solution is $z = b$. Hence

$$z = b + ce^{at}, \qquad z = b + (1 - b)e^{at}$$

where the first result is obtained by adding the general solution of the homogeneous equation and the second is obtained from $z(0) = 1$.

Bernoulli estimated that $a = b = 1/8$ and hence that

$$z(20) = \frac{1}{8} + \frac{7}{8}e^{5/2} \approx 11.$$

The estimate indicates that about 9% of the population of twenty-year-olds would be at risk of smallpox. After studying mortality tables, Bernoulli recommended vaccination.

(d) Pollution

The volume of a lake is G cubic feet. An effluent that is wholly polluted runs into the lake at the rate of a cubic feet per minute and clean water runs in at the rate b cubic feet per minute. The lake drains at the rate $a + b$ cubic feet per minute so that the volume remains constant. (We assume that evaporation is, on the average, canceled by rain.) If the effluent distributes itself more or less uniformly before reaching the exit, it is required to find the concentration $P = E/G$, where $E = E(t)$ is the number of cubic feet of effluent in the lake at time t. Clearly E/G is a better measure of pollution than E itself; a cubic foot of polluted effluent is worse in the bathtub than in Lake Superior.

It is conceptually clearer to assume separate sources of water and pollutant, but the effect is the same if the incoming water itself is polluted. In any case $a/(a + b)$ measures the concentration of pollutant in the water entering the lake.

The first and most important step in the solution of this problem is to write the equation of continuity,

increase equals income minus outgo

for the amount of effluent E. In time Δt the increase is $a\Delta t$ from incoming effluent, and it remains to compute the amount lost due to the outward flow. The number of cubic feet leaving is $(a+b)\Delta t$, the concentration of effluent in the lake is E/G, and hence the $(a+b)\Delta t$ cubic feet leaving the lake contain approximately

$$\frac{E}{G}(a+b)\Delta t$$

cubic feet of effluent. (The result is only approximate because E is not constant.) By the equation of continuity,

$$(5) \qquad \Delta E \doteq a\Delta t - (a+b)\frac{E}{G}\Delta t.$$

Here $\doteq$ denotes *approximate equality* in a special sense, which is discussed in DETA page 55. Suffice it to say that a precise meaning can be given to this symbol in equations such as (5) and it is found, in general, that the error disappears completely when we pass from an equation involving $\doteq$ to a differential equation by taking limits.

In the present case dividing (5) by Δt and letting $\Delta t \to 0$ gives

$$\frac{dE}{dt} = a - (a+b)\frac{E}{G}$$

and with $E = PG$ this becomes

$$(6) \qquad G\frac{dP}{dt} = a - (a+b)P.$$

The equation is easily solved by an integrating factor, but it is more instructive to proceed as follows. Setting $P = $ const. gives the solution

$$P_\infty = \frac{a}{a+b}.$$

Add to this the general solution of the homogeneous equation to get

$$(7) \qquad P = P_\infty + ce^{-(a+b)t/G}$$

where the constant c depends on the initial condition. For any initial condition, that is, for any choice of c, Equation (7) gives

$$\lim_{t\to\infty} P(t) = P_\infty.$$

This agrees with the fact that in the long run the concentration of pollutant in the lake will be the same as the concentration $a/(a+b)$ in the sources feeding the lake. Borrowing from the terminology of electrical engineering, we say that P_∞ is the *steady-state solution* and the remaining term in (7) is a *transient* that depends on the initial conditions.

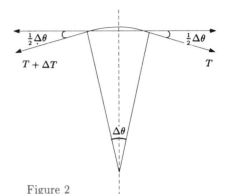

Figure 1   Figure 2

(e) Slipping of a belt

A flexible rope of negligible weight is wrapped around a rotating capstan as shown in Figure 1. A tension T_0 is exerted at one end of the rope in order to raise an anchor at the other end. The more turns around the capstan, the more the capstan helps you raise the anchor. Our problem is to find out how the tension at the anchor end depends on the angle of wrap. The same problem arises in connection with the slipping of a belt on a pully.

Figure 2 shows a small piece of rope with ends at positions specified by the angles θ and $\theta + \Delta\theta$. The statement that the rope is flexible means that the forces due to tension are directed tangentially as in the figure. If these forces have magnitudes T and $T + \Delta T$ the corresponding component of force in a direction normal to the small piece at its center is

$$(8) \qquad \Delta N = (T + \Delta T)\sin\frac{\Delta\theta}{2} + T\sin\frac{\Delta\theta}{2}.$$

L'Hospital's rule

$$\lim_{h \to 0}\frac{\sin h}{h} = \lim_{h \to 0}\frac{\cos h}{1} = 1$$

shows that $\sin h$ is well approximated by h when h is close to 0. We use this with $h = \Delta\theta/2$. We also assume that the tension T is a continuous function of θ, so that $\Delta T \to 0$ as $\Delta\theta \to 0$. These two assumptions in (8) yield the approximate equality

$$\Delta N \doteq 2T\sin\frac{\Delta\theta}{2} \doteq T\Delta\theta, \qquad \Delta\theta \to 0.$$

If the short piece of rope were straight, the frictional force on it would be directed tangentially and would have magnitude $\mu\Delta N$, where μ is the coefficient of static or sliding friction as the case may be. It is an experimental fact that a similar result holds for the curved piece as an approximate equality, namely,

$$\Delta T \doteq \mu\Delta N, \qquad \Delta\theta \to 0.$$

These results yield

$$\Delta T \doteq \mu T \Delta\theta.$$

Dividing by $\Delta\theta$ and letting $\Delta\theta \to 0$ we are led to the differential equation on the left below, with the solution on the right:

$$\frac{dT}{d\theta} = \mu T, \quad T = T_0 e^{\mu\theta}.$$

It is remarkable that the result depends only on the angle of wrap, θ, and is independent of the radius of the capstan.

PROBLEMS

1. According to the square law, if a total force c is divided into two parts a and b, so that $a + b = c$, the effective strength when the parts are used in two separate engagements is $\sqrt{a^2 + b^2}$. Show that this is least when the two parts are equal, and that the effective strength is then $c/\sqrt{2}$.

2. The Battle of Trafalgar in 1805 started an era of British supremacy on the high seas that lasted more than a hundred years and had a profound influence on the science of naval strategy. Nelson's plan of action, written before the battle, assumed 46 ships of the combined French-Spanish fleet against 40 of his own. (The numbers in the actual engagement were somewhat different, but the important thing here is Nelson's intent.) He planned to cut the other fleet exactly in half by use of 16 of his ships, and to attack the two groups of 23 one at a time by means of these 16 ships combined with another 16. The remaining eight ships were to prevent the two groups of 23 from uniting. Note that by Problem 1, a division into equal parts, rather than some other ratio, is optimum.

(a) Check that the 32 ships that Nelson planned to use successively against the two groups of 23 is, within about a half a ship, the correct number to make this part of the battle take place on equal terms.

(b) Check that if Nelson had used all 40 ships against the enemy's 46 ships in a single battle, as was naval practice up to that time, he would have had an effective disadvantage of about 23 ships.

(c) Show, however, that with his plan of $32 + 8$ against $23 + 23$, he had an effective advantage of about $5\frac{1}{2}$ ships.

3. In the pollution problem of the text suppose P has reached its steady-state level P_∞ by time $t = 0$, at which time the source of pollution is removed; in other words, a is replaced by 0. Show that P decays according to the law

$$P = P_\infty e^{-(b/G)t}$$

and hence the time for P to decay to half its initial value is

$$T = (\ln 2)\frac{\text{capacity of lake}}{\text{rate of flow}}$$

These quantities can be measured in any system of compatible units. Note that this problem is mathematically the same as the problem of half-life for radioactive decay we considered in Chapter 1. According to data taken by Rainey in 1967, the ratio multiplying $\ln 2$ is about 185 years for Lake Superior and about 2.6 years for Lake Erie.

4. Suppose a tank contains G gallons of water and that brine containing a pounds of salt per gallon flows into the tank and out again at a constant rate of b gallons per minute, starting at time $t = 0$. The mixture is kept uniform by stirring. Show that the amount $S(t)$ of salt in the tank at time $t \geq 0$ satisfies the differential equation on the left, hence is given by the formula on the right:

$$\frac{dS}{dt} = ab - \frac{b}{G}S, \qquad S(t) = aG(1 - e^{-bt/G})$$

Hint: In the time interval from t to $t + \Delta t$ the number of gallons leaving is $b\Delta t$, the concentration of the mixture in pounds per gallon is S/G, hence the number of pounds leaving is approximately $(S/G)b\Delta t$.

5. In the preceding problem suppose a block of salt in the tank dissolves at the constant rate k pounds per minute. Show that this produces an extra term k on the right side of the differential equation, and hence the effect is the same as if the concentration of the incoming brine were increased from a pounds of salt per gallon to $a + k/b$ pounds of salt per gallon. Do this result and the differential equation of Problem 4 remain valid if a, b, k are continuous functions of t, rather than constants?

6. A sailor raising a 2000 pound anchor with the aid of a capstan finds that a tension of 100 pounds is just enough to keep the rope from slipping. Suppose the rope starts to slip, so that the coefficient of friction changes from its former static-friction value of 0.4 to a value of 0.2 for sliding friction. Show that the sailor must now exert $200\sqrt{5}$ pounds of tension to stop the slipping.

7. Two-link radioactive chain. A radioactive substance A decomposes into a new substance B, which in turn decomposes into a third substance C. Assuming that an amount A_0 of A is present initially, and no B, show that the amount of B at time t is given, in general, by

$$B(t) = aA_0 \frac{e^{-at} - e^{-bt}}{b - a}$$

where a and b are constants giving decay rates of A and B, respectively.

Hint: The rate at which B increases is equal to the rate at which B is formed from A minus the rate at which B decomposes. Thus, using the same letters for the substances and for their amounts,

$$\frac{dB}{dt} = -\frac{dA}{dt} - bB$$

Since $A = A_0 e^{-at}$ you get a linear equation for B. The result for $b = a$ can be obtained by taking the limit as $b \to a$, or also by direct solution, but the probability that $b = a$ in any realistic case is 0.

8. A capacitor of capacity C is discharged through a resistance R as shown in Figure 3. If Q is the charge on the capacitor at time t, v is the voltage across the capacitor, and i is the current in the circuit, show that

$$\frac{dQ}{dt} = -kQ, \quad \frac{dv}{dt} = -kv, \quad \frac{di}{dt} = -ki, \quad k = \frac{1}{RC}$$

Figure 3

Outline of solution: We have $i = dQ/dt$ by the definition of current and $Q = Cv$ by the definition of capacity. The voltage across the resistor is iR by Ohm's law, and $iR + Q/C = 0$ since the voltage drop around the circuit must be 0.

9. The voltage across an inductor is $L\,di/dt$ where L is the inductance, and hence the current i in the RL circuit of Figure 4 satisfies

$$L\frac{di}{dt} + Ri = v_0(t), \qquad i(0) = 0$$

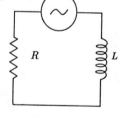

Solve in integral form if $v_0(t)$ is continuous and specialize to the case $v_0 = $ const. The interest of this problem is that the constants R, L, v_0 can be so chosen that the equation exactly duplicates the equations

Figure 4

$$G\frac{dP}{dt} + (a+b)P = a, \qquad G\frac{dS}{dt} = abG - bS$$

obtained in the text and in Problem 4. Before the advent of digital computers, electrical circuits were often used as a means of analog computation.

————————————ANSWERS————————————

5. Yes.   **9.** $i = (1/L)e^{-Rt/L}\int_0^t e^{Rs/L}v_0(s)\,ds, \quad i = (v_0/R)(1 - e^{-Rt/L})$

2. Problems involving geometry

When differential equations are used in scientific investigations, the main ideas sometimes depend on geometry. This is to be expected, since a derivative can be interpreted geometrically as the slope of a curve. Several problems with a geometric flavor are discussed here.

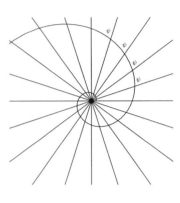

Figure 5

Suppose a curve $r = f(\theta)$ in polar coordinates cuts the radius at a constant angle ψ as shown in Figure 5. We want to get the equation of the curve. If $\psi = 0$ the curve is a ray extending from the origin to ∞ and cannot be represented in the form $r = f(\theta)$. If $\psi = \pi/2$ or $-\pi/2$ the curve is a circle centered at the origin. Setting these cases aside, assume

$$(1) \qquad -\pi/2 < \psi < \pi/2, \qquad \psi \neq 0.$$

Since ψ is constant, $\tan \psi$ is also constant. We write

$$(2) \qquad \tan \psi = \frac{1}{k}$$

Figure 6

where the choice $1/k$ rather than k is made to show the connection with the law of exponential growth discussed in Chapter 1.

The differential triangle in Figure 6 suggests that

$$\tan \psi = \frac{r \, d\theta}{dr}$$

and by elementary calculus this formula is correct whenever the right-hand side is meaningful. Using (2) we get, in succession,

$$\frac{r \, d\theta}{dr} = \frac{1}{k}, \quad \frac{dr}{d\theta} = kr, \quad r = ce^{k\theta}.$$

A curve of this kind is called a *logarithmic spiral* because, apart from constants, $\theta = \ln r$. The steps are reversible and hence a curve $r = f(\theta)$ cuts the radius at a constant angle ψ satisfying (1) if, and only if, it is a logarithmic spiral. Since k is arbitrary, this gives a geometric interpretation of all processes obeying the exponential law of growth.

A logarithmic spiral looks like a snail shell, and this is not just a coincidence. A snail shell has the characteristic that it grows only at one end; the part of the shell already laid down does not change. The growth is specified by its rate in two perpendicular directions, which we take as radial and transverse. Let us suppose that both rates are proportional to the size at time t. The precise meaning of "size" will not matter, and we use the weight W as a quantity that is easily measured. If arc length in the radial direction is s_1 and in the transverse direction is s_2, our hypothesis is

$$(3) \qquad \frac{ds_1}{dt} = k_1 W, \quad \frac{ds_2}{dt} = k_2 W, \qquad k_1, k_2 \text{ const.}$$

By a differential triangle, or also by a familiar formula of calculus, arc length in polar coordinates satisfies

$$ds^2 = dr^2 + (r\,d\theta)^2.$$

Setting $d\theta = 0$ we get $ds_1 = dr$ and setting $dr = 0$ we get $ds_2 = rd\theta$. Thus if the second equation (3) is divided by the first, the result is

$$\frac{r\,d\theta}{dr} = \frac{k_2}{k_1}.$$

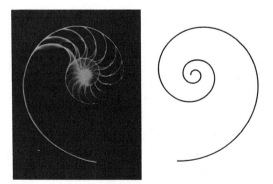

Figure 7

This is the equation we considered in the preceding discussion, with $k = k_1/k_2$. Hence the curve is a logarithmic spiral. Figure 7 shows a photograph of a cross section of a nautilus shell with a logarithmic spiral drawn beside it. The scale and orientation of the spiral are chosen so that the two curves are mirror images of each other. You can check the fit by folding the page.

(b) The parabolic mirror

It is required to find a mirror such that light from a point source at the origin is reflected in a beam parallel to the x axis. The mirror is a surface of revolution whose profile is shown in Figure 8. The ray of light OP strikes the mirror at P and is reflected along PR. If PQ is the tangent at P and $\theta, \phi, \alpha, \beta$ are the angles indicated in the figure, then $\beta = \phi$ because vertical angles are equal and $\beta = \alpha$ by the optical law of reflection. Hence $\alpha = \phi$. Since θ and $\alpha + \phi = 2\phi$ are angles formed by parallel lines meeting a common transversal, we have $\theta = 2\phi$.

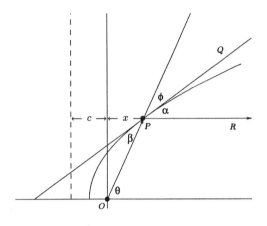

Figure 8

The *signed* angle ψ from the radius to the curve is measured counterclockwise and hence is negative. Thus $\psi = -\phi$. Using the formula

$$\tan \psi = \frac{r \, d\theta}{dr}$$

with $\psi = -\phi$, $\theta = 2\phi$, and $r > 0$, we get

$$-\tan \phi = \frac{2r \, d\phi}{dr}, \quad 2 \cot \phi \, d\phi = -\frac{dr}{r}.$$

If $0 < \phi < \pi/2$ and $r > 0$ as in the figure, integration gives

$$2 \ln \sin \phi = -\ln r + C.$$

The left side is $\ln \sin^2 \phi$ so that, taking exponentials,

$$\sin^2 \phi = \frac{c_1}{r}, \quad c_1 = e^C.$$

By a trigonometric identity $2 \sin^2 \phi = 1 - \cos 2\phi = 1 - \cos \theta$. Hence

(4) $$1 - \cos \theta = \frac{c}{r}, \quad r = \frac{c}{1 - \cos \theta}$$

where $c = 2c_1$. This is the equation of a parabola with focus at the origin and with axis parallel to the x axis. Although we assumed $\theta > 0$ in the derivation, the final result is valid for $|\theta| < \pi/2$, $\theta \neq 0$, and gives the whole parabola. As might be expected, the lower part of the curve, not shown in the figure, is the mirror image of the upper part in the x axis.

The fact that (4) represents a parabola is known from calculus but can be verified directly as follows: Multiplying the left-hand equation (4) by r gives

$$(5) \hspace{4cm} r - x = c$$

since $x = r \cos \theta$. This shows that $r = d$, where $d = x + c$ is the distance from the point (x, y) on the curve to the vertical line $x = -c$. The latter line is the directrix of the parabola and the point $r = 0$ is the focus. Since r is the distance to the focus, we have shown that

distance to focus equals distance to directrix.

Hence the curve is a parabola by definition.

The actual parabolic mirror is obtained by revolving the curve of Figure 8 about the x axis, and we take this point of view from now on. It turns out that the equation $r = x + c$ has an interesting bearing on the relation between *ray optics* and *wave-front optics*. For a point source, the rays are radial lines emanating from the point, and the wave fronts are spherical surfaces centered at the point. The main requirement is that the optical length of the paths leading from the point source to the wave front must be the same on all rays, so the waves arrive in phase. In a uniform medium, which is assumed here, the optical length need not be distinguished from the geometric length.

To apply these ideas to the parabolic mirror, let $x = a$ be a plane perpendicular to the x axis and to the right of the origin as shown in Figure 9. The path length from the focus to this plane is

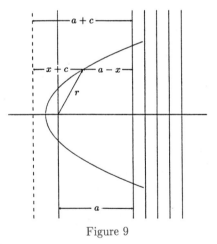

Figure 9

$$r + (a - x) = (x + c) + (a - x) = a + c$$

which is constant. Hence, the plane $x = a$ is a wave front. The parabolic mirror has the effect of converting the *spherical waves* from the point source into *plane waves* parallel to one another.

(c) Orthogonal trajectories

If two families of curves are such that every curve of one family intersects the curves of the other family at right angles, it is said that the two families are *orthogonal trajectories* of each other. For example, the coordinate lines

$$x = c_1, \quad y = c_2$$

in a Cartesian coordinate system form a set of orthogonal trajectories, and so do the circles and radial lines

$$r = c_1, \quad \theta = c_2$$

of a polar coordinate system. In geometric optics the wave fronts are the orthogonal trajectories of the rays, and when heat flows in a uniform plane sheet the lines of flow are orthogonal trajectories of the curves of constant temperature. Further examples of orthogonal trajectories are found in electrostatics, in electromagnetic theory, and in the theory of flow of ideal fluids.

Suppose a given curve in the (x, y) plane is such that the tangent at a point (x, y) on it makes an angle ϕ with the x axis. Then the orthogonal trajectory through the same point (x, y) makes an angle $\phi + \pi/2$ with the x axis. Since

$$\tan(\phi + \pi/2) = -\cot\phi = -\frac{1}{\tan\phi}$$

and since the slope of the curve is $dy/dx = \tan\phi$, we see that we should

$$\text{replace} \quad \frac{dy}{dx} \quad \text{by} \quad -\frac{dx}{dy}$$

in the differential equation for the original family to get the differential equation for the orthogonal trajectories. The process is illustrated in the following example.

EXAMPLE 1. The family of curves

(6) $$x^2 + y^2 = cx$$

represents circles tangent to the y axis as shown in Figure 10. To get the orthogonal trajectories, first differentiate and eliminate c. The result is

$$2x + 2y\frac{dy}{dx} = c$$

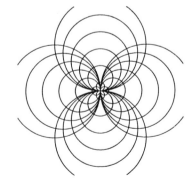

Figure 10

and substituting this value of c into the original equation gives

$$x^2 + y^2 = 2x^2 + 2xy\frac{dy}{dx}$$

which simplifies to

$$y^2 - x^2 = 2xy\,\frac{dy}{dx}.$$

Replace dy/dx by $-dx/dy$ to get the equation of the orthogonal trajectories,

$$y^2 - x^2 = -2xy\,\frac{dx}{dy}.$$

Writing in differential form and putting all terms on the left side gives

$$2xy\,dx - x^2\,dy + y^2\,dy = 0.$$

This equation has the integrating factor $1/y^2$. Multiplying by $1/y^2$ gives

$$d\left(\frac{x^2}{y}\right) + dy = 0, \quad \text{hence} \quad \frac{x^2}{y} + y = c.$$

The result is

$$x^2 + y^2 = cy$$

which represents a family of circles tangent to the x axis as shown in Figure 10. The following comments give useful insight:

Although the analytical steps require

$$x \neq 0, \quad y \neq 0, \quad y^2 \neq x^2$$

the final result is valid without these restrictions.

The quantity c defined by $x^2 + y^2 = cx$ is constant on the original curves of the family, but not constant on the orthogonal trajectories. That is why c must be eliminated in the first step.

Since a horizontal or vertical line that intersects a circle generally intersects it in two points, neither family can be represented in the form

$$y = \phi(x, c) \quad \text{or} \quad x = \psi(y, c).$$

In the study of orthogonal trajectories it is advisable to reinterpret the differential equation by use of differentials, and this was done in the calculation above.

PROBLEMS

1. A family of curves is given by $y = c \cos x$. Compute y', eliminate c by division, and thus verify that the differential equations for the family and for the orthogonal trajectories are respectively

$$\frac{dy}{dx} = -y \tan x, \qquad \frac{dx}{dy} = y \tan x$$

Solve the latter equation by separating variables. Although the analysis assumes $x \neq n\pi/2$ for integral n, these values do no harm. See Figure 11.

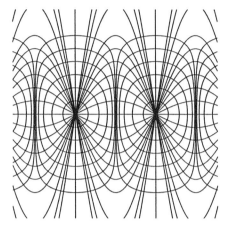

Figure 11

2. Find the orthogonal trajectories:

(a) $x^2 + y^2 = c$ (b) $xy = c$ (c) $y^2 = x + c$ (d) $y = ce^{x^2}$

3. Check that the orthogonal trajectories for $y = ce^x$ and for $y = \ln(cx)$ are given respectively by $2x + y^2 = c$ and $2y + x^2 = c$. Note that one of these families goes into the other when x and y are interchanged, and explain how this property could have been foreseen from the original equations.

4. Verify that the differential equation for orthogonal trajectories of the family of curves $x^2 = 2cy + c^2$ is the same as for the family itself. Such a family is said to be self-orthogonal.

5. Let F_1 be the family of curves whose tangents go through the origin and let F_2 be the family of curves whose normals go through the origin. It is obvious that these families must be orthogonal trajectories of each other. Why is it obvious? Find the families.

6. If the differential equations for a family of curves and for the orthogonal trajectories are written in differential form, it may happen that the one equation goes into the other when x and y are interchanged. Show that this happens for each of the following families, and thus get the orthogonal trajectories by inspection:

(a) $x^3 - 3xy^2 = c$ (b) $y = x + 1 + ce^x$

7. In this problem and the next two, k is a fixed parameter that specifies the family and c varies from one curve to the next in a given family. (a) Show that the orthogonal trajectories of $|y| = c|x|^k$ are ellipses for $k > 0$

and hyperbolas for $k < 0$, with the equation $x^2 + ky^2 = c$. (b) Sketch the original family and the orthogonal trajectories for $k = 2, 1, 0, -2$.

Hint: Assume $x > 0$, $y > 0$ and drop the absolute values. The extension to other quadrants is made by symmetry.

8. If $|k| \neq 1$ show that the following families are orthogonal trajectories of each other:

$$xx^k + yy^k = c, \qquad xx^{-k} - yy^{-k} = c \qquad (x > 0,\ y > 0)$$

When k is an integer, the side condition can be replaced by $x \neq 0$, $y \neq 0$.

9. Find the orthogonal trajectories of $x^k + cy^k = 1$ assuming when necessary that $x > 0$, $y > 0$ or both. What happens if $k = 0$? $k = 1$? $k = 2$?

10. This problem is harder than those above. If a and b are constant and λ is a parameter, show that the family of curves

$$\frac{x^2}{a^2 + \lambda} + \frac{y^2}{b^2 + \lambda} = 1$$

satisfies an equation, free of λ, that is unaltered when y' is replaced by $-1/y'$. This means that the family is self-orthogonal.

─────────────────────ANSWERS─────────────────────

1. $y^2 = \ln(\sin x)^2 + c$
2. (a) $y = cx$ (b) $x^2 - y^2 = c$ (c) $y = ce^{-2x}$ (d) $y^2 + \ln(cx) = 0$
3. It could have been forseen because the original families go into each other if you interchange x and y and replace c by $1/c$.
5. It is obvious because at any point (x, y) common to both curves the tangent to one curve has the same direction as the normal to the other. Hence they are orthogonal. The curves are $y = cx$, $x^2 + y^2 = c$.
6. (a) $y^3 - 3yx^2 = c$ (b) $x = y + 1 + ce^y$
9. $x^2 + y^2 = 2x^{2-k}/(2-k) + c$ in general. If $k = 0$ the problem is nonsensical. If $k = 1$ the family is a set of lines through $(1, 0)$, hence the orthogonal trajectories are circles centered at this point. If $k = 2$ the problem changes character and the trajectories are $x^2 + y^2 = \ln x^2 + c$.
10. $(x + yy')y(y - xy') = (b^2 - a^2)yy'$

───

> Where a mathematical reasoning can be had, it's as great folly to make use of any other, as to grope for a thing in the dark, when you have a candle standing by you.
>
>   —Arbuthnot

Chapter 4

FUNDAMENTAL PRINCIPLES

FOR ANYONE involved in mathematical science, differential equations provide a tool of such power that they are impossible to ignore; and this power increases with each advance in the technology of high-speed computation. Through all the diversity of applications, a few principles and methods are encountered again and again. Among these are the principle of superposition for linear problems, theorems pertaining to existence and uniqueness of solutions, and methods of getting useful information from a differential equation without actually solving it.

1. Linearity

It is a familiar fact that, if u and v are differentiable functions and c is constant, then

$$(cu)' = cu', \quad (u+v)' = u' + v'.$$

In terms of the differential operator $D = d/dx$ the above equations become

$$D(cu) = cDu, \quad D(u+v) = Du + Dv.$$

These results of calculus resemble the algebraic rules

$$p(cu) = cpu, \quad p(u+v) = pu + pv$$

for real numbers p, u, v, c.

An operator T with the two properties described above is said to be *linear*. For example, the operator D is linear. The operation of multiplying by a given function p is also linear.

For precise formulation we assume that T is defined for a class of functions y on an interval I, and that the result Ty of the operation is also a function defined on I. For example, $Dy = y'$ transforms the differentiable function y into another function y'. The class of functions y for which Ty is defined is called the *domain* of T.

Here is the formal definition of linearity:

DEFINITION OF LINEARITY. An operator T is linear if

$$T(cu) = cTu \quad \text{and} \quad T(u+v) = Tu + Tv$$

for all constants c and for all functions u, v in the domain of T.

As an illustration, let us show that the operator T defined by

(1) $$Ty = y' + py$$

is linear, where $p = p(x)$ is a given function, not involving y, defined on an open interval I. The domain of this operator consists of the functions y that are differentiable on I. In terms of T, the linear equation considered in Section 2.4 is $Ty = f$.

It should be emphasized that (1) describes the action of T, not only on y, but on all functions in its domain. For example,

$$Tu = u' + pu, \quad Tv = v' + pv$$

and so on. In view of this, linearity follows from the calculation

$$T(cu) = (cu)' + p(cu) = cu' + pcu = c(u' + pu) = cTu$$

$$T(u + v) = (u + v)' + p(u + v) = u' + v' + pu + pv = Tu + Tv.$$

Here we have used the linearity of the differentiation operator D and of the operation of multiplying by p, as discussed earlier.

Sometimes the fact that an operator is linear is expressed in somewhat different notation. For example, the equations

$$(cu)'' = (cu')' = cu'', \quad (u + v)'' = (u' + v')' = u'' + v''$$

show that the operator T defined by $Ty = y''$ is linear.

The importance of linear operators depends on the following:

GENERAL PRINCIPLE OF SUPERPOSITION. If u satisfies $Tu = f$ and v satisfies $Tv = g$, where T is a linear operator, then

$$y = c_1u + c_2v$$

satisfies $Ty = c_1f + c_2g$ for any choice of the constants c_1 and c_2.

The proof is immediate, since

$$T(c_1u + c_2v) = T(c_1u) + T(c_2v) = c_1Tu + c_2Tv = c_1f + c_2g.$$

The first two equalities follow from linearity and the third from the hypothesis $Tu = f$, $Tv = g$.

We illustrate the concept of linearity by several examples.

EXAMPLE 1. If $p = p(x)$ on I as above, show that the operator T defined by $Ty = y'' + py$ is linear. This follows from

$$T(cu) = (cu)'' + p(cu) = cu'' + cpu = c(u'' + pu) = cTu$$

$$T(u + v) = (u + v)'' + p(u + v) = u'' + v'' + pu + pv = Tu + Tv.$$

EXAMPLE 2. If $Ty = y'' + y$ the functions $u = \sin x$ and $v = \cos x$ obviously satisfy $Ty = 0$. The general principle of superposition shows then that

$$w = c_1 \sin x + c_2 \cos x \quad \text{satisfies} \quad Tw = c_1 0 + c_2 0 = 0.$$

Thus we get a two-parameter family of solutions of $y'' + y = 0$ from the particular solutions $\sin x$ and $\cos x$.

EXAMPLE 3. If $Ty = y' - 2y$ it is seen by inspection that

$$T(1) = -2, \quad T(e^{5x}) = 3e^{5x}, \quad T(e^{2x}) = 0.$$

The general principle of superposition shows then that the function

$$(2) \qquad y = a + be^{5x} + ce^{2x} \quad \text{satisfies} \quad Ty = -2a + 3be^{5x}$$

for any choice of the constants a, b, c. See Example 5 below.

EXAMPLE 4. Using the result of Example 3 find a general solution of

$$(3) \qquad\qquad y' - 2y = 8 + 6e^{5x}.$$

Comparing the right side of (3) with Ty in (2) we set $-2a = 8$ and $3b = 6$. By (2) the desired solution is

$$y = -4 + 2e^{5x} + ce^{2x}.$$

EXAMPLE 5. An expression of the form

$$y = c_1 u_1 + c_2 u_2 + \cdots + c_m u_m,$$

where the c_j are constant, is called a *linear combination* of the functions u_j. If each function u_j satisfies the homogeneous equation $Tu_j = 0$, show that every linear combination of the u_j satisfies the same homogeneous equation.

The result is immediate, since

$$(4) \qquad\qquad Ty = c_1 Tu_1 + c_2 Tu_2 + \cdots + c_m Tu_m$$

by linearity, and each $Tu_j = 0$ by hypothesis. Hence $Ty = 0$, which is what we were to prove. The calculation depends on the fact that the basic linearity property extends to more than two summands by repetition. For example, the property for two summands can be used twice to yield

$$T((c_1u_1+c_2u_2)+c_3u_3) = T(c_1u_1+c_2u_2)+T(c_3u_3) = c_1Tu_1+c_2Tu_2+c_3Tu_3.$$

This gives (4) for $m = 3$. In a like manner

$$T((c_1u_1 + c_2u_2 + c_3u_3) + c_4u_4) = T(c_1u_1 + c_2u_2 + c_3u_3) + T(c_4u_4)$$

and using the result (4) with $m = 3$, we get (4) with $m = 4$. The process continues in an obvious manner and shows that (4) holds whenever the functions u_j are in the domain of the linear operator T.

PROBLEMS

1. In this problem T is defined by $Ty = yy''$.

(a) Check that

$$T(x^2) = 2x^2, \quad T(x^5) = 20x^8, \quad T(\sin x) = -\sin^2 x, \quad T(e^x) = e^{2x}$$

(b) Show that $T(cu) = cTu$ and $T(u + v) = Tu + Tv$ hold for all linear functions u and v, that is, all functions of form $a + bx$ where a and b are constant.

(c) Find a twice-differentiable function u such that $T(2u) \neq 2Tu$. This shows that T is not linear.

2P. Suppose u and v satisfy the equations $Tu = f$ and $Tv = f$, where T is linear. Show that the function $w = u - v$ satisfies $Tw = 0$.

3. If $u' + u\sin x = x^3 \cos 2x$ and $v' + v\sin x = x^3 \cos 2x$ show that

$$u(x) = v(x) + ce^{\cos x}$$

for some constant c. Hint: Use Problem 2 with $Ty = y' + y\sin x$.

4. Let $y = u$ and $y = v$ be two solutions of

$$y' - \frac{e^x}{1 + e^x}y = x^5 \cos x$$

that satisfy $u(0) = v(0) + 2$. Show that $u(x) = v(x) + e^x + 1$.

5. (a) If $Ty = y' + y$ check that $T(1) = 1$, $T(e^{2x}) = 3e^{2x}$, $T(e^{-x}) = 0$, hence

$$T(a + be^{2x} + ce^{-x}) = a + 3be^{2x}$$

for any constants a, b, c. Note that this last equation follows from the general superposition principle; no additional calculation is needed.

(b) If a and b are constant, solve the linear equation

$$y' + y = a + 3be^{2x}$$

by the methods of Chapter 2, and thus confirm the result (a).

6P. If u is a particular solution of $Ty = f$, where T is a linear operator, and if v is any solution of $Tv = 0$, show that $y = u + v$ satisfies $Ty = f$.

REVIEW PROBLEMS————————————————————————————

7. One of the following equations is separable, two are linear in y, and one becomes linear if x and y are interchanged. Decide which is which, and solve:

(a) $y(1 + y^2)\, dx = 2(1 - xy^2)\, dy$ (c) $x(1 + y^2)\, dx = 2e^x y\, dy$

(b) $x\, dy = (8xe^{x^4} - 3y)\, dx$ (d) $x\, dy + (1 - 2x^2)y\, dx = 2x\, dx$

8. One of the equations in Problem 7 admits an integrating factor $1/y$. Which one? Prove your answer, of course.

————————————————————ANSWERS————————————————————

7. (a) $x(1 + y^2) = \log(y^2) + c$ (c) $e^{-x}(1 + x) + \log(1 + y^2) = c$

   (b) $x^3 y = 2e^{x^4} + c$ (d) $xy = ce^{x^2} - 1$ 8. (a)

2. Existence and uniqueness

When a problem from the sciences is formulated in mathematical terms, two demands are usually imposed: The solution should exist, and it should be unique. These demands lead to the subject of existence and uniqueness theorems. The difference between the two kinds of theorems is as follows:

An existence theorem asserts that a problem has *at least* one solution.

A uniqueness theorem asserts that a problem has *at most* one solution.

Although existence theory is not slighted in this text, one can defend the view that existence is less important than uniqueness in many applications. There is nothing wrong with looking for a solution before you know it exists, and if you find it the question of existence is settled then and there.

To bring these remarks into focus, suppose we have a physical system that satisfies

$$(1) \qquad\qquad y'' + y = 0, \qquad y(0) = 0,\ y'(0) = 1.$$

Suppose further that we have noticed that $y = \sin x$ is a solution. Then the question of existence is settled. But as long as uniqueness is in doubt, we cannot say that $y = \sin x$ describes the behavior of the system. The most we can say is: $y = \sin x$ describes a possible behavior of the system.

This example is not artificial. The equation $y'' + y = 0$ describes vibration of a mass-spring system, the current in a circuit having capacitance and inductance, the small oscillations of a pendulum, and small oscillations of an object floating in still water, to mention four examples.

Next, suppose another physical system leads to the problem

$$(2) \qquad\qquad y'' + y = 0, \qquad y(0) = 0,\ y(\pi) = 0.$$

Again $y = \sin x$ is a solution, and again the question of existence is trivial. But there is a spectacular failure of uniqueness, since the system has infinitely many solutions

$$y = c\sin x, \qquad c = \text{const.}$$

Yet this system, like the other, is described by a linear second-order equation with two side conditions. This problem describes small oscillations of a stretched string or the electric field in a microwave resonant cavity. The side conditions $y(0) = 0$, $y(\pi) = 0$ are given on the boundary of the interval in which the differential equation holds, and (2) is an example of a *boundary-value problem*.

Although much of the theory of differential equations is involved with the proof of existence or uniqueness under suitable conditions, in applications a failure of existence or uniqueness can bring important information. For example, the solution $y = \tan x$ of the problem

$$y' = 1 + y^2, \qquad y(0) = 0$$

becomes infinite as $x \to \pi/2$ and does not exist on any interval containing the point $\pi/2$. Such behavior in a chemical system can mean that the reaction leads to an explosion. In the problem of free fall through a resisting medium, the fact that the solution fails to exist precisely when the velocity exceeds a certain critical value v_c means that v_c gives the terminal velocity; that is, the velocity that is approached as the time increases infinitely.

From the point of view of applications, failure of uniqueness is also significant. A point where a presumably unique solution branches into two solutions is called a *bifurcation point*. At such points the system can experience a sudden and sometimes catastrophic change in behavior. For example, when a supporting column in a building is subject to increasing axial pressure, at a certain critical pressure p_c the column suddenly buckles and becomes curved. Mathematically this means that uniqueness fails, and that besides the solution $y = 0$ corresponding to the straight column there is another solution.

When there is a failure of either existence or uniqueness, the trouble is usually caused by the fact that certain functions connected with the differential equation are not bounded. Since the concept of boundedness is important in both the theory and applications of differential equations, we give a formal definition:

DEFINITION OF BOUNDEDNESS. A function f is bounded on a region R if there is a constant M such that $|f(p)| \leq M$ for all p in R.

The precise sense of the word "region" does not matter. For functions $f(x)$ of one variable the region is usually an interval; for functions $f(x, y)$ of two variables with $p = (x, y)$ the region is usually a rectangular region in the plane, and similarly in higher dimensions. What is important is that the inequality $|f(p)| \leq M$ must hold for all values of the variables under consideration. As an example, the function $f(x) = 1/x$ is bounded for $1 \leq x \leq 2$, and also for $1 < x < \infty$, but not for $0 < x < 1$.

Let us now state an existence and uniqueness theorem for the initial-value problem

$$I_1 : \qquad\qquad y' = f(x, y), \qquad y(x_0) = y_0$$

on a two-dimensional rectangular region

$$R_2 : \qquad\qquad a_1 < x < b_1, \quad a_2 < y < b_2.$$

Here are three hypotheses on f:

B: The function f is bounded.
C: The function f is continuous.
D_1 :   The partial derivative $f_y = \partial f/\partial y$ is bounded.

THEOREM 1. *If conditions BCD_1 hold in R_2, the problem I_1 has one, and only one, solution for each point (x_0, y_0) in R_2. The solution can be extended in both directions until it reaches the boundary of R_2.*

The concluding statement about reaching the boundary means that, if $a < x < b$ is the longest interval on which the solution can exist while the point $(x, y(x))$ remains in R_2, then $(x, y(x))$ tends to a definite point on the boundary of R_2 as $x \to a$ through values on (a, b), and also tends to a definite point on the boundary of R_2 as $x \to b$ through values on (a, b). Existence of these limits depends on the hypothesis B that f is bounded. For existence and uniqueness of the solution, conditions CD_1 suffice.

EXAMPLE 1. For the equation $y' = 15y^{2/3}(y-x)^{4/5}$, the function

$$f(x,y) = 15y^{2/3}(y-x)^{4/5}$$

is real and continuous at every point (x, y). Leibniz' rule $(uv)_y = uv_y + vu_y$ gives

$$\frac{\partial f}{\partial y} = 12y^{2/3}(y-x)^{-1/5} + 10(y-x)^{4/5}y^{-1/3}$$

which is undefined when $y = 0$ or $y = x$. The hypothesis of Theorem 1 holds in any rectangular region whose boundary does not intersect the x axis or the line $y = x$. But if $y_0 = 0$ or $y_0 = x_0$, no such region contains (x_0, y_0) and the theorem cannot be applied.

A result similar to Theorem 1 holds for the nth-order initial-value problem. For example, when $n = 2$ the problem is

$$\text{I}_2: \qquad\qquad y'' = f(x, y, y'), \qquad y(x_0) = y_0, \; y'(x_0) = y_1$$

where $f(x, y, z)$ is defined in a three-dimensional rectangular region

$$\text{R}_3: \qquad\qquad a_1 < x < b_1, \quad a_2 < y < b_2, \quad a_3 < z < b_3.$$

The conditions BC are the same as before and D_1 is replaced by

D_2:   The partial derivatives $f_y = \partial f/\partial y$ and $f_z = \partial f/\partial z$ are bounded.

THEOREM 2. *If conditions* BCD_2 *hold, the problem* I_2 *has one, and only one solution for each point* (x_0, y_0, y_1) *in* R_3. *The solution can be extended in both directions until it reaches the boundary of* R_3.

Here the concluding statement means: If $a < x < b$ is the longest interval on which the solution can exist while the point

$$(x, y, z) = (x, y(x), y'(x))$$

remains in R_3, then $(x, y(x), y'(x))$ tends to a definite point on the boundary of R_3 as $x \to a$ through values on (a, b), and also tends to a definite point on the boundary of R_3 as $x \to b$ through values on (a, b). Problem 15 shows that Condition B is essential for existence of these limits. However, existence and uniqueness of the solution follow from CD_2 alone.

Some aspects of the proof of Theorems 1 and 2 are developed in Chapters 14 and A. For a more complete discussion, see DETA, Chapters 14 and F.

EXAMPLE 2. Consider the equation $y'' = 3y(y' - x)^{1/3}$. Setting $y' = z$ we have, in this case,

$$f(x, y, z) = 3y(z - x)^{1/3}.$$

This function is continuous for all (x, y, z) and we check to see if the partial derivatives

$$\frac{\partial f}{\partial y} = 3(z - x)^{1/3}, \quad \frac{\partial f}{\partial z} = y(z - x)^{-2/3} = \frac{y}{(z - x)^{2/3}}$$

are bounded. The first is bounded in every rectangular region R, but the second is unbounded as $z \to x$. Hence, if the initial conditions are of the form

$$y(x_0) = y_0, \quad y'(x_0) = x_0$$

the solution might very well fail to be unique even if it exists.

EXAMPLE 3. If p, q, r are bounded continuous functions of x on an interval (a, b) containing x_0, let

$$y'' + p(x)y' + q(x)y = r(x), \quad y(x_0) = y_0, \ y'(x_0) = y_1.$$

To apply Theorem 2, we solve for y'' and set $z = y'$. The result is

$$f(x, y, z) = r(x) - q(x)y - p(x)z,$$

which is continuous. Since $f_y = -q(x)$ and $f_z = -p(x)$ are assumed bounded, the hypothesis of Theorem 2 holds.

A corresponding result for linear equations of order n is established in DETA, page 341, where it is also shown that the solution can be extended over the whole interval $a < x < b$.

PROBLEMS

1. At what points of the (x, y) plane does the hypothesis Theorem 1 fail for the following equations? Further insight is given by Figure 1 and by Problems 2 and 3.

$$\text{(a) } y' = \sin(xy) \qquad \text{(b) } y' = x^{1/3}y^3 \qquad \text{(c) } y' = (xy)^{1/5}$$

2. In Problem 1(b) verify that $y = 0$ is the only solution that exists on an interval of infinite length. Nevertheless the hypothesis of Theorem 1 holds in every rectangular region, no matter how large.

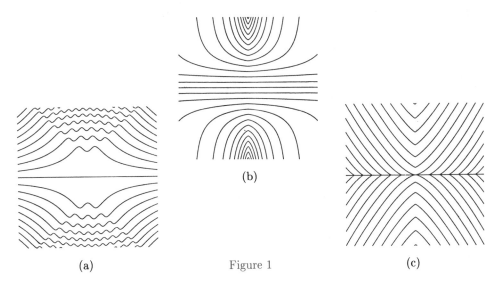

(b)

(a)   Figure 1   (c)

3. This problem requires a hand calculator. The solutions of Problem 1(c) clearly have slope 0 on the axes and yet the graphs in Figure 1(c) seem to have corners there. Near $(0, 1)$ the slope y' is approximately $x^{1/5}$. Compute this for $|x| = .1, .001, .00001$ and thus explain the appearance of the graphs. Similarly show that y' is approximately $\pm(x - 1)^{1/4}$ near $(1, 0)$ and explore numerically. No answer is given.

4. Show that the equation $y' = (y - x)^{1/3}$ does not satisfy the hypothesis of Theorem 1 in any region that intersects the line $y = x$, yet it has a unique solution through every point of that line.

Hint: Set $u = y - x$ and separate variables.

5. At what points of (x, y, z) space does the hypothesis of Theorem 2 fail for the following equations?

(a) $y'' = (xy)^{1/3}$ (b) $y'' = (y' - y)^{1/5}$ (c) $y'' = (yy')^4$ (d) $(\sin x)y'' = e^y$

6. Check that the equation $y' = 3xy^{1/3}$ has two solutions $y = 0$, $y = x^3$ through the origin, although the right-hand side is continuous for all x and y. Why does uniqueness fail?

7. Check that the equation $xy' = y$ has infinitely many solutions through the origin, and find them. Why does uniqueness fail?

8. In what sense does existence fail for $y^2 + (y')^2 = -1$?

9. Show that the equation $y' = 3y^{2/3}$ has a solution $y = (x - c)^3$ where c is any constant, and thus construct a two-parameter family of solutions through the origin.

Hint: Pick $c_1 < 0 < c_2$. Follow a curve of the family to the point $(c_1, 0)$, then follow the x axis to the point $(c_2, 0)$, and finally follow a curve of the family beyond c_2. See Figure 2.

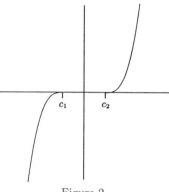

Figure 2

10. Let $f(x) = x/|x|$ for $x \neq 0$ and $f(0) = k$. It will be shown that, no matter how the constant k is chosen, the equation $dy/dx = f(x)$ has no solution on an interval containing the origin. This may serve to explain why, as a rule, a hypothesis of continuity is necessary. Since $f(x) = -1$ for $x < 0$ and $f(x) = 1$ for $x > 0$ we must have

$$y = -x + c_1, \qquad x < 0; \qquad y = x + c_2, \qquad x > 0$$

where c_1 and c_2 are constant. Existence of the derivative $y'(0)$ requires continuity of y at $x = 0$ and hence it requires $c_1 = c_2$. Taking $c_1 = c_2 = c$, show that y is not differentiable at $x = 0$. Hence y does not satisfy the differential equation at $x = 0$, no matter how $f(0)$ is defined.

11. Show that the operator T defined by $Ty = x^2 y'' - 2xy' + 2y$ is linear.

12. Show that the equation $Ty = 0$ in Problem 11 has the two solutions $y = x$, $y = x^2$ and hence the two-parameter family $y = c_1 x + c_2 x^2$. Does the hypothesis of Theorem 2 hold at $x = 0$? Can you use this two-parameter family to get at least the part of the conclusion of Theorem 2 asserting existence at $x = 0$?

13. This problem shows why differential equations are always considered on an interval rather than on a more general set. The equation $xy' + y = 0$ ought to be considered on the interval $(-\infty, 0)$ or on $(0, \infty)$. Suppose, however, that it is considered on the union U of these two intervals. Thus, U is the set of all $x \neq 0$. By separation of variables or by trying $y = x^a$ get the family of solutions $y = \phi(x, c_1, c_2)$ where

$$\phi(x, c_1, c_2) = c_1/x, \qquad x < 0; \qquad \phi(x, c_1, c_2) = c_2/x, \qquad x > 0$$

Thus show that if $x_0 \neq 0$, that is, if $x_0 \in U$, the initial-value problem

$$xy' + y = 0, \quad x \in U; \qquad y(x_0) = y_0$$

has infinitely many solutions.

14. By choosing $1/x = n\pi/2$, where n is an integer, show that the function defined for $x \neq 0$ by $y = \sin(1/x)$ oscillates infinitely often between the values $-1, 0, 1$ as $x \to 0$ and hence has no limit. See Figure 3.

15. For $x \neq 0$, check that the function $y = \sin(1/x)$ of Problem 14 satisfies $(x^2 y')' = -y/x^2$ and hence it satisfies

$$x^4 y'' + 2x^3 y' + y = 0, \qquad x \neq 0$$

This shows that the conclusion asserting existence of a limit at the boundary can fail even for linear equations if the coefficients are unbounded.

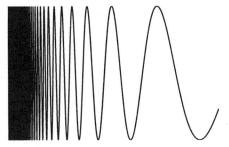

Figure 3

---------------------------------ANSWERS---------------------------------

1. (a) none   (b) none   (c) where $y = 0$

5. (a) where $y = 0$ (c) none

    (b) where $y = z$ (d) where $x = n\pi$, n an integer.

6. Because f_y is undefined when $y \to 0$.

7. Because $y' = y/x$ and the coefficient $1/x$ is unbounded if $x \to 0$.

8. If we solve in the form $y' = f(x, y)$ neither of the possible values for f is real, hence one would not expect real solutions. If complex solutions are allowed, a modified form of the existence theorem would apply.

12. No, since if we solve for y'' the coefficients are unbounded. No, since the only initial values we can satisfy are $y(0) = 0$, $y'(0) = y_1$, which are not general.

---

3. Qualitative behavior

The equation $dy/dx = f(x, y)$ gives the slope of a solution-curve $y = \phi(x)$ at any point of this curve, even before you solve the equation. As an illustration consider the equation

(1) $$\frac{dy}{dx} = (y - 1)(y - 2)(y - 4) = g(y)$$

where $g(y)$ is defined by the product in the center. The sign of $g(y)$ can be determined by use of the following two properties:

(a) If $y > 4$ all factors are positive, hence $g(y) > 0$.

(b) $g(y)$ changes sign at the points 1,2,4 and only there.

The first of these statements is obvious and the second follows from the fact that a continuous function defined on an interval can change sign only by passing through the value 0.

If we plot the points 1,2,4 on the number line the above information is summarized as follows:

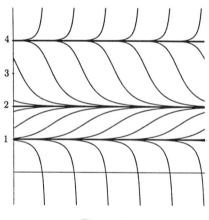

The encircled + indicates that the function is positive for $y > 4$. Starting at the right of 4 we move to the left, changing sign whenever we get to one of the points 4, 2, 1. This produces the three unencircled signs. The arrows below the line indicate that y increases when $dy/dx > 0$. If $dy/dx < 0$ then y decreases and the arrow is directed to the left. Note that the sign of $g(y)$ is also the sign of dy/dx in (1).

Equation (1) has the stationary solutions $y = 1$, $y = 2$, $y = 4$. For example if $y = 2$ then $dy/dx = 0$ and $g(y) = 0$, hence the equation is satisfied. The arrows in the figure indicate that if y is on either of the intervals $(-\infty, 1)$ or $(1, 2)$ adjacent to the stationary value $y = 1$ then y moves away from that value as x increases. For this reason the stationary solution $y = 1$ is said to be *unstable*. The stationary solution $y = 4$ is also unstable. However if y is on either of the intervals $(1, 2)$ or $(2, 4)$ adjacent to the value $y = 2$ then y moves towards the value 2 as x increases, and the stationary solution $y = 2$ is said to be *stable*.

Typical solution-curves for (1) are shown in Figure 4. Note that the stationary solutions $y = 1, 2, 4$ divide the plane into strips, and no solution crosses from one strip into another.

Figure 4

The same sort of thing can be done when the polynomial has factors $(y - y_j)^m$ corresponding to repeated roots. If m is an odd integer such a factor changes sign as y passes through the value y_j, but if m is even, the factor does not change sign. Thus, the first of the factors

$$(y - 3)^5, \quad (y - 3)^6$$

changes sign when y passes from values $y > 3$ to $y < 3$, but the second is positive in either case.

Suppose, for example, that

(2)
$$\frac{dy}{dx} = (y + 1)^3 y^4 (y - 2)^7 (y - 3)^5 (y - 5)^8.$$

The relevant information is summarized as follows:

The small letters e and o above the values $-1, 0, 2, 3, 5$ indicate whether the corresponding zero is of even or odd order. For example, the letter e appears above 5 because the exponent 8 on the factor $y - 5$ is even. For $y > 5$ all factors are positive, hence $g(y) > 0$, and this is indicated by the encircled $+$ sign. Starting with $y > 5$ we move to the left, changing sign whenever we pass a value marked o, but leaving the sign unchanged when we pass a value marked e. This produces the unencircled signs, $+$ or $-$. Note that the results would be the same if $g(y)$ had additional factors corresponding to complex roots, such as

$$(y^2 + 1)(y^2 + 2y + 2)(y^4 + 6).$$

Factors of this kind are always positive, hence they do not alter the pattern of signs.

The arrows below the line show the direction in which y moves on that interval as x increases. The stationary solutions $y = -1$ and $y = 3$ are unstable while $y = 2$ is stable. The stationary solutions $y = 0$ and $y = 5$ are said to be *semistable* because they have one arrow on an adjacent interval pointing toward them and one arrow pointing away from them.

An interval adjacent to a stationary solution, whose arrow points toward that solution, is called an *interval of attraction* for the solution. In this example the stationary solutions $y = 0$ and $y = 5$ each have one interval of attraction, $y = 2$ has two, and $y = -1$ and $y = 3$ have none.

The following theorem supplements the foregoing analysis:

THEOREM 1. *Let $dy/dx = g(y)$ where $g(y)$ is a real polynomial whose real zeros are y_1, y_2, ..., y_m; there may also be complex zeros but these do not affect the result. Then:*

(a) If a solution $y = \phi(x)$ meets a stationary solution $y = y_j$, it coincides with that stationary solution on its whole interval of definition.

(b) If a solution $y = \phi(x)$ starts on an interval of attraction for a stationary solution y_j, then $\lim_{x \to \infty} \phi(x) = y_j$.

The result (a) is easily proved. Since y_j is a zero of $g(y)$ we can write

$$g(y) = (y - y_j)h(y)$$

where $h(y)$ is a polynomial. Hence our solution $y = \phi(x)$ satisfies

$$\frac{dy}{dx} = (y - y_j)h(y)$$

or also

$$\frac{dy}{dx} = (y - y_j)f(x) \quad \text{where} \quad f(x) = h(\phi(x)).$$

Since $f(x)$ is continuous we can solve in the usual way and deduce

$$y - y_j = ce^{F(x)} \quad \text{where} \quad F(x) = \int f(x)dx.$$

Since $y = y_j$ at some value x_0 by hypothesis, the constant c must be 0. Hence $y = y_j$, as was to be shown. This gives Part (a). A result that contains Part (b) as a special case is established in DETA, page 646.

PROBLEMS

1P. An equation $dy/dx = g(y)$ of the type considered in the text is in standard form if each factor is of the form $y - y_j$ rather than $y_j - y$.

(a) Write the following equation in standard form:

$$\frac{dy}{dx} = (8 - y)^4(2 - y)(y - 6)^3(4 - y)^2$$

(b) Plot the points 2,4,6,8 on the number line accompanied by letters e or o to indicate whether the corresponding zero is of even or odd order.

(c) If $y > 8$ all factors $y - y_j$ are positive but the function is prefixed by a minus sign (assuming you did Part (a) correctly) and hence it is negative. Starting with an encircled minus sign at $y > 8$ move to the left, changing sign when you get to a zero of odd order. Thus make a diagram similar to those in the text. In particular, classify the four stationary solutions as stable, unstable, or semistable.

(d) As $x \to \infty$, what happens to solutions $y = \phi(x)$ that pass through the points $(x, y) = (5, 3)$, $(-4, 7)$, $(15, 5)$, $(-7, 9)$? Hint: Use Theorem 1(b).

2. If $y' = dy/dx$, determine the stationary solutions, the nature of their stability, and their intervals of attraction:

(a) $y' = (y-3)(y-5)$ (b) $y' = (1-y)(y-2)^3$ (c) $y' = y(5-y)(y-4)^2$

3. Proceed as in Problem 2:

(a) $y' = (1 + y^2)(y - 3)^3(y - 5)^4$ (c) $y' = (y^2 - 1)(2 - y)(y + 4)^5$
(b) $y' = (1 - y)(3 - y)(4 - y)$ (d) $y' = (y^2 + 3y + 2)(y^4 - 16)$

4. (a) If $y = \phi(x)$ satisfies $dy/dx = g(y)$, show that $y = \phi(x - c)$ satisfies the same equation for any choice of the constant c. This means that any particular solution $y = \phi(x)$ generates a whole family of solutions by horizontal translation. If the original solution $y = \phi(x)$ lies in some strip $y_{j-1} < y < y_j$ determined by the stationary values y_j, the process gives all solutions in that strip.

(b) If x and y are interchanged, the nonlinear equation $dy/dx = g(y)$ becomes a linear equation of a very simple type, namely, $dy/dx = 1/g(x)$. Interpret the result (a) in this context.

5. (abc) Sometimes the function $g(y)$ in $dy/dx = g(y)$ is given as a graph. Since the intervals of positivity can be read from the graph, you can apply the methods of this section. A point y_0 where the graph of g crosses the y axis gives a stationary solution $y = y_0$, and a nonzero maximum or minimum of $g(y)$ gives a point of inflection on the graph of the solution. Using these remarks as a clue, plot y when $g(y)$ is as in Figure 5.

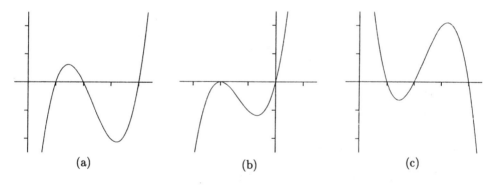

(a)   (b)   (c)

Figure 5

REVIEW PROBLEMS

Here are twelve equations, for which you are to make a list of twelve letters a,b, $\cdots$, l:

(a) $dy + 2xy\,dx = e^{-x^2}\,dx$

(b) $(1 + x)\,dy = (1 + y)\,dx$

(c) $dx + 4y^3x\,dy = y^7\,dy$

(d) $dx + x\,dy = x^3y\,dy$

(e) $y^3d(xy) = x^2\,dx$

(f) $d(xy) = x^6y^3\,dy$

(g) $d(xy) = y^2\sin y\,dy$

(h) $d(xy) = y^2\log x\,dx$

(i) $(\sin x\cos^2 y)dx = \cos^2 x\,dy$

(j) $x\,dy + y\,dx = e^{xy}\,dx$

(k) $dy + y\cos x\,dx = \cos^3 x\,dx$

(l) $x\,dx + y\,dy = \sec(x^2 + y^2)\,dy$

6. Which are separable? Cross them off your list.

7. Of those remaining, which are linear? Cross them off your list.

8. Among those remaining, which are of Bernoulli type? Cross them off.

9. Some of those remaining become linear when x and y are interchanged. Which are they? Cross them off.

10. Some of those remaining become of Bernoulli type when x and y are interchanged. Which? Cross them off.

11. If all has gone well, you should have two equations remaining that do not belong in any of the categories above. They can be solved, however, almost by inspection. Which are they, and what are their solutions?

12. Assuming that $f(x)$ is continuous for all relevant x, solve the following equations with the initial condition $y(0) = y_0$:

$$\text{(a) } y' - (\tan x)y = f(x) \qquad \text{(b) } y' + (\tan x)y = f(x)$$

What, if any, are the limitations on the interval of existence?

13. (a) Show that the solution-curves of the initial-value problem

$$y' = 3y^{2/3}, \quad y(0) = 0$$

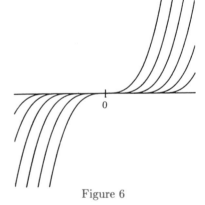

entirely fill up the region between the curve $y = x^3$ and the x axis. That is, if (x_1, y_1) is any point in this region, a solution satisfies $y(x_1) = y_1$. See Figure 6.

Figure 6

(b) Show that if the equation $y' = 3y^{2/3}$ is considered in the whole (x, y) plane, there are infinitely many solutions through each point (x_0, y_0). The two facts illustrated here are typical of what happens when uniqueness fails.

Hint for both parts: See Section 2, Problem 9.

14. (a) Show that the equation $y' = y - y^2$ becomes a linear equation upon the substitution $y = 1/u$, and thus solve it.

(b) Solve the same equation by separation of variables.

(c) Verify the equivalence of the results. No answer is given.

15. Check that the solution of $y'' + 2xy' = 0$, $y(0) = 0$, $y'(0) = 2/\sqrt{\pi}$ is

$$y = \text{erf } x \quad \text{where} \quad \text{erf } x = \frac{2}{\sqrt{\pi}} \int_0^x e^{-t^2} dt$$

The function erf x is called the *error function*; it is important in probability.

————————————ANSWERS————————————

1. (c) 2 unstable, 4 semistable, 6 stable, 8 semistable
   (d) $y \to 4$, $y \to 6$, $y \to 6$, $y \to 8$

2. Partial answer: (a) 3 stable, 5 unstable   (b) 1 unstable, 2 stable
   (c) 0 unstable, 4 semistable, 5 stable

3. Partial answer: (a) 3 unstable, 5 semistable
   (b) 1 stable, 3 unstable, 4 stable
   (c) -4 unstable, -1 stable, 1 unstable, 2 stable
   (d) -2 semistable, -1 stable, 2 unstable

4. (b) The result (a) expresses the fact that all solutions of $dy/dx = 1/g(x)$ on an interval can be obtained by adding a constant to any particular solution.

5. Partial answer: The function $g(y)$ in (a) is the same as that in (1) and the trajectories are shown in Figure 4.

6. b,i   7. a,k   8. e,h   9. c,g   10. d,f

11. j: $x + e^{-xy} = c$,   l: $\sin(x^2 + y^2) - 2y = c$

12. (a) $y = (\sec x) \int_0^x f(t) \cos t \, dt + y_0 \sec x$, $-\pi/2 < x < \pi/2$
    (b) $y = (\cos x) \int_0^x f(t) \sec t \, dt + y_0 \cos x$, $-\pi/2 < x < \pi/2$

Mathematics takes us into the region of absolute necessity, to which not only the actual world but every possible world must conform.

—Bertrand Russell

Chapter 5

APPLICATIONS

THE FACT THAT scientific problems lead to differential equations has already been illustrated in this text, and further examples are given here. Among these are problems of population growth, endangered species and harvesting in ecology, and a number of problems in mathematical physics. Our goal is not to write a treatise on ecology or physics, however, but merely to indicate the diversity of the scientific problems to which differential equations can be applied.

1. The logistic equation

The exponential law for population growth goes back to an essay that appeared anonymously in 1798, but was actually written by the English economist Thomas Malthus. His main thesis was that human misery is inevitable, because population tends to grow geometrically while the means of subsistence grows only arithmetically. From the point of view of differential equations, geometric growth means $dP/dt = kP$ and arithmetic growth means $dP/dt = k$.

In 1845 the Belgian mathematician Pierre Verhulst gave a number of reasons for thinking that the growth constant k should not remain constant indefinitely, but should diminish like $M - P$ where M is the theoretical maximum for the number P of individuals in the population. The Verhulst equation is

$$(1) \qquad \frac{dP}{dt} = k(M - P)P, \qquad P(0) = P_0$$

where k and M are positive constants. Verhulst called the solutions *logistic curves* from a Greek word meaning "skilled in computation" and (1) is often referred to as the *logistic equation*.

Three of the many uses of the logistic equation are discussed in the following problems. The first is the analysis of population growth in the study of demography. The logistic curves involve k, M, and the constant of integration c. Knowing the population P at three times one can determine these constants and thus predict future growth, including the maximum. The computation is easy when the times at which the population is known are equally spaced, and that is the case we consider.

The second application pertains to the problem of endangered species, of which the California condor is an outstanding example. For most species there is a threshold L such that if the population falls below L it cannot reproduce and becomes extinct. This means that the population drops to zero in a finite time. The situation is described by the equation

(2)
$$\frac{dQ}{dt} = k(M - Q)(Q - L), \qquad 0 < L \leq M.$$

If Q satisfies (2) let $P = Q - L$. Then $Q = L + P$ and hence

(3)
$$\frac{dP}{dt} = k(M - L - P)P$$

which is the same as (1) with M replaced by $M - L$. Hence (1) applies to the threshold problem.

The third application is to problems of harvesting. We assume that the population satisfies the logistic equation and we consider harvesting either at a constant rate or at a rate proportional to the population size. The questions are: How much harvesting can be done without driving the population to extinction, or (which is essentially the same thing) what is the maximum rate of harvesting possible in the steady state? To see how these questions lead to the logistic equation, suppose the unharvested population satisfies (1). If the rate of harvesting is proportional to the population size there is an extra term $-kHP$, where the positive constant H measures the vigor of the harvesting effort as a fraction of the growth constant k. Adding this negative term to the right-hand side of (1) gives

$$\frac{dP}{dt} = k(M - H - P)P.$$

Hence the effect of the harvesting is to change M to $M - H$. In the accompanying problems it is seen that similar considerations apply when the harvesting is at a constant rate or when the basic population equation is the threshold equation (2) rather than the logistic equation (1). In all such cases we are led back to (1) with a new choice of parameters.

The rest of this section is devoted to a study of (1). Equation (1) has the stationary solutions $P = 0$ and $P = M$. The graphs of these solutions are horizontal lines that divide the plane into three regions: the half plane $P < 0$, the strip $0 < P < M$, and the half plane $M < P$. No solution can cross either of these horizontal lines, because such a crossing would violate uniqueness; see Section 4.3, Theorem 1.

In Problem 1 it is seen that the substitution $P = 1/p$ leads to a linear equation that can be solved more or less by inspection. The result is

(4)
$$\frac{1}{P} = \frac{1}{M} + \frac{c}{e^{kMt}}.$$

The constant c is given by the initial condition as

(5)
$$c = \frac{1}{P_0} - \frac{1}{M}.$$

If $P_0 > 0$ then $P(t)$ is defined for $t > 0$ and (4) gives

$$\lim_{t \to \infty} P(t) = M.$$

We can get additional information by considering the sign of P' and P'' as obtained from the differential equation

$$P' = k(M - P)P = k(MP - P^2).$$

Differentiating the second form above gives

(6) $$P'' = k(M - 2P)P'$$

by the chain rule. We mark off points $0, M/2, M$ on the number line and construct the following chart:

location of P	$(-\infty, 0)$	$(0, M/2)$	$(M/2, M)$	(M, ∞)
sign of P'	$-$	$+$	$+$	$-$
sign of $M - 2P$	$+$	$+$	$-$	$-$
sign of P''	$-$	$+$	$-$	$+$

The second line is filled in by looking at $(M - P)P$, the third is obvious, and the fourth is the product of the preceding two in accord with (6). From this chart we can read off the sense of monotonicity and concavity of the graph $y = P(t)$ and confirm the behavior shown in Figure 1. In particular, if $0 < P < M$ the chart shows that P is increasing and hence M is approached from below as an asymptotic maximum. If $M < P$, on the other hand, the chart shows that P is decreasing and M is approached from above as an asymptotic minimum.

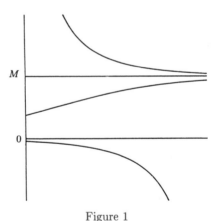

Figure 1

The last line of the table shows that P'' changes sign at $M/2$ and hence the curve has an inflection point. It is most interesting that the value of P there is half the maximum, no matter what k and c may be.

The above remarks apply to the threshold equation with $P = Q - L$ and with M replaced by $M - L$. The main new feature is that the condition $Q_0 < L$ leads to $P_0 < 0$, which we have not considered up to now. We shall show that if $Q_0 < L$ then $Q(t)$ vanishes at a certain time T. Since $Q = 0$ means that the population is extinct, the time T is called the *extinction time*.

We can get a formula for the extinction time as follows. When $Q_0 < L$ and $Q(T) = 0$ the interval

$$Q_0 \geq Q \geq 0 \quad \text{corresponds to} \quad 0 \leq t \leq T.$$

Separate variables in (2) and integrate between corresponding limits to get

$$\int_0^T k\,dt = \int_{Q_0}^0 \frac{dQ}{(M - Q)(Q - L)} = \frac{1}{M - L} \ln \frac{L - Q}{M - Q} \bigg|_{Q_0}^0.$$

The integral was evaluated by expanding the integrand in partial fractions. We summarize as follows:

EXTINCTION THEOREM. Suppose a population Q satisfies the threshold equation

$$\frac{dQ}{dt} = k(M - Q)(Q - L), \qquad Q(0) = Q_0$$

where $k > 0$ and $0 < L < M$. Suppose further that $Q_0 < L$. Then the population becomes extinct at a time T given by

$$kT = \frac{1}{M - L} \ln \left(\frac{M - Q_0}{L - Q_0} \frac{L}{M} \right).$$

Instead of taking $Q_0 = Q(0)$ we could take $Q_0 = Q(t_0)$ and the conclusion would be that the population becomes extinct at the time $t_0 + T$. The gist of the result is that *if the population ever falls below L, it goes to extinction.* That is why L is called the threshold.

The logistic equation applies not only to populations of fish and animals, including man, but also to such things as yeast plants and the growth of sunflowers. A specific numerical equation may be realistic if the population is measured in units of 1000 and the time in days, or in units of 100,000,000 and the time in decades. Changing the units changes the numerical values of the coefficients. With this thought in mind, we have chosen simple coefficients in most of the following problems.

PROBLEMS

1. The logistic equation is a Bernoulli equation with $1/(1 - n) = -1$. Show that the substitution $P = 1/p$ leads to the equation on the left and hence, with $m = 1/M$ and $C = cM$, to the two formulas on the right:

$$p' + kMp = k, \qquad p = m + ce^{-kMt}, \qquad P = \frac{M}{1 + Ce^{-kMt}}$$

2. This problem illustrates essentially everything a logistic curve can do. On a single diagram of large size sketch the solutions of

$$P' = P(1 - P), \qquad t \geq 0; \qquad P(0) = -1, \quad 0, \quad \frac{1}{8}, \quad 1, \quad 2$$

One of these five curves exists only on a finite interval $0 \leq t < t_1$. Which one, and what is the value of t_1?

Suggestion: For all but one case, use the right-hand formula of Problem 1. The value of C is given by (5) as $C = (M/P_0) - 1$.

3. A population P satisfying the logistic equation is known at three equally spaced times t_0, t_0+d, t_0+2d. Without loss of generality $t_0 = 0$; this amounts to choosing, temporarily, a new origin for the time coordinate. Thus

$$P(0) = P_0, \qquad P(d) = P_1, \qquad P(2d) = P_2$$

are known. To get a formula for $P(t)$, let

$$p = \frac{1}{P}, \qquad p_i = \frac{1}{P_i}, \qquad m = \frac{1}{M}, \qquad p(t) = m + ce^{-kMt}$$

where the last equation follows from Problem 1. With $x = e^{-kMd}$ check that

$$p_0 = m + c, \qquad p_1 = m + cx, \qquad p_2 = m + cx^2$$

and hence that

$$x = \frac{p_1 - p_2}{p_0 - p_1}, \qquad c = \frac{p_0 - p_1}{1 - x}, \qquad m = p_0 - c, \qquad kM = \frac{-\ln x}{d}$$

4. Suppose a population measured in units of $10,000,000$ satisfies the logistic equation and has the values $1/7, 1/5, 1/4$ at $1850, 1900, 1950$, respectively.

(a) Show that $x = 1/2$, $c = 4$, $m = 3$, $kM = (\ln 2)/50$ in the notation of Problem 3. Thus get the formula

$$P(t) = \frac{1}{3 + 4 \cdot 2^{-s/50}}, \qquad s = t - 1850$$

(b) Check that the formula gives the right result when $t = 1850, 1900, 1950$.

(c) Show that $P(2000)/P(1800)$ is a good approximation to π.

5. The population of the United States in units of $100,000,000$ is given by the following table (courtesy Department of Commerce, Bureau of Census):

year	1820	1860	1900	1940	1980	1985
pop.	.0964	.3144	.7599	1.317	2.265	2.377

Make a corresponding table of reciprocals $p_i = 1/P_i$ and, assuming that $y = P(t)$ is a logistic curve, get two formulas, one based on $1820, 1900, 1980$, and one on $1900, 1940, 1980$.

(a) How well does the first curve match the values at 1860 and 1940?

(b) Which curve gives a better match at 1985?

(c) What maximum M is predicted by the two curves and what do you think is the cause of this astonishing discrepancy?

Note: Alaska and Hawaii are included in the last two figures but not in the first four. This makes a difference of .014 in 1980 and .016 in 1985.

6. This problem pertains to the harvesting equation $P' = k(M - H - P)P$.

(a) If $H = M$ show that $P(t) = P_0/(1 + kP_0t)$, hence $P(t) \to 0$ as $t \to \infty$.

(b) If $H > M$ show that $P(t) \to 0$ at a much faster rate than in (a), but $P(t)$ does not vanish completely at any finite time T.

(c) If $H < M$ show that $P(t) \to M - H$ as $t \to \infty$, hence the steady-state yield kHP is $kH(M - H)$.

(d) Show that $kH(M - H)$ in Part (c) is maximum when $H = M/2$. The maximum value $kM^2/4$ is called the *maximum sustainable yield*.

(e) Show that the maximum in (d) is the same as the maximum value of P' for the unharvested population when $0 < P < M$.

Hint for (b) and (c): Use the result of Problem 1 with $M - H$ instead of M.

7. This problem pertains to the threshold equation (2) with $L = M$, which is a condition of zero probability but leads to simple solutions.

(a) If $Q(0) > M$ show that $Q(t) \to M$ as $t \to \infty$.

(b) If $Q(0) < M$ find the time T at which the population becomes extinct.

──────────────────────────────ANSWERS──────────────────────────────

2. The first, $t_1 = \ln 2$.   4. (c) $22/7$

5. (a) 0.289, 1.54 (b) The value 2.32 is not quite as good as 2.42 but the average of these two values is nearly exact.   (c) Partial answer: 2.86, 52.6

7. (b) $kMT = Q_0/(M - Q_0)$ where $Q_0 = Q(0)$

2. Physical problems

The first of the topics considered here is an elaboration of the following question: If a stone is dropped from height h, with what speed does it hit the ground? One could find the position at any time t and differentiate to get the speed, but it is more efficient to reason as follows: Let m be the mass of the stone and g the acceleration of gravity. Then the weight of the stone is mg and the work needed to raise it through the height h is mgh. If it drops from rest and hits with velocity v_0, the gain in kinetic energy is $(1/2)mv_0^2$. Equating the gain in kinetic energy to the work needed to raise the stone gives

$$\frac{1}{2}mv_0^2 = mgh \quad \text{or} \quad |v_0| = \sqrt{2gh}.$$

(a) Conservation of energy

Suppose an object of mass m moves in a straight line under a force $F(s)$, where s is the displacement from the origin and F is a continuous function. It is required to deduce a first-order equation for the velocity $v = ds/dt$ as a function of s. Newton's law

$$(\text{mass})(\text{acceleration}) = \text{force}$$

gives

(1) $$ms'' = F(s)$$

where the primes denote differentiation with respect to the time t.

We shall use the transformation $y' = v$, $y'' = v\,dv/dy$ that was introduced in Chapter 2. In the present case $y = s$, $v = ds/dt$, and the result is

$$mv\frac{dv}{ds} = F(s) \quad \text{or} \quad mv\,dv = F(s)\,ds.$$

Let us suppose that this holds on an interval $s_0 \le s \le s_1$, that $v = v_0$ when $s = s_0$, and that $v = v_1$ when $s = s_1$. Integrating between corresponding limits yields

(2) $$\frac{1}{2}mv^2\Big|_{v_0}^{v_1} = \int_{s_0}^{s_1} F(s)\,ds.$$

This is the *principle of conservation of energy* as applied to the equation (1).

One can also go in the opposite direction, if (2) is given for variable upper limits $v_1 = v$ and $s_1 = s$, and deduce (1) at any point where $v \ne 0$. Thus Newton's law is equivalent to the principle of conservation of energy on any interval on which the velocity does not vanish and the force F is a continuous function of displacement s.

To get a physical interpretation, recall that in moving a distance against a constant force the work is given by

$$\text{work} = (\text{force})(\text{distance}).$$

Hence $F(s)\Delta s$ is the work done in moving through a distance Δs against the force $F(s)$, and by a limit process with which you are probably familiar, the integral on the right of (2) represents the work done in moving from s_0 to s_1 against the variable force F. To interpret the left side, $(1/2)mv^2$ represents the *kinetic energy* due to the fact that the object is in motion. Then the principle of conservation of energy says:

$$(\text{change in kinetic energy}) = (\text{work done by the force}).$$

A rocket is fired vertically upwards and its motor is turned off as soon as it is high enough so that air resistance can be neglected — about 70 miles. If its velocity when the motor is turned off is v_0, what value v_0 just ensures that the rocket will not return to earth? This problem will be solved here by use of the principle of conservation of energy. The velocity needed to escape the earth's gravitational field is called the *escape velocity*.

Let m be the mass of the rocket and m_e the mass of the earth. By Newton's law of gravitation the force on the rocket due to gravity is

$$F(r) = -\frac{Gmm_e}{r^2}$$

where r is the distance to the earth's center and G is a constant, called the gravitational constant, whose value need not concern us in this problem. The minus sign is used because the positive direction is upwards, away from the earth's center.

The force at the earth's surface is $-mg$ and hence, if r_e is the radius of the earth,

$$\frac{-Gmm_e}{r_e^2} = -mg.$$

This gives G and shows that the force at a distance $r \geq r_e$ from the earth's center is

$$F = -mg\left(\frac{r_e}{r}\right)^2.$$

As a check and an aid to memory, note that the expression is of the form k/r^2, as it should be, and it gives the right value when $r = r_e$.

Although the rocket's coasting trip starts somewhat above the earth's surface, so air resistance can be neglected, we do the problem as if it starts at $r = r_e$. The work needed to raise the rocket to an unlimited height is

$$\int_{r_e}^{\infty} mg \left(\frac{r_e}{r}\right)^2 dr = \lim_{r \to \infty} mgr_e^2 \left(-\frac{1}{r} + \frac{1}{r_e}\right) = mgr_e.$$

By the principle of conservation of energy, this should be just enough to equal the kinetic energy $(1/2)mv_0^2$. Hence

$$\frac{1}{2}mv_0^2 = mgr_e \quad \text{or} \quad v_0 = \sqrt{2gr_e}.$$

Comparing with the formula $v_0 = \sqrt{2gh}$ mentioned at the beginning of this section, we see that the escape velocity is the same as would be attained in a free fall through a height $h = r_e$ in a constant gravitational field equal to that at the earth's surface.

For numerical approximation let us ignore the extra height needed to get above the atmosphere and let us take $g \approx 32$ ft/sec^2, $r_e \approx 4000$ miles. These values give

$$v_0^2 = 2gr_e \approx \left(\frac{64 \text{ ft}}{\text{sec}^2}\right)(4000 \text{ mi})\left(\frac{1 \text{ mi}}{5280 \text{ ft}}\right).$$

Hence the escape velocity is about 7 miles per second.

(c) Maximum height

Suppose that a stone is thrown vertically upward with velocity v_0 and that air resistance is not neglected. What is the maximum height attained? The point of this problem is that besides the force $-mg$ on the stone due to gravity there is a retarding force that depends on the velocity v. For viscous damping the force is $k|v|$ where k is constant. For power-law damping it is $k|v|^a$ where k and a are constant. The special case kv^2 is called *Newtonian damping*.

Here we assume, more generally, that the retarding force is a continuous function $k(v) \geq 0$. If s is the height of the stone measured positively upward and m is its mass, Newton's law gives

$$ms'' = -mg - k(v).$$

The minus sign is used on $-k(v)$ because v is positive as long as the stone is moving upward, and the resistance force has a sign opposite to the sign of v. This equation is of the form $s'' = g(s')$, and not $s'' = g(s)$ as in the previous discussion. Nevertheless, we shall use the same substitution $s'' = v \, dv/ds$. The result is

$$m v \frac{dv}{ds} = -mg - k(v), \quad \text{hence} \quad \frac{mv \, dv}{mg + k(v)} = -ds.$$

Since the stone is thrown upward with velocity v_0 we have $v = v_0$ when $s = 0$. Also $v = 0$ at the highest point. Denoting the maximum height by h and integrating between corresponding limits, we get

$$(3) \qquad\qquad \int_0^{v_0} \frac{mv \, dv}{mg + k(v)} = \int_h^0 -ds = h.$$

This is an explicit formula for the maximum height h.

It is of some interest to compare this result with the formula $v_0 = \sqrt{2gh}$ for free fall through a distance h. If $k(v) = 0$ the formula (3) gives $v_0^2/(2g) = h$, which is consistent with the former result. Thus, in the absence of air resistance, we get the same pair of values (h, v_0) if a stone is dropped from height h and hits the ground with speed v_0, or if it is thrown upward with speed v_0 and attains the maximum height h.

Most people would consider this equivalence to be rather obvious. However, such a result is never to be expected in the presence of air resistance. To see why, suppose you make a parachute jump from 10,000 feet and land with the same speed $|v_0|$ you would acquire in jumping off a ten-foot building. If you now reverse the process, and jump upwards with speed $|v_0|$, you would only go back the 10 feet to the top of the building (and not the 10,000 feet to your starting point). Still less, if you are dragging the parachute behind you.

(d) The Torricelli-Borda law

Evangelista Torricelli is known mainly for his invention of the barometer in 1643. However he also stated a law of fluid flow under gravity when a tank is drained through a hole in its bottom. Namely, the exit speed from the hole is the same as would be attained by a free fall through a vertical distance equal to the head. Here the "head" means the depth y of the hole beneath the surface of the water. Since a stone dropped from height h hits the ground with speed $\sqrt{2gh}$, the Torricelli law gives

$$(4) \qquad\qquad \text{(exit speed at depth } y) = \sqrt{2gy}.$$

It was observed by Torricelli's contemporary Borda that the fluid stream has a smaller diameter than the aperture, so that the effective size of the latter is diminished. Specifically, if the cross-section area of the exit pipe is

a the effective area is ba where the *Borda constant* b is usually between $1/2$ and 1. For water, b is about 0.6. We summarize as follows:

(5)   (effective aperture size) $= ba$, $0 < b \leq 1$.

It is a general principle of physics that the rate of flow in a pipe is

(6)   (rate of flow)=(effective cross section)(speed of fluid flow).

The units depend on the units used to measure speed and cross-section area. For example, if the speed is in feet per second and the area in square feet, the rate of flow is in cubic feet per second. To see why (6) is reasonable, let the speed of flow be S and the effective cross-section area A. In time Δt a cylinder of water of length $S\Delta t$ and of base-area A exits from the pipe, and hence the change in volume is $\Delta V = AS\Delta t$. This gives $dV/dt = AS$, which is what the principle says. When the principle is combined with (4) and (5) the result is the *Torricelli-Borda law*

(7)   (rate of flow) $= (ba)\sqrt{2gy} = k\sqrt{y}$, $k = ab\sqrt{2g}$.

Note that a single constant k takes account of the size of the drain pipe as measured by a, of the nature of the fluid as measured by b, and even of the acceleration of gravity, g. (If the tank were drained on the moon, g and hence k would change accordingly.)

(e) Draining a tank

Figure 2

Figure 2 shows a water tank that is to be drained by a pipe at its bottom. We assume that the tank has no pits or depressions, so that all the water has access to the drain. Also the diameter of the exit pipe is negligible compared to the size of the tank. As the depth of the water decreases, the rate of flow through the pipe also decreases, and the rate approaches zero when the depth approaches zero. Hence it is not immediately obvious that the tank will drain in finite time. Our problem is to show that the time is finite, and to get a formula for the time in terms of the geometry of the tank. The key to the solution is provided by the Torricelli-Borda law.

Let $A(x)$ denote the cross-section area of the tank at height x above the bottom. We assume that $A(x)$ is positive and continuous. By elementary calculus (volume by slicing) the volume of water when the depth is y is

$$V(y) = \int_0^y A(x)\,dx.$$

This gives

$$\frac{d}{dt}V(y) = \frac{dV}{dy}\frac{dy}{dt} = A(y)\frac{dy}{dt}$$

where the first equality follows from the chain rule and the second from the fundamental theorem of calculus. Since the volume is decreasing, its rate of increase dV/dt is negative and is given by (7) as $-k\sqrt{y}$. Hence

$$-k\sqrt{y} = A(y)\frac{dy}{dt} \quad \text{or} \quad -k\,dt = \frac{A(y)}{\sqrt{y}}\,dy.$$

The value $t = 0$ corresponds to the initial depth h and $y = 0$ corresponds to the time T for emptying the tank. Integrating between corresponding limits gives a formula of remarkable simplicity and generality,

(8)
$$kT = \int_0^h \frac{A(y)}{\sqrt{y}}\,dy.$$

PROBLEMS

If nothing is said about air resistance, air resistance is to be neglected. For numerical evaluation $g = 32\,\text{ft/sec}^2$.

1. A stone is thrown vertically upward with velocity 8 feet per second at time $t = 0$. Find the velocity and position at any time t. Find the time at which the velocity is 0 and show that the height is then maximum. Verify that the maximum height agrees with the principle of conservation of energy, that is, with the formula $mgh = mv_0^2/2$.

2. An iceboat weighing 500 pounds is driven by a wind that exerts a force of 25 pounds. Five pounds of this force are expended in overcoming frictional resistance. What speed will the boat acquire at the end of 30 seconds if it starts from rest?

Hint: The relation between weight w and mass m is $w = mg$.

3. (a) What constant acceleration is needed to run the hundred-yard dash in 10 seconds?

(b) If you and your frictionless rocket-powered skateboard together weigh 160 pounds, what constant thrust of the rocket will do the same job?

4. A brick is set moving over ice with initial speed v_0. If the coefficient of friction between the brick and the ice is μ, how long will it be before the brick stops? By integrating $ds = v\,dt$ obtain the total distance s traveled and verify that the work μmgs done by frictional forces equals the loss in kinetic energy. This illustrates the fact that conservation of energy can apply to dissipative systems.

5. A particle of mass m slides down an inclined plane making an angle θ with the horizontal. If friction is neglected the component of force in the direction of motion is $F = mg \sin \theta$. Hence the equation of motion is $m\, d^2r/dt^2 = mg \sin \theta$ where r is the distance along the inclined plane. If a number of particles are released from rest at the same point and time on a number of inclined planes as suggested in Figure 3, show that at any later time T the particles all lie on a circle. This result was known to Galileo before Newton's laws were available.

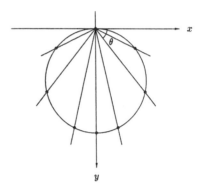

Figure 3

Suggestion: Regard (r, θ) as polar coordinates, multiply by r and convert to Cartesian coordinates.

6. If the coefficient of friction in the above problem is μ there is an extra term $-mg\mu \cos \theta$ that tends to retard the motion. Find the locus of the various particles at any subsequent time.

7. In this problem x is measured horizontally and y vertically with the positive direction upward. At time $t = 0$ a golf ball is hit with speed v_0 in a direction making an angle $\theta > 0$ with the horizontal. Thus the horizontal and vertical components of the initial velocity are respectively

$$v_1 = v_0 \cos \theta, \qquad v_2 = v_0 \sin \theta$$

Figure 4

It is an important principle of physics that x and y act independently of each other in situations such as this. Hence if (x, y) gives the position of the golf ball at time t and m is its mass, the equations of motion are

$$mx'' = 0, \qquad my'' = -mg$$

(a) If the golf ball starts at the origin show that

$$x = v_1 t \qquad y = v_2 t - (1/2)gt^2 \qquad y = x \tan \theta - \frac{g}{2v_0^2} x^2 \sec^2 \theta$$

The third equation holds for $\theta \neq \pi/2$ and shows that the orbit is a parabola in that case.

(b) Show that y is maximum when $gt = v_2$. Conclude that the vertex of the parabola is at (a, b) where $ag = v_1 v_2$, $2bg = v_2^2$.

(c) By setting $y = 0$ check that the distance to the point where the ball lands is twice the value a, which equals $(v_0^2/g) \sin 2\theta$. See Figure 4.

8. A building with vertical walls stands on a horizontal plain. A stone is thrown horizontally out of a window at height H feet above the plain and is D feet from the building when it lands; see Figure 5. Show that the speed with which it was thrown is approximately $4D/\sqrt{H}$ feet per second.

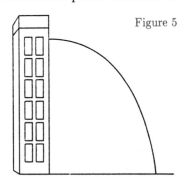

Figure 5

9. Water is flowing from a 2-inch horizontal pipe running full. Find the rate of discharge in cubic feet per minute if the jet of water strikes the ground 4 feet beyond the end of the pipe when the pipe is 2 feet above the ground.

10. An ancient water clock called a clepsydra is so designed that the surface descends at a constant rate as the water runs out through a hole in the bottom. This design allows use of a uniform scale to mark the hours. Find the cross-section area $A(y)$ of the tank as a function of y, and conclude that, if the tank is a volume of revolution with vertical axis, its profile is given by $y = cx^4$.

Hint: In the notation of Part 4(e), $T(h) = (\text{const.})h$. Differentiate.

11. For the tank of Part 4(e) let $A(y) = cy^n$ where c and n are positive constants. Compute kT, V, and kT/V and thus show that

$$kT = \frac{V}{\sqrt{h}} \frac{2n+2}{2n+1}$$

where V is the volume of the tank. Find n for the four tanks illustrated in Figure 6, given that the curve in the last two figures is a parabola.

   (a)   (b)   (c)   (d)

Figure 6

12. Two tanks are both right circular cylinders of height L and diameter L and both are initially full. The first stands on end and drains through a hole in its flat base. The second lies on its side, with its axis horizontal, and drains through a hole in the lowest part of the curved side. If the holes are the same size, which tank do you think will drain faster? Check your conjecture by finding the ratio of the two times, first/second.

13. Show that the time T for maximum height in the problem leading to (3) is given by

$$T = \int_0^{v_0} \frac{m\,dv}{mg + k(v)}$$

14. You drop a wrench into an abandoned mine shaft and 16/3 seconds later you hear the noise it made when it hit the bottom.

(a) If $g = 32$ feet per second per second and the speed of sound is 1200 feet per second how deep is the shaft?

(b) Does it make a significant difference if you use the more accurate values 32.17, 1125 for the acceleration of gravity and the speed of sound, respectively?

15. This problem requires knowledge of the theory of envelopes. An anti-aircraft gun standing on a broad horizontal plane has the constant muzzle velocity v_0. Its angle of elevation θ can be set at any value from 0 to $\pi/2$, and the gun swivels so that it can point in any direction. Describe the region that is within the range of the gun. See Figure 7.

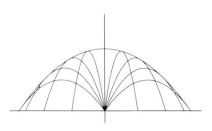

Figure 7

Hint: The Cartesian equation of the trajectories is given in Problem 7. Write in terms of $c = \tan\theta$ and find the envelope.

———————————————ANSWERS———————————————

1. $v = 8(1 - 4t)$, $s = 8t(1 - 2t)$, $v = \pm 8\sqrt{1 - s}$ 2. 192/5 ft/sec
3. (a) 6 ft/sec^2 (b) Since this is $(3/16)g$, and since 160 pounds is the force associated with g, the rocket force is $(3/16)(160) = 30$ lbs.
4. $t_0 = v_0/(\mu g)$, $s_0 = v_0^2/(2\mu g)$
5. $r = 2c\sin\theta$ where $4c = gT^2$. Hence $r^2 = 2cy$, which is a circle.
6. $(|x| + \mu c)^2 + (y - c)^2 = c^2(1 + \mu^2)$ where $4c = gT^2$. This represents two circular arcs through the points $(0,0)$ and $(0,2c)$. The arcs make an

angle θ with the horizontal at $(0,0)$ where $\tan\theta = \mu$. At smaller angles the particle does not move.

9. About 15 cubic feet per minute. "Running full" means that the effective pipe size is 2 inches after taking account of the Borda constant.

10. $A(y) = (\text{const})\sqrt{y}$ 11. $n = 2, 1, 1/2, 1$ 12. $3\pi/8$

14. (a) 400 feet (b) No. The new answer is 398.7 feet, which corresponds to an error of about 0.006 seconds in the time. This is well beyond the probable accuracy of the measurement. Similar considerations motivate our use of simple values, such as $g = 32$, in other problems.

15. The envelope is the parabola through the three points $(0, v_0^2/2g)$ and $(\pm v_0^2/g, 0)$ that describe the maximum height and the maximum horizontal range, respectively. If you revolve the envelope about a vertical axis, the region is bounded by the parabolic surface so obtained and by the plane $y = 0$.

Not only the movements of the heavenly host must be determined by observation and elucidated by mathematics, but whatsoever else can be expressed by a number and defined by natural law.

—d'Arcy Wentworth Thompson

Nature first made things in her own way, and then made human reason skillful enough to be able to understand, but only by hard work, some part of her secrets.

—Galileo

As far as the laws of mathematics refer to reality they are not certain, and as far as they are certain, they do not refer to reality.

—Albert Einstein

Chapter 6

LINEAR SECOND ORDER, CONSTANT

SECOND-ORDER linear equations with constant coefficients are discussed here by means of general principles that are valid in a broader context. Among these are the principle of superposition and the principle of the complementary function. An interesting feature of the analysis is that many problems are best solved by use of complex-valued functions, especially when the solution exhibits an oscillatory behavior. The bridge between real and complex solutions is provided by a third general principle, the principle of equating real and imaginary parts.

1. Basic principles

Numerous problems in mechanical and electrical engineering lead to an equation of the form

$$(1) \qquad\qquad y'' + py' + qy = f(t), \qquad t \in I$$

where f is continuous on a given interval I, the coefficients p and q are constant, and the primes denote differentiation with respect to t. If we define an operator T by

$$(2) \qquad\qquad Ty = y'' + py' + qy$$

then T is linear; that is,

$$(3) \qquad\qquad T(cu) = cTu, \quad T(u+v) = Tu + Tv$$

for any constant c and for all functions u and v in the domain of T. In the present case the domain of T consists of functions twice differentiable on I. The equation $T(cu) = cTu$ follows from

$$(cu)'' + p(cu)' + q(cu) = cu'' + pcu' + qcu = c(u'' + pu' + qu).$$

The proof that $T(u+v) = Tu + Tv$ is similar.

Since T is linear, we can use the following principle to get a family of solutions of $Ty = 0$ from two solutions:

PRINCIPLE OF SUPERPOSITION. Let T be any linear operator. If $Tu = 0$ and $Tv = 0$, then $T(c_1 u + c_2 v) = 0$ for any constants c_1 and c_2.

The proof is immediate because $T(c_1u + c_2v) = c_1Tu + c_2Tv$ by linearity and Tu and Tv are 0 by hypothesis.

In order to get the particular solutions u and v, note that, if s is any constant, the function e^{st} satisfies

$$(e^{st})' = se^{st}, \quad (e^{st})'' = s^2 e^{st},$$

and so on. Hence substitution in (2) gives

$$T(e^{st}) = e^{st}(s^2 + ps + q)$$

when the common factor e^{st} is factored out. This equation is used in the following example.

EXAMPLE 1. By setting $y = e^{st}$ find two solutions of $y'' + 3y' + 2y = 0$. The suggested substitution gives

$$Ty = e^{st}(s^2 + 3s + 2) = e^{st}(s + 1)(s + 2)$$

and $Ty = 0$ if $s = -1$ or $s = -2$. The solutions are e^{-t}, e^{-2t}.

The equation
$$s^2 + ps + q = 0$$

is called the *characteristic equation* of the operator T defined by (2). The gist of the above method is that the homogeneous equation $Ty = 0$ has a solution $y = e^{st}$ if, and only if, s satisfies the characteristic equation.

Most people would think that two solutions e^{-t} and $2e^{-t}$ are more or less the same, while e^{-t} and e^{-2t} are really different. This intuitive feeling is given a mathematical significance by the following definition:

DEFINITION. On a given interval two functions u and v are linearly dependent if one of them is a constant multiple of the other, and they are linearly independent if neither is a constant multiple of the other.

According to this definition e^{-t} and $2e^{-t}$ are linearly dependent on every interval and e^{-t} and e^{-2t} are linearly independent on every interval. The concept of linear dependence is discussed in Section 10.1, where it is seen that linearly independent solutions of the homogeneous equation $Ty = 0$ can be used to satisfy arbitrary initial conditions. Instead of developing theory at this point, we give an example.

EXAMPLE 2. Find the solution in Example 1 satisfying arbitrary initial conditions

$$y(0) = y_0, \quad y'(0) = y_1.$$

Since the independent solutions e^{-t} and e^{-2t} were found in Example 1, the superposition principle gives the solution

$$y = c_1 e^{-t} + c_2 e^{-2t}$$

for any constants c_1 and c_2. The initial conditions hold if

$$c_1 + c_2 = y_0, \quad -c_1 - 2c_2 = y_1.$$

Adding the two equations we get $c_2 = -(y_0 + y_1)$ and the first equation then gives c_1. The result is

$$y = (2y_0 + y_1)e^{-t} - (y_0 + y_1)e^{-2t}.$$

So far we have considered only the homogeneous equation, that is, the equation with $f = 0$. The case of general f can often be handled by a trial solution, much as we used the trial solution $y = e^{st}$ to discuss the homogeneous equation. Since the trial solution involves coefficients that are initially undetermined, this method is called the *method of undetermined coefficients*.

EXAMPLE 3. If the constant a does not have one of two exceptional values, show that the equation

$$y'' + 3y' + 2y = 7e^{at}$$

has a solution of the form $y = Ae^{at}$ where A is constant. The function Ae^{at} satisfies the equation if

$$Ae^{at}(a^2 + 3a + 2) = 7e^{at}.$$

Dividing by e^{at} shows that the equation holds if, and only if,

$$A(a^2 + 3a + 2) = 7.$$

This can be solved for A unless a has one of the values -1 or -2 for which the expression in parentheses vanishes. In the following section we shall see what to do about that case.

EXAMPLE 4. If a, b, c are constant and $a^2 \neq b^2$ find a particular solution of

$$y'' + a^2 y = c \sin bt.$$

Here we try $y = A \sin bt$ where A is constant. The result is

$$(-b^2 + a^2)A \sin bt = c \sin bt, \quad \text{hence} \quad A = \frac{c}{a^2 - b^2}.$$

In Section 4 we shall see what to do if $a^2 = b^2$.

The following result enables us to get a family of solutions of the inhomogeneous equation $Ty = f(t)$ from a particular solution:

PRINCIPLE OF THE COMPLEMENTARY FUNCTION. Let u be a particular solution of $Tu = f$, where T is a linear operator, and let v be any solution of $Tv = 0$. Then $T(u + v) = f$, and every solution of $Ty = f$ can be obtained in this way.

That $T(u + v) = f$ is evident because $T(u + v) = Tu + Tv$ by linearity and $Tu = f$ and $Tv = 0$ by hypothesis. The meaning of the second assertion is that if $Ty = f$, then y can be written in the form $y = u + v$, where v satisfies the homogeneous equation $Tv = 0$. To see this, define $v = y - u$. Then $y = u + v$ and $Tv = T(y - u) = Ty - Tu = f - f = 0$.

EXAMPLE 5. In three steps, we shall solve the initial-value problem

(4) $y'' - 4y = 8t - 16, \qquad y(0) = 4, \; y'(0) = 2.$

First step: Solve the homogeneous equation $Ty = 0$. The substitution $y = e^{st}$ into the equation $y'' - 4y = 0$ gives $s^2 - 4 = 0$. Hence $s = 2$ or -2, and the superposition principle gives the family of solutions

$$v = c_1 e^{2t} + c_2 e^{-2t}, \quad Tv = 0.$$

Second step: Find a particular solution u of the inhomogeneous equation $Ty = f$. Since the right-hand side has the form $at + b$ we try $y = At + B$. Substitution into the differential equation (4) gives

$$-4(At + B) = 8t - 16.$$

Hence, by equating coefficients, $-4A = 8, -4B = -16$. The result is

$$u = 4 - 2t, \quad Tu = f.$$

Third step: Use the principle of the complementary function to form the family of solutions

(5) $y = 4 - 2t + c_1 e^{2t} + c_2 e^{-2t}, \quad Ty = f.$

The constants are determined by the initial conditions as follows:

$$
\begin{array}{lll}
y(0) = 4 : & 4 + c_1 + c_2 = 4, & c_1 + c_2 = 0 \\
y'(0) = 2 : & -2 + 2c_1 - 2c_2 = 2, & c_1 - c_2 = 2.
\end{array}
$$

The values $c_1 = 1$, $c_2 = -1$ can be read from the last two equations and (5) gives the solution,

$$y = 4 - 2t + e^{2t} - e^{-2t}.$$

We conclude with some statements that have not been proved at this point but give useful insight. On an interval I let u and v be linearly independent solutions of $Ty = 0$ and let w be a particular solution of $Ty = f$. Then the family of functions

$$y = c_1 u + c_2 v + w$$

is both the *general solution* and the *complete solution* of $Ty = f$. It is a general solution because the constants c_i can be chosen so as to satisfy arbitrary initial conditions

$$y(t_0) = y_0, \quad y'(t_0) = y_1$$

at any specified point $t_0 \in I$. It is the complete solution because every solution of $Ty = f$ is contained in the family. Since any general solution coincides with the complete solution, one can speak of "the general solution" rather than "a general solution." The function $c_1 u + c_2 v$ is called the *complementary function* because when added to any particular solution of $Ty = f$, it produces the complete solution of this equation.

PROBLEMS

1P. Show that the functions u and v are linearly dependent on an interval I if, and only if, there are constants c_1 and c_2, not both 0, such that

$$c_1 u(t) + c_2 v(t) = 0, \qquad t \in I$$

2P. In this problem a and b are constant and I is an interval. Show that e^{at} and e^{bt} are linearly independent on I if $a \neq b$, and e^{at} and te^{at} are linearly independent on I for all a.

3. Obtain two linearly independent solutions of the form e^{st} where s is constant:

(a) $y'' = 9y$ (b) $y'' = 4y'$ (c) $y'' - y' = 2y$ (d) $y'' + 5y' + 6y = 0$

4P. This problem pertains to the equation $y'' + 2y = 3y'$.

(a) Obtain two linearly independent solutions by trying $y = e^{st}$.
(b) Write the general solution by use of the superposition principle.
(c) Thus find the solution satisfying $y(0) = y'(0) = 1$.

5. Solve the equation subject to the initial conditions below it:

(a) $y'' - y = 0$ (b) $y'' + 7y' + 12y = 0$ (c) $y'' + 7y' + 10y = 0$

$y(0) = 1, \; y'(0) = 0$ $y(0) = -1, \; y'(0) = 4$ $y(0) = 12, \; y'(0) = -3$

6. Obtain a particular solution by use of a trial function of the form Ae^{st}, $A \sin st$, $A \cos st$, or $At + B$, whichever is appropriate:

(a) $y'' + 3y = 4 \sin 2t$ (c) $y'' - y = t + 1$ (e) $y'' = y' + y + 10e^{3t}$

(b) $y'' + y' + y = 9e^t$ (d) $y'' - y = 10 \cos t$ (f) $y'' + t = y' + y + 4$

7. Obtain a particular solution under the assumption that the constant a does not have certain exceptional values. What are these exceptional values?

(a) $y'' + y = \sin at$ (c) $y'' + y = e^{-at}$ (e) $y'' + 9y' + 14y = e^{at}$

(b) $y'' - y' = e^{at}$ (d) $y'' - 9y = \cos at$ (f) $y'' + 8y' + 12y = e^{-at}$

8P. The *rest solution* of a differential equation is the solution that satisfies null initial conditions at $t = 0$. We are going to find the rest solution of

$$Ty = y'' + y = 3 \sin 2t + 3 + 4e^t$$

(a) Solve $Ty = 3 \sin 2t$, $Ty = 3$, $Ty = 4e^t$ by use of appropriate trial solutions. The general superposition principle of Section 4.1 then yields a particular solution of the given equation.

(b) Adding the complementary function gives the two-parameter family

$$y = c_1 \cos t + c_2 \sin t - \sin 2t + 3 + 2e^t$$

Determine the constants so that $y(0) = y'(0) = 0$.

9. Obtain the rest solution:

(a) $y'' + y = 2e^t$ (c) $y'' + y' = 2e^{-2t} + 4e^t$

(b) $y'' - y = 4 \sin t$ (d) $y'' + 9y = 15 \sin 2t + 36e^{-3t}$

10. Solve, subject to the given initial conditions:

(a) $y'' + 3y' + 2y = 24e^{-5t} - 4$, $y(0) = 0$, $y'(0) = -9$

(b) $y'' - y' - 2y = -8e^{-2t} - 2t - 3$, $y(0) = 1$, $y'(0) = 9$

11P. The operator $D = d/dt$ denotes differentiation; for example,

$$D \sin t = \cos t, \qquad De^{2t} = 2e^{2t}$$

Give a reasonable interpretation of each of the following expressions, and decide whether they are equal. No answer is provided:

$$(D + 2)(D - 3)t^3, \qquad (D - 3)(D + 2)t^3$$

what is this   idiocy?

─────────────────────────ANSWERS─────────────────────────

1. If $u = cv$ then $c_1 u + c_2 v = 0$ with $c_1 = 1 \neq 0$ and $c_2 = -c$.
   If $c_1 u + c_2 v = 0$ and $c_1 \neq 0$ then $u = cv$ with $c = -c_2/c_1$.
3. (a) e^{3t}, e^{-3t} (b) 1, e^{4t} (c) e^{-t}, e^{2t} (d) e^{-2t}, e^{-3t} 4. e^t
5. (a) $\cosh t$ (b) $-e^{-4t}$ (c) $19e^{-2t} - 7e^{-5t}$
6. (a) $-4\sin 2t$ (b) $3e^t$ (c) $-t - 1$ (d) $-5\cos t$ (e) $2e^{3t}$ (f) $t - 5$
7. (a) $(\sin at)/(1 - a^2)$, $a \neq 1, -1$ (d) $-(\cos at)/(a^2 + 9)$, no exceptions
   (b) $e^{at}/(a^2 - a)$, $a \neq 1, 0$ (e) $e^{at}/(a^2 + 9a + 14)$, $a \neq -7, -2$
   (c) $e^{-at}/(a^2 + 1)$, no exceptions (f) $e^{-at}/(a^2 - 8a + 12)$, $a \neq 6, 2$
8. $c_1 = -5$,   9. (a) $e^t - \cos t - \sin t$ (c) $e^{-2t} + 2e^t - 3$
   $c_2 = 0$ (b) $-2\sin t + 2\sinh t$ (d) $3\sin 2t + 2e^{-3t} - 2\cos 3t$
10. (a) $2e^{-5t} + e^{-t} - e^{-2t} - 2$ (b) $1 + t + 4\sinh 2t$

──

2. Operator notation

The study of linear differential equations becomes easier if we introduce a special symbol $D = d/dt$ for the operation of differentiation. This symbol should be distinguished from the symbol d used in connection with differentials. Thus

$$D(t^3) = 3t^2, \quad d(t^3) = 3t^2 dt.$$

Since

(1) $$D(cu) = cDu, \quad D(u + v) = Du + Dv$$

for all differentiable functions u, v and for any constant c, the operator D is linear. Its usefulness stems from additional properties that indicate that D behaves much like a number. For example, if we define powers by

$$D^0 u = u, \quad D^m u = u^{(m)} = \frac{d^m u}{dt^m} \quad (m = 1, 2, \cdots)$$

whenever u has m derivatives, then

(2) $$D^m D^n = D^{m+n}, \quad (D^m)^n = D^{mn} \quad (m, n = 0, 1, 2, \cdots).$$

Operator equations such as these mean that the operators on each side have the same domain and do the same thing to functions in their domain. Since

(2) is an immediate consequence of the definitions, we do not give a formal proof.

An operator expression such as

$$P(D) = D^2 + pD + q$$

is defined by prescribing its domain and stating what it does to any function in its domain. In this case the domain consists of all functions twice differentiable on some interval I. For such functions, by definition,

(3) $$(D^2 + pD + q)u = D^2u + pDu + qu = u'' + pu' + qu.$$

Although (3) resembles the linearity property (1), it is merely a definition while (1) is a theorem about differentiation.

If a and b are constant a short calculation based on (1) and (3) gives

$$(D + a)(D + b)u = D^2u + (a + b)Du + abu$$

for any twice-differentiable function u. Since the right-hand side is unchanged when a and b are permuted, that is, when (a, b) is replaced by (b, a), the left-hand side must have the same property. This shows that

$$(D + a)(D + b) = (D + b)(D + a), \qquad a, b \text{ const.}$$

The gist of these remarks is that polynomials $P(D)$ can be manipulated much as if D were a number, provided the coefficients are constant. That this provision is essential is shown in Problem 1.

There is an interesting connection between the polynomial $P(D)$ and the characteristic equation introduced in Section 1. If s is constant,

$$De^{st} = se^{st}, \qquad D^2e^{st} = s^2e^{st},$$

and so on. Hence

$$(D^2 + pD + q)e^{st} = (s^2 + ps + q)e^{st}.$$

The result is summarized by the equation

(4) $$P(D)e^{st} = P(s)e^{st},$$

which shows that the characteristic equation for the operator $T = P(D)$ can be obtained by writing s instead of D.

EXAMPLE 1. Find two linearly independent solutions by inspection:

(5) $$y = e^{st} \qquad (D - 4)(D + 5)y = 0.$$

The characteristic equation is $(s-4)(s+5)=0$ with roots $s=4$, $s=-5$. This yields the solutions

$$e^{4t}, \quad e^{-5t}.$$

In numerous investigations one requires trial solutions of the form $e^{kt}u$ where u is a sufficiently differentiable function and k is constant. By the rule for differentiating a product

$$D(e^{kt}u) = e^{kt}Du + ke^{kt}u = e^{kt}(D+k)u.$$

Hence if b is constant

$$(D-b)e^{kt}u = e^{kt}(D+k-b)u.$$

Thus, the factor e^{kt} can be moved to the left of the operator $D-b$ provided D is replaced by $D+k$. Repetition gives a similar result for two factors,

$$(D-a)(D-b)e^{kt}u = (D-a)e^{kt}(D+k-b)u = e^{kt}(D+k-a)(D+k-b)u.$$

This is important enough to state as a theorem:

EXPONENTIAL SHIFT THEOREM. Let k be constant and let

$$P(D)y = (D-a)(D-b)y$$

where a and b are constant. Then a factor e^{kt} can be moved across the operator $P(D)$ provided D is replaced by $D+k$. In symbols

$$P(D)e^{kt}u = e^{kt}P(D+k)u$$

for all u in the domain of $P(D)$.

As our first application of the shift theorem let us see what to do about the case of repeated roots. Here the differential equation is

$$(D-a)^2 y = 0, \quad a = \text{const.}$$

and the trial solution $y = e^{st}$ yields only the single solution e^{at} with a corresponding family

$$y = e^{at}c.$$

Using a method suggested by the French-Italian mathematician Lagrange, we replace c by a function u and try

$$y = e^{at}u.$$

By the shift theorem with $k = a$

$$(D-a)^2 e^{at} u = e^{at}(D+a-a)^2 u = e^{at} D^2 u.$$

The desired equation $(D-a)^2 y = 0$ holds if $D^2 u = 0$ and hence if u is a linear function of t. Thus we get the family of solutions

$$y = (c_1 + c_2 t)e^{at}.$$

In particular, the choices $(c_1, c_2) = (1, 0)$ and $(0, 1)$ yield two linearly independent solutions

$$e^{at}, \quad te^{at}.$$

As a second illustration, the shift theorem shows how to get a particular solution of an equation such as

(6) $$(D-3)^2 y = 16e^{3t}.$$

The trial solution $y = Ae^{3t}$ does not work because 3 is a root of the characteristic equation and hence this choice of y gives $Ty = 0$. Instead let us try $y = e^{3t} u$ and use the shift theorem with $k = 3$. The result is

$$(D-3)^2 e^{3t} u = e^{3t}(D+3-3)^2 u = e^{3t} D^2 u = 16e^{3t}$$

where the last equality is imposed because we want a solution of (6). Clearly $D^2 u = 16$ does the job. Integrating twice gives $u = 8t^2$ and hence

$$y = 8t^2 e^{3t}.$$

We drop the constants of integration because only a particular solution is wanted.

As a third illustration let us find a particular solution of

$$(D-1)(D+2)y = 12e^t.$$

The trial solution $y = Ae^t$ is not appropriate because $s = 1$ is a root of the characteristic equation, so that this choice makes the left-hand side vanish. Instead let us set $y = e^t u$ and use the shift theorem with $k = 1$. The result is

$$(D-1)(D+2)e^t u = e^t(D+1-1)(D+1+2)u = 12e^t.$$

This reduces to

$$D(D+3)u = 12 \quad \text{or} \quad (D+3)Du = 12.$$

The second form is preferable because with $v = Du$ it becomes

$$(D+3)v = 12.$$

This has the obvious solution $v = 4$. Since $v = Du$ an integration gives u and the final result is

$$v = 4, \quad u = 4t, \quad y = 4te^t.$$

Although we assumed $P(D)$ to be in factored form when proving the theorem, the following example shows that $P(D)$ need not be factored when the theorem is used.

EXAMPLE 2. By the trial function $y = e^{-2t}u$ obtain a particular solution:

$$(D^2 + 2D + 8)y = 4(1 + 4t)e^{-2t}.$$

The suggested substitution gives

$$(D^2 + 2D + 8)e^{-2t}u = 4(1 + 4t)e^{-2t}$$

or, by the shift theorem with $k = -2$,

$$e^{-2t}((D - 2)^2 + 2(D - 2) + 8)u = 4(1 + 4t)e^{-2t}.$$

Canceling the factor e^{-2t} and simplifying, we get

$$(D^2 - 2D + 8)u = 4(1 + 4t).$$

The trial function $u = at + b$ gives

$$-2a + 8(at + b) = 4 + 16t, \quad \text{hence} \quad 8a = 16, \quad -2a + 8b = 4.$$

The final result is $u = 2t + 1$, $y = (2t + 1)e^{-2t}$.

PROBLEMS

1. Show that the equation $(D+t)(D+1) = (D+1)(D+t)$ fails if both sides are applied to the function $u(t) = 1$.

2. Write the general solution by inspection:

(a) $(D - 1)^2 y = 0$ (b) $(D + 1)(D - 2)y = 0$ (c) $(D + 4)^2 y = 0$

3P. We are going to find the general solution of

$$(D + 1)(D - 2)y = (9t^2 + 6t - 3)e^{2t}$$

(a) Set $y = e^{2t}u$ and use the shift theorem to get the equivalent equation

$$(D + 3)Du = 9t^2 + 6t - 3$$

(b) Set $Du = At^2 + Bt + C$ and find first A, then B, and finally C.

(c) Finish the job.

4. Find the general solution:

(a) $(D-1)^2 y = 80(t^3 + 1)e^t$ (c) $(D-1)(D+2)y = (18t^2 - 12t + 3)e^{-2t}$

(b) $(D^2 - 9)y = (2 + 12t)e^{3t}$ (d) $(D-\pi)^2 y = 120t^4 e^{\pi t}$

5. Solve $Dy = t^3 e^t$ by the shift theorem and thus integrate $t^3 e^t$ without using the formula for integration by parts.

6. Obtain a particular solution without factoring $P(D)$, as in Example 2:

(a) $(D^2 + D + 1)y = te^t$ (b) $(D^2 - 2D + 8)y = e^t \sin 2t$ (c) $e^t(D^2 + 1)y = t^2$

7P. If a, b, c, k are constant and u is a three-times differentiable function, explain why

$$(D-a)(D-b)(D-c)e^{kt}u = e^{kt}(D+k-a)(D+k-b)(D+k-c)u$$

8P. With a, b, c as in Problem 7, explain why

$$(D-a)(D-b)(D-c) = (D-c)(D-b)(D-a)$$

---------------------------------ANSWERS---------------------------------

2. (a) $(c_1 + c_2 t)e^t$ (b) $c_1 e^{-t} + c_2 e^{2t}$ (c) $(c_1 + c_2 t)e^{-4t}$

3. $y = (t^3 - t + c_1)e^{2t} + c_2 e^{-t}$

4. (a) $(4t^5 + 40t^2 + c_1 t + c_2)e^t$ (c) $(c_1 - t - 2t^3)e^{-2t} + c_2 e^t$

   (b) $(t^2 + c_1)e^{3t} + c_2 e^{-3t}$ (d) $(4t^6 + c_1 t + c_2)e^{\pi t}$

5. $\int t^3 e^t dt = (t^3 - 3t^2 + 6t - 6)e^t + c$

6. (a) $(t-1)e^t/3$ (b) $(e^t \sin 2t)/3$ (c) $e^{-t}(t+1)^2/2$

7. Because you can move e^{kt} to the left of each factor one at a time, as was done for two factors in the text.

8. Because an interchange of two adjacent factors does no harm, and you can get from $(D-a)(D-b)(D-c)$ to $(D-c)(D-b)(D-a)$ by three interchanges of adjacent factors, as follows:

$$(D-b)(D-a)(D-c), \quad (D-b)(D-c)(D-a), \quad (D-c)(D-b)(D-a)$$

3. Complex numbers

If you try to solve $y'' + y = 0$ by setting $y = e^{st}$ you are led to

(1) $$s^2 + 1 = 0.$$

This equation has no solution in real numbers because the square of a real number cannot be negative. But if you adjoin a symbol i to the real number system, which satisfies the equation

$$i^2 = -1$$

by definition, you can construct a system of *complex numbers* in which (1) has a solution.

Complex numbers are an indispensable aid in the study of oscillations of mechanical and electrical systems, in electrostatics, in the theory of fluid flow, and in other branches of physics and engineering. They greatly increase the scope of the methods we have introduced up to now and, as seen presently, they lead to a complete solution of the general linear equation with constant coefficients. In the next section we indicate some of the reasons why complex numbers are so useful. Here we explain what a complex number is.

A complex number is a symbol $a + ib$ where a and b are real, and this condition (that a and b are real) is always understood when we write $a + ib$ without explanation. The number a is called the *real part* and b the *imaginary part* of $a + ib$, in symbols

$$\mathrm{Re}(a + ib) = a, \quad \mathrm{Im}(a + ib) = b.$$

One could define the imaginary part to be ib but this is not as convenient.

With two exceptions, the algebra of complex numbers is just like the algebra of real numbers, provided we agree to replace i^2 by -1 whenever the occasion arises. The first exception involves the notion of equality . By definition,

$$a + ib = c + id \quad \text{means} \quad a = c \text{ and } b = d.$$

Thus, a *single* complex equation leads to *two* real equations. We emphasize this point because it might cause confusion if you are seeing complex numbers for the first time.

The second exceptional feature of complex numbers involves the notion of inequality. Since the square of a complex number can be negative, there is no relation of inequality between complex numbers that has the familiar properties of this relation as applied to real numbers. However, inequalities can be expressed by means of the symbol for *absolute value*,

$$|a + ib| = \sqrt{a^2 + b^2}.$$

If the complex number $A = a + ib$ is associated with the point (a, b) in the (x, y) plane then A is the distance from the representative point of A to the origin and $|A - B|$ is the distance from the representative point for A to that for B. The geometric interpretation suggests the inequality

$$(2) \qquad\qquad |A + B| \leq |A| + |B|$$

and an algebraic proof is outlined in Problem 12.

A *complex-valued function* is defined by $w = u + iv$ where u and v are real-valued functions of the real variable t. The domain of w is the set of points t where both $u(t)$ and $v(t)$ are defined. Familiar concepts such as the notions of limit, continuity, derivative, and integral extend in a natural way from u and v to w. For example, w is differentiable if, and only if, both u and v are, and in that case the derivative is $w' = u' + iv'$. The phrase "complex function" means the same as "complex-valued function."

With the aid of (2) one can extend the theory of limits from real to complex functions and the proofs remain virtually unchanged. As a consequence, familiar formulas for differentiating sums, products, and so on, continue to hold for complex-valued functions of a real variable t. One can also develop the calculus of complex functions from that of real functions without repeating the proofs, as illustrated in Problems 13 and 14. These results are not deep and are taken for granted in the sequel.

PROBLEMS

1P. To solve the equation $A(2 + 3i) = 6 - 7i$ you can do this:

$$A = \frac{6 - 7i}{2 + 3i}\frac{2 - 3i}{2 - 3i} = \frac{12 - 14i - 18i - 21}{4 + 9}$$

Express A in the form $x + yi$ by completing the calculation and verify that it satisfies the given equation.

2. Assuming $|a + ib| \neq 0$ in (c), express A in the form $x + yi$ and check:

(a) $A(1 - i) = 4 + 6i$ (b) $A(5 - 2i) = 29(3 + 4i)$ (c) $A(a + bi) = c + di$

3P. Write the real and imaginary parts by inspection:

(a) $(2 + 3i)(4 - i)$ (b) $(1 + i)(c + is)$ (c) $(a + ib)(c + is)$

4P. Check that $|\cos t + i \sin t| = 1$, $-\infty < t < \infty$.

5P. In this problem we assume that e^{it} can be so defined that its derivative is ie^{it}, and also $e^{i0} = e^0 = 1$.

(a) Using the quotient rule show that the derivative of the following function is 0 for all t. How do you know the denominator does not vanish?

$$\phi(t) = \frac{e^{it}}{\cos t + i \sin t}$$

(b) Hence $\phi(t) = c$, a constant; see Problem 14. Determine c by setting $t = 0$ and thus establish the *Euler-de Moivre formula*

$$e^{it} = \cos t + i \sin t$$

(c) What does the formula say if $t = \pi$?

6P. If the complex number $z = x + iy$ is correlated with the point (x, y) in the Cartesian plane, show that polar coordinates

$$x = r \cos \theta, \qquad y = r \sin \theta$$

give $z = re^{i\theta}$ when you use the result of Problem 5. This is called the *polar form* of the complex number z.

7P. Let ω be real. Writing $C = c_1 - ic_2 = Ae^{-i\phi}$, show that

$$c_1 \cos \omega t + c_2 \sin \omega t = \text{Re } (Ce^{i\omega t}) = A \cos(\omega t - \phi)$$

8. Using the result of Problem 5, show that the equation $e^{i(a+b)} = e^{ia}e^{ib}$ is equivalent to the familiar formulas for $\sin(a + b)$ and $\cos(a + b)$.

9P. For $z = x+iy$ we define $e^z = e^{x+iy} = e^x e^{iy}$ where e^{iy} is given by Problem 5 with $t = y$. Check that $|e^z| = e^x$ and

$$e^{z_1}e^{z_2} = e^{z_1+z_2}$$

Hint: $e^{x_1}e^{iy_1}e^{x_2}e^{iy_2} = e^{x_1}e^{x_2}e^{iy_1}e^{iy_2}$. Use properties of the real exponential function together with Problem 8.

10P. The *conjugate* of a complex number z is denoted by $\bar{z}$ and is obtained by changing this sign of its imaginary part. Thus,

$$z = x + iy \iff \bar{z} = x - iy$$

(a) Check that $\overline{z_1 + z_2} = \bar{z}_1 + \bar{z}_2$, $\quad \overline{z_1 z_2} = \bar{z}_1 \bar{z}_2$, $\quad |z|^2 = z\bar{z}$.

(b) Explain the following proof that $|z_1 z_2| = |z_1||z_2|$:

$$|z_1 z_2|^2 = z_1 z_2 \overline{z_1 z_2} = z_1 z_2 \bar{z}_1 \bar{z}_2 = z_1 \bar{z}_1 z_2 \bar{z}_2 = |z_1|^2 |z_2|^2$$

11. The Schwarz inequality, special case. Apply the inequality $|\operatorname{Re} z| \leq |z|$ to $z = (a + ib)(c - id)$ and deduce

$$|ac + bd| \leq \sqrt{a^2 + b^2}\sqrt{c^2 + d^2}$$

12. Fill in the details and deduce $|A + B| \leq |A| + |B|$:

$$|A + B|^2 = |A|^2 + 2\operatorname{Re}A\overline{B} + |B|^2 \leq (|A| + |B|)^2$$

13. At any point where w_1' and w_2' exist, use familiar results from the calculus of real functions to get the first equation, and from this get the second:

$$(w_1 w_2)' = w_1 w_2' + w_2 w_1', \qquad (e^{at}e^{ibt})' = (a + ib)e^{(a+ib)t}$$

14. Let w be continuous for $a \leq t \leq b$ and let $w'(t) = 0$ for $a < t < b$. Show that w is constant for $a \leq t \leq b$.

Hint: By definition, the hypothesis on $w = u + iv$ gives corresponding conditions on u and v. Use a familiar result from the calculus of real functions.

————————————ANSWERS————————————

2. Self-checking. **3.** (a) 11, 10 (b) $c - s, c + s$ (c) $ac - bs, bc + as$
5. (a) By Problem 4. (c) $e^{i\pi} = -1$

4. Complex solutions

Since the equation $y'' + y = 0$ leads to the values $s = i$ or $-i$ upon the substitution $y = e^{st}$, it is natural to ask whether the corresponding solutions

$$e^{it}, \quad e^{-it}$$

can be given an interpretation that justifies this substitution. Such an interpretation is possible and has the utmost importance for differential equations.

In Section 10.1 it is shown that every solution of $y'' + y = 0$ must have the form

$$y = c_1 \cos t + c_2 \sin t$$

and this result is taken for granted here. (If complex-valued solutions are in question the constants c_1 and c_2 are complex.) Since we want the function e^{it} to be a solution, it follows that

(1) $$e^{it} = c_1 \cos t + c_2 \sin t.$$

Let us suppose that e^{it} can be differentiated by the usual rules, so that by differentiation of (1)

(2) $ie^{it} = -c_1 \sin t + c_2 \cos t.$

Setting $t = 0$ in (1) and (2) gives $c_1 = 1$, $c_2 = i$. Hence the only possibility, subject to our requirements, is

(3) $e^{it} = \cos t + i \sin t.$

It is not hard to show, conversely, that if e^{it} satisfies (3), then e^{it} has all the properties used in the derivation. Hence we accept (3) as a definition, which is completed by setting

$$e^{s+it} = e^s e^{it} = e^s(\cos t + i \sin t).$$

That z in e^z acts like an exponent was shown in Section 3, Problem 9.

Equation (3) is known as the *Euler-de Moivre formula* after the British mathematician de Moivre and the Swiss mathematician Euler (pronounced "oiler") who discovered it independently. If you know about Taylor series you will recognize that the substitution $x = it$ in the series for e^x produces (3) at once. However, this argument is not a proof of (3), but is only another way of motivating the definition.

EXAMPLE 1. By setting $y = e^{st}$ obtain two linearly independent solutions of

$$y'' + 4y' + 13y = 0.$$

The suggested substitution leads to the equation

$$s^2 + 4s + 13 = 0$$

which has the roots

$$s = \frac{-4 \pm \sqrt{16 - 52}}{2} = -2 \pm 3i.$$

Using (3) with t replaced by $3t$ or $-3t$, we get the solutions

(4) $e^{(-2\pm 3i)t} = e^{-2t} e^{\pm 3it} = e^{-2t}(\cos 3t \pm i \sin 3t).$

The center expression shows that the ratio of the solutions is e^{6it}, which is not constant. Hence the solutions are independent.

Since the coefficients of the equation leading to (4) are real, it will be seen shortly that the real and imaginary parts of any complex solution are again solutions. Hence we get the linearly independent real solutions

$$e^{-2t} \cos 3t, \quad e^{-2t} \sin 3t$$

by taking real and imaginary parts of the expression with the plus sign on the right of (4), and the pair

$$e^{-2t} \cos 3t, \quad -e^{-2t} \sin 3t$$

from the expression with the minus sign. Thus the minus sign gives nothing new.

When a linear operator T is applied to a complex function $x + iy$ it is assumed that Tx and Ty are meaningful and, by linearity,

$$T(x + iy) = Tx + iTy.$$

However, this is not necessarily a decomposition into real and imaginary parts, because Tx and Ty on the right might not be real. If Tu is real for every real function u in the domain of T the operator T is said to be *real*. As an illustration, the operator T defined by

$$Tz = z'' + pz' + qz$$

is real if the constants p and q are real.

Much of the usefulness of complex solutions depends on the following:

PRINCIPLE OF EQUATING REAL PARTS. Let $z = x + iy$ be a complex solution of $Tz = f$, where T is a real linear operator. Then $Tx = \operatorname{Re} f$ and $Ty = \operatorname{Im} f$.

The proof is simple. By linearity $Tz = Tx + iTy$ and this equals f by hypothesis. Hence the real part of f is Tx and the imaginary part is Ty, as stated in the theorem. In the special case in which $Tz = 0$ the principle says that x and y satisfy the homogeneous equation, a fact that was used in Example 1.

EXAMPLE 2. Find a solution of each equation:

(5) $$x'' + 3x' + 2x = \cos t, \quad y'' + 3y' + 2y = \sin t.$$

Since $\cos t$ and $\sin t$ are the real and imaginary parts of e^{it} we consider, instead of (5), the equation

$$z'' + 3z' + 2z = e^{it}.$$

The trial solution $z = Ae^{it}$ satisfies this equation if

$$Ae^{it}(i^2 + 3i + 2) = e^{it}$$

or $A(1 + 3i) = 1$. This gives

$$A = \frac{1}{1 + 3i} \frac{1 - 3i}{1 - 3i} = \frac{1 - 3i}{10}$$

and $z = Ae^{it}$ becomes $z = (.1 - .3i)(\cos t + i \sin t)$. The real and imaginary parts of $z = x + iy$ can be found by inspection with the result

$$x = .1 \cos t + .3 \sin t, \quad y = .1 \sin t - .3 \cos t.$$

These functions satisfy (5).

The problem could be solved without using complex numbers by means of the trial solution

$$y = A \cos t + B \sin t.$$

However this procedure leads to two pairs of simultaneous equations for the constants A and B and the details are harder.

EXAMPLE 3.   Obtain a solution of

(6) $$y'' + y = 5e^t \sin t.$$

Since the right-hand side is the imaginary part of $5e^t e^{it} = 5e^{(1+i)t}$ we consider, instead of (6),

(7) $$z'' + z = 5e^{(1+i)t}.$$

The trial solution $z = Ae^{(1+i)t}$ leads to the equation

$$A((1 + i)^2 + 1) = 5 \quad \text{or} \quad (1 + 2i)A = 5.$$

By a brief calculation $A = 1 - 2i$ and the solution of (7) is

$$z = (1 - 2i)e^{(1+i)t} = (1 - 2i)e^t(\cos t + i \sin t).$$

The imaginary part $y = e^t(\sin t - 2 \cos t)$ satisfies (6).

This problem could be solved by use of the trial solution

(8) $$y = e^t(A \cos t + B \sin t)$$

but the calculations would be harder.

For an equation of the form $Ty = e^{at} \cos bt$ or $e^{at} \sin bt$, one can use the shift theorem to eliminate the factor e^{at} and thus avoid the complex exponent $a + bi$. This approach is illustrated in the following example.

EXAMPLE 4. By the trial function $y = e^t u$ obtain a particular solution:

$$(D^2 + 4D + 7)y = e^t(17\cos t + 5\sin t).$$

By the shift theorem with $k = 1$ it is found that

$$(D^2 + 6D + 12)u = 17\cos t + 5\sin t.$$

This equation can be solved with equal ease by trying $u = A\cos t + B\sin t$ or by consideration of the complex equation

$$(D^2 + 6D + 12)z = e^{it}.$$

It is left for you to verify that either method gives

$$u = \cos t + \sin t, \quad \text{hence} \quad y = e^t(\cos t + \sin t).$$

Complex numbers shed light on an important class of functions, called *hyperbolic functions*, that are related to the equation $y'' - y = 0$ in much the same way as the trigonometric functions are related to the equation $y'' + y = 0$. Namely, instead of satisfying $y'' + y = 0$ with the initial values

$$y(0) = 0, \quad y'(0) = 1; \qquad y(0) = 1, \quad y'(0) = 0$$

the functions $\sinh x$ and $\cosh x$ satisfy $y'' - y = 0$ with these same initial values. They are called the *hyperbolic sine* and *hyperbolic cosine*, respectively.

If you solve the two initial-value problems, you will find that

(9) $$\sinh x = \frac{e^x - e^{-x}}{2}, \quad \cosh x = \frac{e^x + e^{-x}}{2}$$

and hence that

$$(\sinh x)' = \cosh x, \quad (\cosh x)' = \sinh x, \quad \cosh^2 x - \sinh^2 x = 1.$$

The close resemblance of these properties to those of $\sin x$ and $\cos x$ is explained, in part, by the substitution $x = it$ in (9). In view of (3) this gives

$$\sinh it = i\sin t, \quad \cosh it = \cos t.$$

Many integrals and differential equations lead to trigonometric functions for some values of the parameters and to exponential functions for other values. Hyperbolic functions often allow a unified approach to such problems by which two investigations are, in effect, reduced to one.

PROBLEMS

1P. For the equation $z'' - 10z' + 26z = 0$ obtain the complex solution

$$z = e^{(5+i)t} = e^{5t}e^{it} = e^{5t}(\cos t + i\sin t)$$

by trying $z = e^{st}$. Thus get two linearly independent real solutions, and from this, get the general solution in real form. What if you use the root $5 - i$ instead of $5 + i$?

2. Find the general solution in real form:

(a) $y'' + 9y = 0$ (b) $y'' - 2y' + 2y = 0$ (c) $(D^2 - 8D + 25)y = 0$

3. (a) For the equation $z'' + 4z' + 3z = 130e^{2it}$ show that the trial solution $z = Ae^{2it}$ leads to $(-1 + 8i)A = 130$. Solve for A and thus obtain

$$z = (-2 - 16i)(\cos 2t + i\sin 2t)$$

(b) Using the result (a) obtain particular solutions of

$$x'' + 4x' + 3x = 130\cos 2t, \qquad y'' + 4y' + 3y = 130\sin 2t$$

4. By use of an equivalent complex equation or by a trial solution of the form $A\sin st + B\cos st$, as you prefer, obtain particular solutions of:

(a) $x'' - 3x' + 2x = 20\cos 2t, \quad y'' - 3y' + 2y = 20\sin 2t$

(b) $x'' + x' + 17x = 17\cos 4t, \quad y'' + y' + 17y = 17\sin 4t$

5. As in Example 3 or 4, or by a trial function of the form suggested by (8), obtain a particular solution:

(a) $x'' + 2x' + 5x = 20e^t\cos 2t, \quad y'' + 2y' + 5y = 20e^t\sin 2t$

(b) $x'' - x' - 6x = 102e^{3t}\cos 3t, \quad y'' - y' - 6y = 102e^{3t}\sin 3t$

6. Solve $z' = e^{(a+ib)t}$, separate into real and imaginary parts, obtain the formula on the left, and answer the question on the right:

$$\int e^{at}\cos bt\, dt = e^{at}\frac{a\cos bt + b\sin bt}{a^2 + b^2} + c, \qquad \int e^{at}\sin bt\, dt = \ ?$$

7P. Let ω be a nonzero real constant. Solve $(D + i\omega)(D - i\omega)z = e^{i\omega t}$ and thus show that the equations below have the indicated solutions:

$$x'' + \omega^2 x = \cos\omega t, \quad x = \frac{t\sin\omega t}{2\omega}; \qquad y'' + \omega^2 y = \sin\omega t, \quad y = -\frac{t\cos\omega t}{2\omega}$$

8. Write two linearly independent real solutions:

(a) $y'' + 3y' = 54y$ (d) $y'' + 6y = 5y'$ (g) $y'' + 6y' + 10y = 0$

(b) $y'' + y = 2y'$ (e) $y'' + 4y = 4y'$ (h) $y'' + 10y' + 29y = 0$

(c) $y'' + 5y = 4y'$ (f) $y'' + 50y = 2y'$ (i) $y'' + 12y' + 35y = 0$

9. This problem pertains to the equation $y'' + 4y' + ay = 0$ where a is a real constant. For $a = 3, 4, 5$, respectively, obtain the particular solutions

$$e^{-t} \qquad te^{-2t} \quad . \quad e^{-2t} \cos t$$

10. This problem pertains to the equation $y'' + 2by' + 2y = 0$ where b is a real constant. For $b^2 = 1, 2, 3$, respectively, obtain the particular solutions

$$e^{-bt} \cos t \qquad e^{-bt}t \qquad e^{-bt}e^t$$

11. You are to get a particular solution for both the equations

$$x'' + 2x' + 2x = 2e^{-t} \cos t, \qquad y'' + 2y' + 2y = 2e^{-t} \sin t$$

Do (a) or (b), whichever you prefer: (a) Try $te^{-t}(A \cos t + B \sin t)$.

(b) Set $z = e^{(-1+i)t}u$ in the equivalent complex equation

$$(D + 1 + i)(D + 1 - i)z = 2e^{(-1+i)t}$$

——————————————ANSWERS——————————————

1. $e^{5t}(c_1 \cos t + c_2 \sin t)$. The root $5 - i$ gives $e^{5t}(c_1 \cos t - c_2 \sin t)$, which is the same family in different notation.
2. (a) $c_1 \cos 3t + c_2 \sin 3t$ (c) $e^{4t}(c_1 \cos 3t + c_2 \sin 3t)$
   (b) $e^t(c_1 \cos t + c_2 \sin t)$
3. $-2 \cos 2t + 16 \sin 2t, \quad -16 \cos 2t - 2 \sin 2t$
4. (a) $-3 \sin 2t - \cos 2t, \ 3 \cos 2t - \sin 2t$
   (b) $\cos 4t + 4 \sin 4t, \quad -4 \cos 4t + \sin 4t$
5. (a) $e^t(\cos 2t + 2 \sin 2t), \quad e^t(-2 \cos 2t + \sin 2t)$
   (b) $e^{3t}(-3 \cos 3t + 5 \sin 3t), \quad e^{3t}(-5 \cos 3t - 3 \sin 3t)$
8. (a) e^{-9t}, e^{6t} (d) e^{3t}, e^{2t} (g) $e^{-3t} \cos t, e^{-3t} \sin t$
   (b) e^t, te^t (e) e^{2t}, te^{2t} (h) $e^{-5t} \cos 2t, e^{-5t} \sin 2t$
   (c) $e^{2t} \cos t, e^{2t} \sin t$ (f) $e^t \cos 7t, e^t \sin 7t$ (i) e^{-5t}, e^{-7t}
11. $te^{-t} \sin t, \quad -te^{-t} \cos t$

Chapter 7

APPLICATIONS

THE DISPLACEMENT of a mass suspended from a spring and the voltage across the condenser in a series circuit both satisfy a second-order linear differential equation with constant coefficients. Analysis of either problem leads to the concepts of underdamped and overdamped oscillation, resonance and beats, and phase lag and gain. A series of curves showing the transition from one mode of resonance to another gives insight into the concept of a filter as used in electrical engineering. Further insight is given by the phase plane, in which the response is described by a graph of y' versus y rather than y versus t.

1. Mechanical and electrical systems

If a mass m is attached to one end of a suspended spring, it will produce an elongation s which, according to Hooke's law, is proportional to the force mg. Thus

$$F = ks = mg$$

where k is the stiffness constant for the spring. If an additional force is applied to produce an additional extension y, and is then removed, the mass-spring system will start to oscillate; see Figure 1. The problem is to determine the subsequent motion. Here, and throughout this chapter, we assume that the spring does not go slack or exceed its elastic limit.

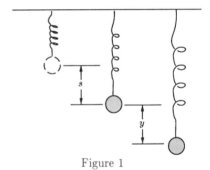

Figure 1

The forces acting on the mass are the force of gravity mg downward and the tension T in the spring that acts in a direction opposite to the force of gravity. We take the positive direction as downward. Since T is the tension when the elongation is $s + y$, Hooke's law gives $T = k(s + y)$. The resultant downward force due to the combined action of the spring and gravity is therefore

$$F = mg - k(s + y) = -ky$$

when we recall that $mg = ks$. Here y is the displacement from the equilibrium position. By Newton's second law

$$(\text{mass})(\text{acceleration}) = \text{force}$$

we get $my'' = -ky$ where, as throughout this chapter, primes denote differentiation with respect to t. Thus

$$my'' + ky = 0.$$

This differential equation describes free oscillations of the undamped mass-spring system.

Suppose next that the mass is in a resisting medium in which the damping is proportional to the velocity y'. Since the resistance opposes the motion, the damping acts in a direction opposite to the velocity. Hence the resistive force is $-ry'$, where r is a positive constant. By Newton's law $my'' = -ky - ry'$ and hence

$$my'' + ry' + ky = 0.$$

This is the equation for free oscillations of the damped mass-spring system.

Next, let the point of support vibrate in accordance with some law that gives the displacement of the top of the spring as a function of time, say $x = x(t)$ where x is measured positively downward. Since the elongation of the spring is diminished by the amount x, the term ky in the foregoing discussion must be replaced by $k(y - x)$. The result is

(1) $$my'' + ry' + ky = kx$$

when the term involving x is transferred to the right-hand side. This equation describes forced oscillation of the damped mass-spring system. The result for the undamped spring is obtained by setting $r = 0$ and for free oscillations by setting $x = 0$. Thus (1) summarizes the preceding discussion.

We shall exhibit a striking analogy between the mass-spring system and a corresponding electrical circuit. This analogy makes it possible to apply the methods of electrical engineering to a variety of mechanical problems.

Let a source of electromotive force be connected in series with a capacitor, an inductor, and a resistor, as shown in Figure 2. If the current is i the voltage drop across the resistance R is iR by Ohm's law and the drop across the inductance L is

$$L\frac{di}{dt}$$

by the law of Faraday. The voltage equation is therefore

(2) $$Li' + Ri + v = v_0(t)$$

where v is the voltage across the capacitor and $v_0(t)$ is the voltage at the source.

Figure 2

The charge Q on a capacitor plate satisfies $Q = Cv$ where C is a constant called the *capacitance*. Since the current i is the rate of flow of charge by definition, we have $i = Q' = Cv'$. Substitution into (2) gives

$$LCv'' + RCv' + v = v_0(t)$$

or, dividing by C and setting $S = 1/C$,

$$(3) \qquad\qquad Lv'' + Rv' + Sv = Sv_0(t).$$

The constant $S = 1/C$ is called the *elastance*.

Upon comparing (1) and (3) we see an exact correspondence between the variables

$$v, \ L, \ R, \ S$$

of the circuit problem and the variables

$$y, \ m, \ r, \ k$$

of the mechanical problem. The displacement $x(t)$ of the top of the spring corresponds to the voltage $v_0(t)$ of the source. Correspondences of this kind form the basis for analog computers in which mechanical systems are simulated by electrical circuits. More important is the fact that the whole technology of transmission-line theory, with its concepts of impedance, transfer function, and so on can be applied to mechanical systems. In fact, from the point of view of communications engineering the mass-spring system is a *transducer* that transforms an input x into an output y.

2. Undamped unforced oscillations

If the equation

$$(1) \qquad\qquad my'' + ky = kx$$

for undamped oscillations is divided by m the result is

$$(2) \qquad\qquad y'' + \omega^2 y = \omega^2 x(t)$$

where $\omega^2 = k/m$. This equation is considered here under the assumption that $\omega > 0$. We begin with the case of unforced oscillations, for which $x = 0$.

(a) Simple harmonic motion

According to Chapter 6 the general solution of the equation

$$(3) \qquad\qquad y'' + \omega^2 y = 0$$

is given by

$$(4) \qquad\qquad y = c_1 \cos \omega t + c_2 \sin \omega t$$

or also by

$$(5) \qquad\qquad y = A \cos(\omega t - \phi), \qquad A \geq 0.$$

The equivalence of these two forms is seen most easily by writing

$$y = \operatorname{Re} C e^{i\omega t}, \qquad C = c_1 - ic_2 = Ae^{-i\phi}$$

but can also be verified by real-variable methods. The expression (4) is convenient for satisfying prescribed initial conditions while (5) makes it easy to plot the graph of the solution.

Motion described by any one of the equivalent equations (3), (4), and (5) is called *simple harmonic motion*. The *amplitude* is A, the *phase* is ϕ, and the *period* of the motion is

$$T = \frac{2\pi}{\omega}.$$

This is the smallest positive constant such that, for all t,

$$y(t + T) = y(t).$$

In other words, T is the smallest time for one complete oscillation. The *frequency*

$$f = \frac{1}{T} = \frac{\omega}{2\pi}$$

gives the number of oscillations per second and ω is the *angular frequency* in radians per second. The notation T for period and f for frequency is standard. One must not, however, confuse this use of the letters T, f with their use to denote an operator T or a forcing function $f(t)$ in a differential equation.

(b) The phase plane

If we set $v = y'$, $y'' = v\, dv/dy$ as in Chapter 2, the equation

$$my'' + ky = 0$$

becomes $mv\, dv + ky\, dy = 0$ after multiplication by dy. An integration gives

$$(6) \qquad\qquad \frac{1}{2}mv^2 + \frac{1}{2}ky^2 = E$$

where E is constant. This equation expresses the principle of conservation of energy. The term $(1/2)\,mv^2$ represents the kinetic energy in the moving mass and the term $(1/2)\,ky^2$ gives the work

$$W = \int_0^y F\,dy = \int_0^y ky\,dy$$

needed to stretch the spring by the amount y. Hence this term represents the potential energy. Since there is no damping and no forcing, one would expect the total energy, kinetic plus potential, to remain constant. That expectation is confirmed by the analysis.

The (y, v) plane is called the *phase plane* and a plot of (y, v) in this plane as t varies is called an *orbit*. By equation (6) the orbits in simple harmonic motion are ellipses. A given ellipse corresponds to a given energy-level E. As shown in Figure 3, the orbits obtained for equal increments of E, such as $E = 1, 2, 3, \cdots$, are not equally spaced, but crowd together as the orbits get larger. When $\omega = 1$ the ellipses are circles, and the configuration has the same mathematical structure as a diagram of Fresnel zones in geometrical optics.

Figure 3

PROBLEMS

1P. Review Section 6.4, including the P problems.

2P. (a) The general solution of $y'' + \omega^2 y = 0$ is $c_1 \cos \omega t + c_2 \sin \omega t$. Determine c_1 and c_2 from the initial conditions $y(0) = y_0$, $y'(0) = y_1$ and thus get

$$y = y_0 \cos \omega t + y_1 \frac{\sin \omega t}{\omega}$$

(b) Let I be an interval containing the point t_0. A *fundamental solution* of a second-order homogeneous linear equation $Ty = 0$ at t_0 is a pair of functions u, v such that $Tu = Tv = 0$ on I and

$$u(t_0) = 1, \quad u'(t_0) = 0; \qquad v(t_0) = 0, \quad v'(t_0) = 1$$

If (u, v) is a fundamental solution show that $y = y_0 u + y_1 v$ satisfies

$$Ty = 0, \qquad y(t_0) = y_0, \qquad y'(t_0) = y_1$$

3. A force of 900 dynes stretches a spring 1 cm. A mass of 100 grams is suspended from the spring and allowed to oscillate.

(a) Find the differential equation of the motion.

(b) Using Problem 2, obtain displacement versus time if the mass is pulled down 2 cm from its equilibrium position and gently released.

(c) Obtain displacement versus time if the mass is projected downward from its equilibrium position with speed 9 centimeters per second.

4. (a) A mass M is hung on a spring of stiffness k and allowed to come to equilibrium. The total length of the spring is now y_0. If an additional mass m increases the length to y_1, show that $mg = k(y_1 - y_0)$ independently of the mass M.

(b) Two equal masses m are suspended from a spring of stiffness k. After equilibrium is reached one of the masses falls off. Describe the motion of the remaining mass.

5. A concrete block suspended by a cable oscillates vertically with frequency f. When a mass m_1 is added, the frequency drops to a new value f_1. Both f and f_1 are determined by observation of the oscillating mass. Show that the mass m of the block and the stiffness constant k of the cable can be calculated as follows:

$$m = \frac{f_1^2}{f^2 - f_1^2} m_1, \qquad k = \frac{(2\pi f f_1)^2}{f^2 - f_1^2} m_1$$

6. (a) Explain how you can determine the phase-plane ellipse for a solution of $y'' + \omega^2 y = 0$ from the initial conditions $y(0)$ and $y'(0)$.

(b) If y and z are two solutions satisfying

$$y(t_0) = a, \quad y'(t_0) = b, \quad z(t_1) = c, \quad z'(t_1) = d$$

what conditions on a, b, c, d ensure that both solutions have the same orbit in phase space?

(c) Give the equation of the phase-plane orbit for $y'' + 9y = 0$, $y(0) = 2$, $y'(0) = -3$, and sketch.

(d) In an artificial mass-spring system in which the spring has a negative stiffness constant $k = -1$, the displacement satisfies $y'' = y$. Sketch the phase-plane orbits. What if $|y(t_0)| = |y'(t_0)|$ at some value t_0?

7P. An important generalization of $y'' + y = 0$ is $y'' + g(y) = 0$ where g is continuous. As in the text, set $v = y'$ and obtain the equation

$$\frac{1}{2}v^2 + G(y) = E, \qquad G(y) = \int_0^y g(\eta)\, d\eta$$

Hence the orbits in the phase plane can be found by intersecting the surface

$$z = \frac{1}{2}v^2 + G(y)$$

with the plane $z = E$ and then projecting onto the (y, v) plane.

8. In many cases the orbits of Problem 7 form closed curves and the corresponding solutions are periodic in t; the period is the time needed to go once around the orbit. Given a closed orbit, let α be the minimum and β the maximum of y on it. Check that the period is given, in general, by

$$T = \sqrt{2} \int_\alpha^\beta \frac{dy}{\sqrt{E - G(y)}}$$

Hint: Set $v = dy/dt$ and separate variables. The time to traverse half the orbit (from $y = \alpha$ to $y = \beta$) is $T/2$. If the integral diverges it takes infinitely long to traverse the orbit and the period is infinite.

9. In the special case $y'' + \omega^2 y = 0$ check that the formula of Problem 7 yields the correct result, $T = 2\pi/\omega$.

10. This is difficult. (a) For $y'' + y = y^3$ and (b) for $y'' + y^3 = y$, find the phase-plane orbits. Solutions with $v = 0$ are called stationary. Check that each equation has the stationary solutions $y = 0$, $y = -1$, $y = 1$. Choose E so that the stationary point is on the orbit and plot these three orbits; use a hand calculator if you wish. Also plot orbits for two other well-chosen values of E and thus confirm the behavior shown in Figure 4. A systematic method of analyzing orbits near a stationary point is given in Chapter 20.

(a)

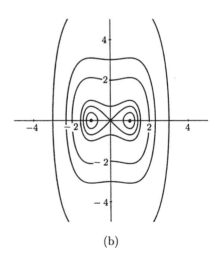

(b)

Figure 4

—ANSWERS—
---

3. (a) $y'' + 9y = 0$ (b) $y = 2\cos 3t$ (c) $y = 3\sin 3t$
4. (b) $y = (mg/k)\cos t\sqrt{k/m}$
6. (a) If $v = y'$ then $v^2 + \omega^2 y^2$ has the same value at t as at $t = 0$, since it is independent of t. Hence $v^2 + \omega^2 y^2 = y'(0)^2 + \omega^2 y(0)^2$.
    (b) $b^2 - d^2 = \omega^2(c^2 - a^2)$ (c) $v^2 + 9y^2 = 45$ (d) $v^2 - y^2 = 2E$, $v = \pm y$
10. (a) $2v^2 + 2y^2 - y^4 = 4E$ (b) $2v^2 - 2y^2 + y^4 = 4E$

---

3. Undamped forced oscillations

In the equation

$$(1) \qquad y'' + \omega^2 y = \omega^2 x(t), \qquad \omega > 0$$

the forcing-function $x(t)$ is often a sinusoidal input,

$$(2) \qquad x(t) = A_0 \cos \omega_0 t \quad \text{or} \quad x(t) = B_0 \sin \omega_0 t,$$

where A_0, B_0, and ω_0 are real constants with $\omega_0 > 0$. Part of the importance of this choice stems from the fact that the output of an alternating-current generator is, as a rule, sinusoidal. But more fundamental is the fact that an arbitrary periodic input of period T can generally be approximated by sums of the form

$$(3) \qquad \frac{1}{2}a_0 + a_1 \cos \omega_1 t + b_1 \sin \omega_1 t + \cdots + a_n \cos \omega_n t + b_n \sin \omega_n t$$

where $T\omega_k = 2\pi k$, $k = 1, 2, 3, \cdots$. If we know the response to the particular inputs (2), the general principle of superposition gives the response to (3) and hence, approximately, to an arbitrary periodic input $f(t)$. A discussion of the underlying theory can be found in DETA, Chapter 18.

(a) Sinusoidal input

Suppose, then, that $x(t)$ in (1) is given by either expression (2). If $\omega_0 \neq \omega$ as we now assume, the trial solution

$$y = A \cos \omega_0 t \quad \text{or} \quad y = B \sin \omega_0 t$$

gives $A(-\omega_0^2 + \omega^2) = \omega^2 A_0$. By this and by a similar result for B,

$$(4) \qquad A = \frac{\omega^2}{\omega^2 - \omega_0^2} A_0, \quad B = \frac{\omega^2}{\omega^2 - \omega_0^2} B_0.$$

The ratio A/A_0 of output to input amplitude is called the *gain* of the system. This ratio can be read off from (4), where it is seen that the same value is obtained for B/B_0. Note that the gain is large if ω_0 is close to ω.

When ω_k is different from ω for all relevant k, the response to the input (3) is given by a series of the same form with coefficients

$$\frac{a_0}{2}, \quad \frac{\omega^2 a_k}{\omega^2 - \omega_k^2}, \quad \frac{\omega^2 b_k}{\omega^2 - \omega_k^2}.$$

This follows from (4) together with the general superposition principle. Addition of the complementary function

$$c_1 \cos \omega t + c_2 \sin \omega t$$

gives the complete solution, but in realistic problems there is always at least a small amount of dissipation, and the complementary function tends to zero as $t \to \infty$. The effect of dissipation is discussed later. Anticipating the result of this discussion, we omit the complementary function here.

If a particular frequency ω_k is close to ω the corresponding term in the series is strongly emphasized and the output may look entirely different from the input. This behavior is now illustrated numerically.

(b) Example of a filter

Let $x(t)$ be the function of period 2π defined for $|t| \leq \pi$ by

$$x(t) = \frac{\pi}{2} - |t|.$$

Figure 5

The graph of this function is the saw-tooth wave illustrated in Figure 5. In the theory of Fourier series it is shown that

$$x(t) = \frac{4}{\pi} \left(\frac{\cos t}{1^2} + \frac{\cos 3t}{3^2} + \frac{\cos 5t}{5^2} + \cdots \right)$$

and that if the series is carried only to n terms the error is at most $1/(4n-2)$. The above discussion indicates that if $x(t)$ is approximated by such a finite sum, the steady-state response to the approximation is given by the same number of terms in

$$y(t) = \frac{4\omega^2}{\pi} \left(\frac{\cos t}{1^2(\omega^2 - 1^2)} + \frac{\cos 3t}{3^2(\omega^2 - 3^2)} + \frac{\cos 5t}{5^2(\omega^2 - 5^2)} + \cdots \right).$$

When ω is close to an odd integer the corresponding term strongly dominates and the other frequencies are, effectively, filtered out. Similar behavior is found whenever a system resonates at one or more frequencies, and the general theory of this effect leads to an important class of engineering systems known as *filters*. The response for several values of ω is shown in Figure 6.

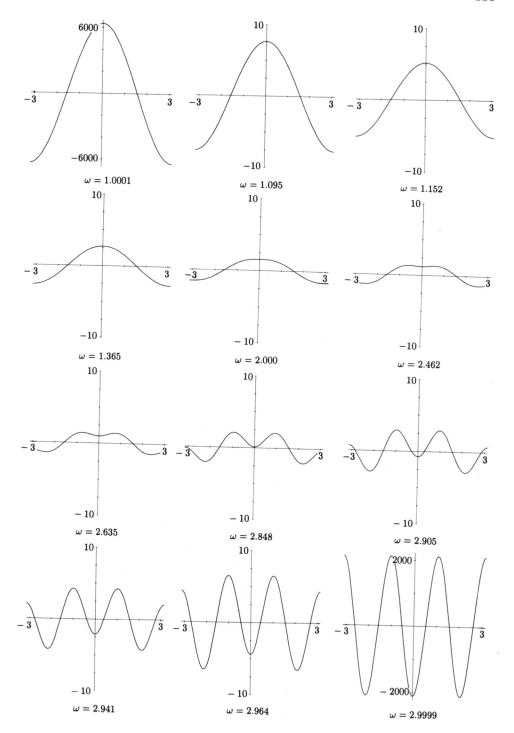

Figure 6

Assuming $\omega_0 \neq \omega$, we are going to find the rest solution of

$$y'' + \omega^2 y = A \cos \omega_0 t \quad \text{where} \quad A = \omega_0^2 - \omega^2.$$

Since any solution can be multiplied by a constant, the choice of A involves no loss of generality. The general solution is

$$y = c_1 \cos \omega t + c_2 \sin \omega t - \cos \omega_0 t.$$

The initial conditions $y(0) = y'(0) = 0$ give $c_1 = 1$ and $c_2 = 0$, so that

(5) $$y = \cos \omega t - \cos \omega_0 t.$$

Physically, this represents the response of a system with the natural frequency ω to a forcing function of frequency ω_0, if the system is initially at rest.

By a trigonometric identity the function (5) can be written

$$y = 2 \sin \frac{\omega_0 + \omega}{2} t \, \sin \frac{\omega_0 - \omega}{2} t.$$

If ω_0 is close to ω, and both are large, the second sine term is slowly varying in comparison to the first and the curve has the general appearance shown in Figure 7. The low-frequency factor $\sin(\omega_0 - \omega)t/2$ induces an *amplitude modulation* on the high-frequency factor $\sin(\omega_0 + \omega)t/2$. The resulting modulated oscillation is an example of the phenomenon of *beats*.

When a tuning fork is sounded simultaneously with a piano string at nearly the same frequency, the beats allow you to tell how far apart the two notes are even if you are tone-deaf; that is, if your ear is sensitive only to

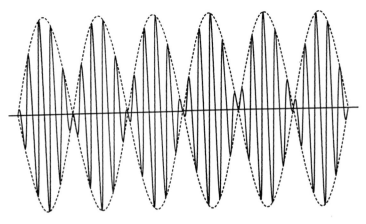

Figure 7

intensity and not to pitch. The slower the beats, the more nearly the string is in tune. This technique is used by piano tuners.

The phenomenon of beats has an interesting bearing on electronic communication and detection. The year 1902 saw the invention of the *heterodyne method* in which a circuit at the receiver generates a signal that differs only slightly in frequency from the high-frequency signal of the sender. The difference produces a frequency in the audible range. As a spin-off from radar technology, in 1946 the heterodyne method was used to detect one-centimeter microwave radiation from the sun. That experiment marked a major step in radio astronomy and microwave spectroscopy.

The phenomenon of amplitude modulation illustrated in Figure 7 is important in radio transmission. A carrier wave of high frequency can be modulated symmetrically by one signal as shown in Figure 8, or by two different signals as suggested in Figure 9. Further development of this technique makes it possible to transmit several messages with a single carrier. A mathematical description is given by the product of a high-frequency carrier wave

$$c(t) = A \cos(\omega t - \phi)$$

with a comparatively slowly varying signal $s(t)$. The signal changes A to $s(t)A$, and that is the reason for the term "amplitude modulation."

In *frequency modulation* it is ω that is altered and in *phase modulation*, ϕ is altered. The point we want to make is that a good part of this highly developed and sophisticated technology has its roots in the simple phenomenon illustrated in Figure 7.

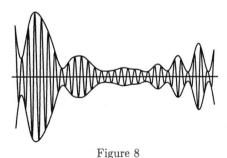

Figure 8

Figure 9

(e) Resonance

When $\omega_0 = \omega$ it is said that the system is *resonant*. Thus, resonance means that the angular frequency ω_0 of the forcing function agrees with the natural frequency ω of the oscillating system. From results obtained in Chapter 6 the two equations

$$x'' + \omega^2 x = \omega^2 A_0 \cos \omega t, \quad y'' + \omega^2 y = \omega^2 B_0 \sin \omega t$$

have the particular solutions

(6)
$$x = \frac{1}{2} A_0 \omega t \sin \omega t, \quad y = -\frac{1}{2} B_0 \omega t \cos \omega t$$

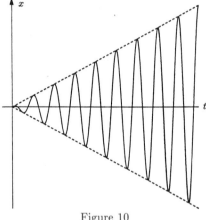

respectively. Here it is convenient to let both x and y denote responses, in agreement with the notation $z = x + iy$. Thus, x must not be confused with the input-function x introduced earlier. The oscillatory factor $\sin \omega t$ has the expected behavior but the amplitude is proportional to t and is unbounded as $t \to \infty$. See Figure 10.

When present at all, resonance is essentially unaffected by the initial conditions. For this reason it has profound importance in engineering and physical situations. The failure of airplane wings,

Figure 10

propellor shafts, and the Tacoma bridge has been attributed, in part, to forced vibrations. Accidents have been caused by resonance in connection with power steering of automobiles, and there are instances of the collapse of buildings induced by the rhythmic swaying of dancing couples. See also, Joshua 6:20.

PROBLEMS

1. (a) If $|\omega|$ does not equal 1, 2, or 3 obtain the following particular solution for the equation $y'' + \omega^2 y = \sin t + \sin 2t + \sin 3t$:

$$y = \frac{\sin t}{\omega^2 - 1} + \frac{\sin 2t}{\omega^2 - 4} + \frac{\sin 3t}{\omega^2 - 9}$$

(b) This problem is best done with a hand calculator. Sketch a graph showing the approximate behavior of the input and of the particular solution found in Part (a) when $\omega = 1.01$, 2.01, and 2.99. (Four graphs.)

2. When placed on a rubber pad, a motor compresses the pad $1/24$ inch. Find the frequency of natural vibrations of the system and thus show that there is a risk of uncontrolled oscillations when the motor reaches a speed of about 900 revolutions per minute.

Hint: $mg = k(1/24)(1/12)$ where $g \approx 32$, and $\omega^2 = k/m$.

3. As you and your new Ferrari together are being lowered by a crane, your friend notices that it sinks 6 inches from the time the tires just touch the ground.

(a) Check that the frequency of free oscillations when you're on the road is about $4/\pi$ cycles per second.

(b) On your way to school you traverse a washboard section of road in which the pavement forms a sinusoid of low amplitude and period 70; that is, the distance from one maximum to the next is 70 feet. At what speed will you be shaken so badly that you might lose control of the car?

Note: 70 is about 22π and 30 miles per hour is 44 feet per second.

4. This problem pertains to the equations

$$y'' + \omega^2 y = \cos \omega_0 t, \qquad y'' + \omega^2 y = \sin \omega_0 t$$

where ω_0 is close to but different from ω. The choice $A_0 = B_0 = 1/\omega^2$ in (4) yields the particular solutions

$$\frac{\cos \omega_0 t}{\omega^2 - \omega_0^2}, \qquad \frac{\sin \omega_0 t}{\omega^2 - \omega_0^2}$$

(a) As $\omega_0 \to \omega$ show that one of the initial conditions $y(0)$ or $y'(0)$ becomes infinite. (The one that becomes infinite is different for the two solutions.)

(b) Check that the following are particular solutions for which the initial conditions remain finite as $\omega_0 \to \omega$:

$$\frac{\cos \omega_0 t - \cos \omega t}{\omega^2 - \omega_0^2}, \qquad \frac{\sin \omega_0 t - \sin \omega t}{\omega^2 - \omega_0^2}$$

(c) By l'Hospital's rule show that the limit as $\omega_0 \to \omega$ in (b) gives the resonant solutions (6) with $A_0 = B_0 = 1/\omega^2$.

————————————————ANSWERS————————————————

2. $f = 48/\pi$ cycles per second.   **3.** (b) About 60 miles per hour.

4. Viscous damping

The equation obtained for free damped oscillations in Section 1 can be written

(1) $$y'' + 2by' + a^2y = 0$$

where a and b are positive constants. In the case of a mass-spring system

$$2b = \frac{r}{m}, \quad a^2 = \frac{k}{m}$$

whereas if the equation refers to an electrical circuit then

$$2b = \frac{R}{L}, \quad a^2 = \frac{1}{LC}.$$

The following discussion is worded in terms of the mass-spring system but the analysis applies to both cases.

The characteristic polynomial $P(s) = s^2 + 2bs + a^2$ associated with (1) has roots

$$r_1 = -b + \beta, \quad r_2 = -b - \beta \quad \text{where} \quad \beta = \sqrt{b^2 - a^2}.$$

It will prove instructive to interpret the behavior of the solution in the three cases $b > a$, $b = a$, $b < a$ for which β is real and nonzero, or $\beta = 0$, or β is purely imaginary, respectively.

(a) Nonoscillatory behavior

When $\beta \neq 0$ the roots are distinct and the general solution is

(2) $$y = c_1 e^{(-b+\beta)t} + c_2 e^{(-b-\beta)t}.$$

If $b > a$ then β is real and the equation $\beta^2 = b^2 - a^2$ shows that $|\beta| < b$. Hence both of the quantities $-(b - \beta)$ and $-(b + \beta)$ in the exponents of (2) are negative. This is the _overdamped_ case. The retarding force is so great that no oscillation can occur and the solution decays exponentially as $t \to \infty$. Figure 11 shows the behavior of several curves with a common initial value but several initial velocities. The latter is the slope at the left-hand endpoint of the curves.

When $a = b$ the characteristic polynomial is

$$s^2 + 2bs + b^2 = (s + b)^2$$

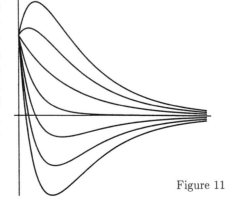

Figure 11

and the two roots are both equal to $-b$. By Chapter 6 the solution is

$$y = e^{-bt}(c_1 + c_2 t), \qquad a = b.$$

This is the *critically damped* case, so called because the motion becomes oscillatory if the damping is reduced by an arbitrarily small amount. In the design of instruments such as a voltmeter one wants the response to be as fast as possible without oscillation. This leads to critical damping as a useful design feature. Aside from that application, however, the case $a = b$ has little practical importance, because it is unstable with respect to small changes of the coefficients. A minute change of a or b moves the system out of the critically-damped case and into one of the other two cases. The general appearance of the curves when $a = b$ is similar to that when $a < b$.

(b) Oscillatory behavior

When $b < a$ the roots of the characteristic equation are complex and the solution is oscillatory. If ω is defined by

$$\omega^2 = a^2 - b^2, \qquad \omega > 0,$$

comparison with $\beta^2 = b^2 - a^2$ shows that $\beta = \pm i\omega$. Hence the solution can be written in the form

$$y = e^{-bt}(c_1 \cos \omega t + c_2 \sin \omega t)$$

or also in the form

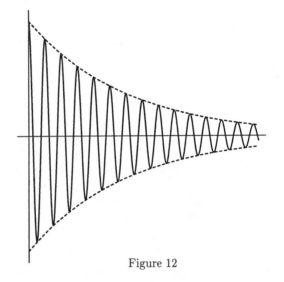

Figure 12

$$(3) \qquad\qquad y = e^{-bt} A \cos(\omega t - \phi)$$

where $C = c_1 - ic_2 = Ae^{-i\phi}$. This is the *underdamped* case. The first expression is useful for satisfying initial conditions and the second for plotting the graph of the solution. See Figure 12.

According to (3) the motion consists of a cosine factor with period $2\pi/\omega$ multiplied by the damping factor e^{-bt}. Since the cosine factor oscillates between -1 and 1, the graph of y versus t oscillates between the curves

$$y = -Ae^{-bt} \quad \text{and} \quad y = Ae^{-bt}.$$

It touches the lower curve at the values

$$\omega t = \phi + \pi, \quad \phi + 3\pi, \quad \phi + 5\pi, \quad \cdots$$

where the cosine is -1 and it touches the upper curve at the values halfway between these.

(c) Forced oscillations

The sinusoidal input discussed in Section 2 leads to the two equations

(4) $\qquad x'' + 2bx' + a^2 x = a^2 \cos \omega_0 t, \quad y'' + 2by' + a^2 y = a^2 \sin \omega_0 t$

where, without loss of generality, the amplitudes A_0 and B_0 have been taken equal to 1. As in the previous section, here too x denotes a response rather than an input. These equations can be solved by use of

(5) $\qquad\qquad\qquad z'' + 2bz' + a^2 z = a^2 e^{i\omega_0 t}.$

The trial solution $z = Ce^{i\omega_0 t}$ in (5) leads to $C(-\omega_0^2 + 2bi\omega_0 + a^2) = a^2$, which gives the particular solution

(6) $\qquad\qquad\qquad z = \dfrac{a^2}{a^2 - \omega_0^2 + 2ib\omega_0} e^{i\omega_0 t}.$

Since the roots of the characteristic equation have negative real parts, as seen above, the complementary function tends to 0 as $t \to \infty$. This means that every other solution has the same behavior as the solution found above, in the long run. It is said, briefly, that the equation has a well-defined *steady-state* behavior, and that the effect of the initial conditions is a *transient*.

When $b = 0$ the complementary function does not tend to 0 and the steady-state response to an input of period T is interpreted as follows:

(a) Introduce a small dissipation $b > 0$.

(b) Keep only the terms of the response that do not tend to 0 as $t \to \infty$.

(c) Now let $b \to 0$ in the result (b).

Further discussion of these matters can be found in Section 8.3.

(d) Phase lag and gain

Let the denominator in (6) be written in polar form $Re^{i\phi}$. The equation

(7) $\qquad\qquad a^2 - \omega_0^2 + 2bi\omega_0 = R(\cos \phi + i \sin \phi)$

holds if R is the magnitude of the expression on the left and ϕ is the phase angle. By a short calculation

(8) $\qquad\qquad R^2 = (a^2 - \omega_0^2)^2 + 4b^2\omega_0^2, \quad \tan \phi = \dfrac{2b\omega_0}{a^2 - \omega_0^2}$

where $R \geq 0$ and $0 < \phi < \pi$. The inequality $0 < \phi < \pi$ follows from the equation $2b\omega_0 = R \sin \phi$, which shows that $\sin \phi > 0$.

Let us define G by

(9) $$G = \frac{a^2}{R} = \frac{a^2}{\sqrt{(a^2 - \omega_0^2)^2 + 4b^2\omega_0^2}}.$$

Equations (6), (7), and (9) give $z = Ge^{i(\omega_0 t - \phi)}$ and the corresponding real solutions are

$$x = G\cos(\omega t - \phi), \quad y = G\sin(\omega t - \phi).$$

The steady-state response has the same angular frequency ω as the impressed oscillation, but there is a phase lag ϕ given by (8). When $\omega_0 < a$ the angle ϕ is between 0 and $\pi/2$, and it approaches 0 as $\omega_0 \to 0$. Thus, the system tends to follow a slow oscillation in phase. However if $\omega_0 > a$ then ϕ is between $\pi/2$ and π, and ϕ tends to π as $\omega_0 \to \infty$. This shows that when the impressed oscillation is rapid, the response is almost out of phase with it. As $\omega_0 \to a$ the phase lag tends to $\pi/2$. This corresponds to the condition for near-resonance, as seen next.

Since the input amplitudes are 1 and the response amplitudes are G the coefficient G is a measure of the gain of the system. When ω_0 is large G is approximately a^2/ω_0^2, which is small. Thus a rapid impressed oscillation does not produce much response. But if $\omega_0 = a$ then $G = a/(2b)$, and this can be dangerously large when the damping term b is close to 0. As $b \to 0$ we are led to the condition of resonance discussed in Section 3, in which the amplitude of the response tends to ∞ with t.

PROBLEMS

1. In the overdamped case check that the conditions $y(0) = y_0$, $y'(0) = y_1$ lead to the equations

$$c_1 + c_2 = y_0, \qquad c_1 - c_2 = (y_1 + by_0)\beta^{-1}$$

Thus obtain the solution $y = y_0 u + y_1 v$ where

$$u = e^{-bt}\left(\cosh \beta t + b\frac{\sinh \beta t}{\beta}\right), \qquad v = e^{-bt}\frac{\sinh \beta t}{\beta}$$

Note that (u, v) gives a fundamental solution at $t = 0$. The same applies to u and v in Problem 2.

2. (a) Using $e^{i\omega t} = \cos \omega t + i \sin \omega t$, check that

$$\cosh i\omega t = \cos \omega t, \qquad \frac{\sinh i\omega t}{i\omega} = \frac{\sin \omega t}{\omega}$$

(b) Setting $\beta = i\omega$ in the result of Problem 1, obtain the following formulas for the underdamped case:

$$u = e^{-bt}\left(\cos\omega t + b\frac{\sin\omega t}{\omega}\right), \qquad v = e^{-bt}\frac{\sin\omega t}{\omega}$$

3. (a) For the critically damped case $a = b$, obtain the fundamental solution

$$u = e^{-bt}(1 + bt), \quad v = e^{-bt}t$$

(b) Show that the result of Part (a) can be obtained by letting $\beta \to 0$ in the result of Problem 1, and also by letting $\omega \to 0$ in the result of Problem 2.

4. (a) A force of 3380 dynes extends a spring 2 cm. A mass of 10 grams is suspended from one end of the spring, pulled down 5 cm from its equilibrium position, and gently released. Find a differential equation and a formula for the motion if resistance is neglected.

(b) Solve Part (a) under the hypothesis that the motion is viscously damped. It is given that a resistance of 100 dynes corresponds to a speed of 1 cm per second.

5. What value of the resistance in Problem 4 would make the motion critically damped, and what is the motion in that case?

6. (a) A force of 120 pounds extends a spring one inch. A weight of 320 pounds is suspended from the spring, pulled down 1 foot below its equilibrium position, and gently released. Find a differential equation and a formula for the motion when resistance is neglected.

Hint: One inch is 1/12 foot. The weight in pounds is related to the mass in newtons by the formula $w = mg$. Take $g = 32$ feet per second per second.

(b) Solve the problem of Part (a) if there is viscous damping of 260 pounds corresponding to 1 foot per second.

7. What value of the resistance in Problem 6 would give critical damping and what is the motion in that case?

8. Logarithmic decrement. Let $y(t) = Ae^{-bt}\cos(\omega t - \phi)$ where A, b, and ω are positive constants. If $T = 2\pi/\omega$ is the period of the cosine factor show that

$$\frac{y(t + T)}{y(t)} = e^{-bT}$$

independently of t. Show that successive maxima y_k, y_{k+1} of the curve y versus t are spaced by a time interval T and hence the foregoing equation also gives y_{k+1}/y_k. You need not compute the maximum values y_k. The quantity $\delta = \log(y_k/y_{k+1}) = bT$ is called the *logarithmic decrement*. It allows estimation of the damping constant b from observation of the motion. Since b depends on the medium in which the motion takes place, whether liquid or gaseous, the logarithmic decrement is useful for measuring viscosity.

9. A body suspended from the end of a viscously damped spring makes 60 oscillations per minute and after one minute the amplitude is halved. Find the differential equation of the motion.

10. Show that $G > 1$ if, and only if, $\omega_0^2 + 4b^2 < 2a^2$. This is the condition for amplification, that is, for the steady-state response to have a larger amplitude than the input.

11. If $s = \omega_0 - a$ and $|s|$ is small, show that $4G^2 \approx a^2/(s^2 + b^2)$. Sketch the corresponding approximate graph of G^2 versus s for fixed $b > 0$ and show that G^2 assumes half its maximum value when $s = b$. Hence if G^2 is plotted versus ω_0, the *full width at half maximum* is approximately $2b$. The quantity $2b$ has been given the misleading name of "half width." It embodies a relation betweeen frequency response and dissipation that is useful in the analysis of microwave and acoustic resonators and in the measurement of dielectric materials.

──────────────────────────────ANSWERS──────────────────────────────

4. (a) $y'' + 169y = 0$, $\quad y = 5\cos 13t$
   (b) $y'' + 10y' + 169y = 0$, $\quad y = (5/12)(12\cos 12t + 5\sin 12t)e^{-5t}$
5. $r = 260$, $\quad y = 5(1 + 13t)e^{-13t}$
6. (a) $y'' + 144y = 0$, $\quad y = \cos 12t$
   (b) $y'' + 26y' + 144y = 0$, $\quad y = (1/5)(9e^{-8t} - 4e^{-18t})$
7. $r = 240$, $\quad y = (1 + 12t)e^{-12t}$
9. $y'' + y'\ln 4 + ((120\pi)^2 + (\ln 2)^2)y = 0$ if t is measured in minutes

──

One cannot escape the feeling that mathematical formulas
have an independent existence and intelligence of their own,
that they are wiser than we are.

— Heinrich Hertz

Chapter 8

LINEAR HIGHER ORDER, CONSTANT

THIS CHAPTER is devoted to n th-order linear differential equations with constant coefficients. One application is to coupled systems, which can lead to equations of high order even when the system itself is of low order. Another is to the method of undetermined coefficients. For broad classes of forcing-functions the form of particular solutions can be precisely specified, and this is done here by means of techniques that are computationally effective. The chapter concludes with an introduction to the concept of asymptotic stability, which pertains to the behavior of physical systems in the remote future.

1. Techniques of solution

Consider the equation

$$Ty = y^{(n)} + p_{n-1}y^{(n-1)} + \cdots + p_1 y' + p_0 y = f$$

on an open interval I, where the coefficients p_j are real or complex constants. The domain of the operator T defined by the first equality consists of the functions that are n times differentiable on I. We write, as in Chapter 6,

$$Ty = (D^n + p_{n-1}D^{n-1} + \cdots + p_1 D + p_0)y = P(D)y.$$

Let us show that T is linear. Since D is linear it is readily checked that

$$D^j(u + v) = D^j u + D^j v, \qquad j = 0, 1, 2, \cdots$$

when u and v are j-times differentiable. Multiplying by p_j, we get

$$p_j D^j(u + v) = p_j D^j u + p_j D^j v.$$

If this is summed on j from $j = 0$ to $j = n$ the result is

$$T(u + v) = Tu + Tv.$$

The proof that $T(cu) = cTu$ is similar.

Since T is linear, all the principles established for general linear operators in Chapter 6 hold for it, and we need not repeat the proofs. The action of T on exponential functions e^{st} is also similar to that in the earlier discussion, as seen next.

If s is constant then

$$D^j e^{st} = s^j e^{st}, \qquad j = 0, 1, 2, \cdots$$

If this equation is multiplied by p_j and the results added we get

$$P(D)e^{st} = P(s)e^{st}$$

where $P(s)$ is the *characteristic polynomial,*

$$P(s) = s^n + p_{n-1}s^{n-1} + \cdots + p_1 s + p_0.$$

Hence $y = e^{st}$ satisfies the homogeneous equation $Ty = 0$ if, and only if, s is a root of the *characteristic equation* $P(s) = 0$.

Just as one can solve the characteristic equation by factoring $P(s)$, one can solve the differential equation by factoring $P(D)$. If

$$a_1, a_2, \cdots, a_n$$

are the roots of $P(s) = 0$ counted with their multiplicities, then

$$P(D) = (D - a_1)(D - a_2) \cdots (D - a_n).$$

The action of $P(D)$ on functions in its domain is independent of the order in which the factors are written. To see this, note that the equation

$$(D - a)(D - b) = (D - b)(D - a)$$

obtained in Section 6.2 shows that an interchange of adjacent factors does not affect the result. However, any permutation of the factors can be obtained by successive interchanges of adjacent factors, as is easily proved. Hence the permutation leaves the operator unaltered.

The following example illustrates the correspondence between $P(D)$, $P(s)$, and Te^{st}.

EXAMPLE 1. Find all solutions of the form $y = e^{st}$:

(1) $$(D - 4)(D^2 - 1)(D^2 + 1)y = 0.$$

The characteristic equation is

$$(s - 4)(s^2 - 1)(s^2 + 1) = 0$$

with roots $4, 1, -1, i, -i$. These roots give the solutions

$$e^{4t}, \quad e^t, \quad e^{-t}, \quad e^{it}, \quad e^{-it}.$$

The last two functions could be replaced by

$$\cos t, \quad \sin t.$$

This follows from the principle of equating real parts, or also from the equations

$$\cos t = \frac{e^{it} + e^{-it}}{2}, \quad \sin t = \frac{e^{it} - e^{-it}}{2i}.$$

By the principle of superposition

$$y = c_1 e^{4t} + c_2 e^t + c_3 e^{-t} + c_4 \cos t + c_5 \sin t$$

is a solution for any choice of the constants c_j. We see later that this is both the complete solution and the general solution.

As in Chapter 6, here too the problem of multiple roots is taken care of by the following:

EXPONENTIAL SHIFT THEOREM. If u is in the domain of $P(D)$ and k is constant, then

(2) $$P(D)e^{kt}u = e^{kt}P(D+k)u.$$

The proof is simple. In the equation

$$P(D)e^{kt}u = (D - a_1)(D - a_2) \cdots (D - a_n)e^{kt}u$$

you can move the factor e^{kt} to the left of $(D - a_n)$ if D in this factor is replaced by $D + k$. Similarly you can move it to the left of the next factor, and so on. After n steps e^{kt} is at the left of every factor $(D - a_j)$ and the result is (2).

If some particular root $a = a_j$ is of multiplicity m, so that $P(D)$ has a factor

$$(D - a)^m,$$

the exponential shift theorem shows that the corresponding solutions of the homogeneous equation are

$$e^{at}, \quad te^{at}, \quad t^2 e^{at}, \quad \cdots, \quad t^{m-1}e^{at}.$$

A formal proof is given in Section 3. Here we illustrate by an example.

EXAMPLE 2. Obtain an appropriate set of solutions for each root of the characteristic equation:

$$(3) \qquad (D-1)^3(D+1)(D^2+1)^2 y = 0.$$

The trial solution $y = e^t u$ and the shift theorem with $k=1$ lead to

$$e^t D^3(D+2)((D+1)^2+1)^2 u = 0.$$

The equation can be written

$$(D+2)(D^2+2D+2)^2 D^3 u = 0$$

and holds if $D^3 u = 0$. This gives

$$u = c_0 + c_1 t + c_2 t^2, \qquad y = (c_0 + c_1 t + c_2 t^2)e^t.$$

The functions y obtained for various choices of c_j are solutions corresponding to the root $s = 1$ of multiplicity 3. Note that y is a linear combination of the three functions

$$e^t, \quad te^t, \quad t^2 e^t.$$

This family of three functions is an equally good description of the solutions corresponding to $s = 1$.

Once the logic of the method is understood, it is no longer necessary to use the shift theorem. Instead, you can get the answer by inspection. Since the characteristic equation for (3) has the root -1 with multiplicity 1 and the roots i, $-i$ each with multiplicity 2, the corresponding solutions

$$e^{-t}, \quad e^{it}, \quad te^{it}, \quad e^{-it}, \quad te^{-it}$$

can be written with ease. If a real form is desired one can replace the pair e^{it}, e^{-it} by the pair $\cos t$, $\sin t$ to get

$$\cos t, \quad t\cos t, \quad \sin t, \quad t\sin t.$$

By the principle of superposition any linear combination

$$(c_0 + c_1 t + c_2 t^2)e^t + c_3 e^{-t} + (c_4 + c_5 t)\cos t + (c_6 + c_7 t)\sin t$$

is also a solution. Note that there are eight constants c_j, corresponding to the fact that the equation is of order eight. It can be shown that this expression gives both the complete and the general solution.

So far we have considered only the homogeneous equation, for which $f = 0$. Techniques for solving the inhomogeneous equation are similar to those in Chapter 6, as illustrated next.

EXAMPLE 3. If the constant a is not a root of the characteristic equation, obtain a particular solution of

(4) $$P(D)y = e^{at}.$$

The substitution $y = Ae^{at}$ with A constant gives

$$P(D)Ae^{at} = AP(D)e^{at} = AP(a)e^{at} = e^{at}$$

where the last equality is stated because we want to solve (4). Hence $A = 1/P(a)$ and

$$y = \frac{e^{at}}{P(a)}.$$

EXAMPLE 4. Obtain a particular solution of each equation:

(5) $$(D^2 + 3D + 2)^2 x = \cos t, \quad (D^2 + 3D + 2)^2 y = \sin t.$$

Since $\cos t$ and $\sin t$ are the real and imaginary parts of e^{it} we set $z = x + iy$ and consider, instead of (5),

$$(D^2 + 3D + 2)^2 z = e^{it}.$$

By Example 3 with $a = i$ a solution is

$$z = \frac{e^{it}}{(-1 + 3i + 2)^2} = \frac{e^{it}}{(1 + 3i)^2} = \frac{e^{it}}{-8 + 6i}.$$

Since $64 + 36 = 100$ this gives

$$z = \frac{e^{it}}{-8 + 6i}\frac{-8 - 6i}{-8 - 6i} = (-.08 - .06i)(\cos t + i \sin t).$$

The solutions of (5) are the real and imaginary parts,

$$x = -.08 \cos t + .06 \sin t, \quad y = -.08 \cos t - .06 \sin t.$$

EXAMPLE 5. Obtain a particular solution of

$$(D^2 + 1)(D - 3)^2 y = 12e^{3t}(10t + 1).$$

Substituting $y = e^{3t}u$ and using the shift theorem with $k = 3$, we get

$$e^{3t}((D + 3)^2 + 1)D^2 u = 12e^{3t}(10t + 1)$$

or, setting $v = D^2 u$,

$$(D^2 + 6D + 10)v = 120t + 12.$$

The trial solution $v = At + B$ gives v, integrating twice gives u, and the result is

$$v = 12t - 6, \quad u = 2t^3 - 3t^2, \quad y = (2t^3 - 3t^2)e^{3t}.$$

EXAMPLE 6.   If ω is a nonzero real constant, obtain a solution of each equation:

(6) $(D^2 + \omega^2)^4 x = \cos \omega t, \quad (D^2 + \omega^2)^4 y = \sin \omega t$

Results of this kind are useful in the study of resonance in oscillating systems. We set $z = x + iy$ and consider, instead of (6), the equation

(7) $(D + i\omega)^4 (D - i\omega)^4 z = e^{i\omega t}.$

If $z = e^{i\omega t} u$ the shift theorem with $k = i\omega$ yields

$$e^{i\omega t}(D + 2i\omega)^4 D^4 u = e^{i\omega t}.$$

We set $D^4 u = A$, where A is constant, and are led to

$$(D + 2i\omega)^4 A = 1.$$

Since $DA = 0$, $D^2 A = 0$, and so on, the equation reduces without detailed calculation to

$$(2i\omega)^4 A = 1 \quad \text{or} \quad 16\omega^4 A = 1.$$

This gives A, the equation $D^4 u = A$ gives u, and the result is

$$z = \frac{1}{384} \left(\frac{t}{\omega} \right)^4 (\cos \omega t + i \sin \omega t).$$

The real and imaginary parts satisfy the two equations (6).

Initial conditions are introduced with the aid of the complementary function as in Chapter 6. Initial conditions will not be emphasized here, however, because they are handled more easily by the Laplace transform, and because they are often irrelevant to the steady-state behavior. These matters are taken up in Section 3 of this chapter and in Chapter 21.

PROBLEMS

1. This problem and Problem 2 are to be done, as far as possible, by inspection. No proof is required. Obtain n distinct solutions of the form e^{kt} where n is the order of the equation:

(a) $(D - 1)(D - 2)(D - 3)y = 0$ (c) $(D^3 + 9D)(D^4 - 16)y = 0$

(b) $(D^2 - 9)(D^2 + 3D + 2)y = 0$ (d) $(D^2 + 4D + 5)(D^2 + 2D + 5)y = 0$

2. Obtain n distinct solutions of the form $t^j e^{kt}$ where n is the order of the equation:

(a) $(D - 1)^3(D - 2)^2(D - 3)y = 0$

(b) $(D^2 - 1)^3(D^2 - 7D + 10)^2 y = 0$

(c) $(D^2 - D)^2(D^2 + 4D + 13)^3 y = 0$

(d) $(D^2 + 1)^4 y = 0$

(e) $(D^3 + 4D^2)^3 y = 0$

(f) $(D^2 + 2D + 2)^4 y = 0$

3. Obtain a particular solution by trying $y = Ae^{st}$ with $s = -2$ or $s = 1$:

$\quad$ (a) $(D^4 + D^3 + D^2 + D + 1)y = 33e^{-2t}$ $\qquad$ (b) $(D^4 - 2)^5 y = 5e^t$

4. Obtain a real solution by use of $Ae^{i\omega t}$ or $A\cos\omega t + B\sin\omega t$, whichever you prefer, with $\omega = 1$ or $\omega = 2$:

(a) $(D^2 + 2D + 2)(D^2 + 2)^4 x = \cos t$, $\quad$ $(D^2 + 2D + 2)(D^2 + 2)^4 y = \sin t$

(b) $(D^2 + D + 1)^2(D^2 + 1)x = \cos 2t$, $\quad$ $(D^2 + D + 1)^2(D^2 + 1)y = \sin 2t$

5. Obtain a solution of $(D^2 - 1)^4(D^3 + 1)^5 y = 3e^t$ by the trial solution $At^4 e^t$ or by writing the equation in the form

$$(D^3 + 1)^5(D + 1)^4(D - 1)^4 y = 3e^t$$

and using the shift theorem with $k = 1$. Hint for second method: If $D^4 u = A$ is constant then $D^j A = 0$ for $j \geq 1$ and hence

$$((D + 1)^3 + 1)^5(D + 2)^4 A = (1 + 1)^5 2^4 A$$

6. Obtain a real solution of each of the two equations

$$(D^2 + 1)^2(D^2 + D + 1)x = \cos t, \qquad (D^2 + 1)^2(D^2 + D + 1)y = \sin t$$

by trying
$$y = t^2(A \cos t + B \sin t)$$

or by consideration of the complex equation

$$(D^2 + D + 1)(D + i)^2(D - i)^2 z = e^{it}$$

7. Check that the functions te^{st}, $t^2 e^{st}$, $t^3 e^{st}$, $\cdots$ associated with a multiple root s can be obtained by differentiating e^{st} with respect to s. This illustrates a phenomenon that pervades the theory of differential equations. It will be seen again in connection with the Frobenius method for handling multiple roots in power-series solutions.

————————————————ANSWERS————————————————

1. (a) $k = 1, 2, 3$ (c) $k = 0, 3i, -3i, 2, -2, 2i, -2i$
   (b) $k = 3, -3, -1, -2$ (d) $k = -2 + i, -2 - i, -1 + 2i, -1 - 2i$
2. (a) $(j, k) = (0, 1), (1, 1), (2, 1), (0, 2), (1, 2), (0, 3)$
   (b) $(j, k) = (0, \pm 1), (1, \pm 1), (2, \pm 1), (0, 2), (1, 2), (0, 5), (1, 5)$
   (c) $(j, k) = (0, 0), (1, 0), (0, 1), (1, 1), (0, -2 \pm 3i), (1, -2 \pm 3i), (2, -2 \pm 3i)$
   (d) $(j, k) = (0, \pm i), (1, \pm i), (2, \pm i), (3, \pm i)$
   (e) $(j, k) = (0, 0), (1, 0), (2, 0), (3, 0), (4, 0), (5, 0), (0, -4), (1, -4), (2, -4)$
   (f) $(j, k) = (0, -1 \pm i), (1, -1 \pm i), (2, -1 \pm i), (3, -1 \pm i)$
3. (a) $3e^{-2t}$ (b) $-5e^t$
4. (a) $x = .2 \cos t + .4 \sin t, \quad y = -.4 \cos t + .2 \sin t$
   (b) $507x = -5 \cos 2t + 12 \sin 2t, \quad 507y = -12 \cos 2t - 5 \sin 2t$
5. $y = 2^{-12} t^4 e^t$ 6. $8x = -t^2 \sin t, \quad 8y = t^2 \cos t$

2. A class of linear systems

Let S and T be two linear differential operators of orders m and n, respectively, with constant coefficients. This means

$$S = D^m + p_{m-1} D^{m-1} + \cdots + p_1 D + p_0$$

$$T = D^n + q_{n-1} D^{n-1} + \cdots + q_1 D + q_0$$

where the p_j and q_j are constant. The domain of ST consists of the functions that have $m + n$ derivatives on some interval I, and this is also the domain of TS. Furthermore

$$ST = TS$$

in the sense that each side does the same thing to any function in the common domain. This is true because S and T can be written in factored form. As an illustration, suppose

$$S = (D - a_1)(D - a_2), \quad T = (D - b_1)(D - b_2)(D - b_3).$$

Then ST is the product of these linear factors in one order and TS is the product of the same linear factors in another order. It was seen in Section 1 that the order does not affect the result, hence $ST = TS$.

These remarks lead to a powerful method of solving simultaneous differential equations. We illustrate in the special case

(1) $$Sx = ay, \quad Ty = bx$$

where a and b are nonzero constants. If at least one of the operators S or T actually involves differentiation, it can be shown that all solutions of (1) have derivatives of all orders, and that fact is taken for granted here.

To solve (1), operate on the first equation with T and use the second equation to get

$$TSx = T(ay) = aTy = abx.$$

Thus the system becomes

(2) $$TSx = abx, \quad ay = Sx.$$

Since the first equation (2) involves x alone we can get x by the methods of Section 1 and the second equation then gives y.

Conversely, if x and y satisfy (2), then

$$aTy = T(ay) = TSx = abx$$

and, dividing by a, we get $Ty = bx$. This shows that any solution of (2) is also a solution of (1).

EXAMPLE 1. Solve the system

(3) $$x' = x + y, \quad y' = 6x$$

where, as throughout this section, the primes denote differentiation with respect to t. When the system is written in the form

$$(D - 1)x = y, \quad Dy = 6x$$

the general procedure gives

$$D(D - 1)x = Dy = 6x, \quad \text{hence} \quad (D^2 - D - 6)x = 0.$$

Since $D^2 - D - 6 = (D - 3)(D + 2)$ the result is

$$x = c_1 e^{3t} + c_2 e^{-2t}.$$

The value of y is now found from the first equation (3):

$$y = x' - x = 2c_1 e^{3t} - 3c_2 e^{-2t}.$$

EXAMPLE 2. Solve

(4) $$x'' = 2x + 3y, \quad y'' = 2x + y.$$

Here the standard form is

$$(D^2 - 2)x = 3y, \quad (D^2 - 1)y = 2x$$

and the general procedure gives

$$(D^2 - 1)(D^2 - 2)x = 3(D^2 - 1)y = 6x.$$

Hence

$$(D^4 - 3D^2 - 4)x = 0 \quad \text{or} \quad (D^2 - 4)(D^2 + 1)x = 0.$$

This yields

$$x = c_1 e^{2t} + c_2 e^{-2t} + c_3 \cos t + c_4 \sin t$$

and y is obtained from the first equation (4).

EXAMPLE 3. The solution of Example 2 has four arbitrary constants. Using this fact, try to impose the four initial conditions

$$x(0) = x'(0) = x''(0) = 0, \qquad y(0) = y_0.$$

Setting $t = 0$ in the first equation (4) yields

$$x''(0) = 2x(0) + 3y(0)$$

and hence the initial values of x'', x, and y cannot all be assigned arbitrarily. In the present case the initial values for x give $y(0) = 0$. Hence there is no solution unless $y_0 = 0$. If $y_0 = 0$ a short calculation shows that there are infinitely many solutions,

$$x = c(e^{2t} - e^{-2t} - 4\sin t), \quad 3y = 2c(e^{2t} - e^{-2t} + 6\sin t).$$

The following theorem gives conditions under which the initial conditions determine the solution uniquely, so that the phenomenon noted in the preceding example does not arise:

THEOREM 1. *Let S and T be linear operators, with constant coefficients, of orders m and n respectively. Suppose $m \geq 1$ and $n \geq 1$ and suppose a and b are nonzero constants. Then the system*

$$Sx = ay, \quad Ty = bx$$

has one, and only one, solution satisfying the initial conditions

$$x(t_0) = x_0, \ x'(t_0) = x_1, \ \cdots \ , x^{(m-1)}(t_0) = x_{m-1}$$

$$y(t_0) = y_0, \ y'(t_0) = y_1, \ \cdots \ , y^{(n-1)}(t_0) = y_{n-1}.$$

Theorem 1 is a special case of results in DETA, Chapter 14.

In conclusion, we mention that the foregoing techniques apply to the system

(5) $$Px = ay + f, \quad Qx + Ry = g.$$

Here P, Q, R denote linear operators with constant coefficients, a is a nonzero constant, and f and g are sufficiently differentiable functions on a given interval I. If $RP + aQ$ is not identically 0 we can solve as follows: Operate on the first equation with R and eliminate Ry by the second equation to get

$$(RP + aQ)x = ag + Rf, \quad ay = Px - f.$$

We find x from the first equation and y is then given uniquely by the second.

PROBLEMS

1P. (a) Check that the system $x' = 2(y - x)$, $y' = 3x - y$ can be written

$$(D + 2)x = 2y, \qquad (D + 1)y = 3x$$

(b) Operate on the first equation with $D + 1$ and use the second to get

$$(D + 1)(D + 2)x = 6x, \quad \text{hence} \quad (D^2 + 3D - 4)x = 0$$

(c) Find x and then determine $2y$ from the first of the original equations.

2. Solve as in Problem 1:

  (a) $x' = y$, $y' = -x$ (b) $x' = y$, $y' = x$ (c) $x'' = y$, $y'' = x$

3. Solve as in Problem 1:

  (a) $x' = 3x - 2y$, $y' = 2x - y$ (d) $x' + y = 6x$, $y' - 5x = 4y$

  (b) $x' = 3x + 4y$, $y' = 2x + y$ (e) $x'' = 3x + 2y$, $y'' = -4x - 3y$

  (c) $x' = 5x + 9y$, $y' = -x + 11y$ (f) $x'' = x + 2y$, $y'' = -2x - 3y$

4. Solve the following system subject to the initial conditions $x'(0) = y'(0) = 0$ and to the additional conditions given below:

$$x'' = 4y - 7x, \qquad y'' = -3y + 3x$$

(a) $x(0) = 2$, $y(0) = 3$ (b) $x(0) = -2$, $y(0) = 1$ (c) $x(0) = 0$, $y(0) = 0$

5. We are going to solve the inhomogeneous initial-value problem

$$x' + 2x - 2y = 16t, \qquad y' - 3x + y = 5e^t, \quad x(0) = -1, \quad y(0) = -14$$

(a) Check that $(D+2)x = 2y + 16t$, $(D+1)y = 3x + 5e^t$.

(b) Operate on the first of these equations with $D+1$ and use the second to get an equation in x alone:

$$(D^2 + 3D - 4)x = 10e^t + 16t + 16$$

(c) Finish the job.

6. Let $x' + 2x - 2y = 16t$, $y' - 3x + y = 5e^t$, $x(0) = -3$, and $x'(0) = -2$.

(a) Find $y(0)$, $y'(0)$, and $y''(0)$ without solving either differential equation.

(b) Check by solving completely.

7. The system $(D+1)x+(D+6)y = 1$, $(2D+4)x+(D+15)y = 4$ does not fall into the pattern (5), because neither x nor y is given explicitly in terms of the other. However, if the first equation is subtracted from the second the result is $(D+3)x + 9y = 3$. Thus solve completely.

---------------------------------ANSWERS---------------------------------

1. $x = c_1 e^{-4t} + c_2 e^t$, $\quad 2y = -2c_1 e^{-4t} + 3c_2 e^t$
2. (a) $x = c_1 \sin t + c_2 \cos t$, $\quad y = c_1 \cos t - c_2 \sin t$
   (b) $x = c_1 e^t + c_2 e^{-t}$, $\quad y = c_1 e^t - c_2 e^{-t}$
   (c) $x = c_1 e^t + c_2 e^{-t} + c_3 \cos t + c_4 \sin t$, $\quad y = c_1 e^t + c_2 e^{-t} - c_3 \cos t - c_4 \sin t$
3. (a) $x = (c_1 + c_2 t)e^t$, $\quad 2y = (2c_1 - c_2 + 2c_2 t)e^t$
   (b) $x = c_1 e^{5t} + c_2 e^{-t}$, $\quad 2y = c_1 e^{5t} - 2c_2 e^{-t}$
   (c) $x = (c_1 + c_2 t)e^{8t}$, $\quad 9y = (3c_1 + c_2 + 3c_2 t)e^{8t}$
   (d) $x = e^{5t}(c_1 \cos 2t + c_2 \sin 2t)$, $\quad y = e^{5t}((c_1 - 2c_2) \cos 2t + (2c_1 + c_2) \sin 2t)$
   (e) $x = c_1 e^t + c_2 e^{-t} + c_3 \sin t + c_4 \cos t$
       $y = -c_1 e^t - c_2 e^{-t} - 2c_3 \sin t - 2c_4 \cos t$
   (f) $x = (c_1 + c_2 t) \cos t + (c_3 + c_4 t) \sin t$
       $y = (c_4 - c_1 - c_2 t) \cos t - (c_2 + c_3 + c_4 t) \sin t$
4. (a) $x = 2 \cos t$, $y = 3 \cos t$ $\quad$ (b) $x = -2 \cos 3t$, $y = \cos 3t$ $\quad$ (c) $x = y = 0$
5. $x = 2te^t + 6e^{-4t} - 4t - 7$, $y = (1 + 3t)e^t - 6e^{-4t} - 12t - 9$
6. (a) $-4, 0, -1$
7. $x = 1 + c_1 \sin 3t + c_2 \cos 3t$, $3y = (c_2 - c_1) \sin 3t - (c_2 + c_1) \cos 3t$

3. Supplementary topics

Throughout this section an operator T is defined by

$$T = (D^n + p_{n-1}D^{n-1} + \cdots + p_1 D + p_0)$$

or, in factored form,

$$T = (D - a_1)(D - a_2) \cdots (D - a_n).$$

Here the p_j and a_j are real or complex constants.

It is not assumed that the a_j are distinct. If $P(s)$ has a zero $a_j = a$ of multiplicity m, then

$$P(s) = Q(s)(s - a)^m$$

where $Q(s)$ is a polynomial that does not vanish when $s = a$. The corresponding operator equation is

$$P(D) = Q(D)(D - a)^m.$$

To solve $P(D)y = 0$ we set $y = e^{at}u$ and use the shift theorem with $k = a$, getting

(1) $$P(D)y = Q(D)(D - a)^m e^{at}u = e^{at}Q(D + a)D^m u.$$

The right-hand side is 0 if $D^m u = 0$, hence if u is a polynomial of degree at most $m - 1$. This shows that the expression

$$y = c(t)e^{at}$$

satisfies $Ty = 0$ when $c(t)$ is any polynomial of the form

$$c(t) = c_0 + c_1 t + \cdots + c_{m-1}t^{m-1}.$$

The same sort of thing can be done for each root a_j of the characteristic equation. Let the roots be so numbered that the *distinct* roots are

$$a_1, \quad a_2, \quad \cdots, \quad a_r$$

with multiplicities $m_1, m_2, \cdots, m_r$ respectively. Thus

(2) $$m_1 + m_2 + \cdots + m_r = n$$

where n is the degree of $P(s)$. The above method shows that the homogeneous equation $Ty = 0$ has the solution

(3) $$y = c_1(t)e^{a_1 t} + c_2(t)e^{a_2 t} + \cdots + c_r(t)e^{a_r t}$$

where $c_j(t)$ is an arbitrary polynomial of degree $m_j - 1$. Such a polynomial has m_j coefficients and hence, by (2), the number of arbitrary constants in the solution (3) is n. In Section 11.2 we show that (3) is both the complete solution and the general solution of $Ty = 0$.

(a) Undetermined coefficients

Let $A(t)$ be a polynomial. By substituting and equating coefficients, it is found that the equation

$$(D - a_1)y = A(t), \qquad a_1 \neq 0$$

has one, and only one, polynomial solution $B(t)$, and the degree of B is the same as that of A. Repetition gives a similar result for

$$(D - a_1)(D - a_2)y = A(t), \quad (D - a_1)(D - a_2)(D - a_3)y = A(t),$$

and so on, provided the constants a_j are all different from 0. These remarks are summarized as follows:

THEOREM 1. *Let $A(t)$ be a polynomial and suppose $P(0) \neq 0$. Then the equation $Ty = A(t)$ has one, and only one, polynomial solution $B(t)$, and the degree of B is the same as that of A.*

We now consider the equation $Tz = e^{kt}A(t)$ where k is a real or complex constant and A is a polynomial. The substitution $z = e^{kt}u$ and the shift theorem give the equivalent equation

$$P(D + k)u = A(t).$$

The condition $P(0) \neq 0$ of Theorem 1 holds for this equation if $P(k) \neq 0$. Thus we get:

THEOREM 2. *Let k be a real or complex constant, let $A(t)$ be a polynomial, and suppose $P(k) \neq 0$. Then the equation $Tz = e^{kt}A(t)$ has one, and only one, solution of the form $z = e^{kt}B(t)$ where B is a polynomial. The degree of B is the same as that of A.*

Assume next that k is a zero of $P(s)$ of multiplicity m, so that

$$P(s) = Q(s)(s - k)^m$$

where Q is a polynomial. Taking $P(D) = Q(D)(D - k)^m$ and using the shift theorem as above, we are led to the equation

$$Q(D + k)D^m u = A(t).$$

Since m is the multiplicity of k we know that $Q(k) \neq 0$. Hence Theorem 1 shows that there is one, and only one, solution such that $D^m u = B(t)$, a polynomial. The function u is found from B by integrating m times. Exactly one of the functions obtained by integrating B has the form $t^m C(t)$, where $C(t)$ is a polynomial of the same degree as B. Namely, such a function $t^m C(t)$ is obtained when, and only when, the constants of integration are set equal to 0. These considerations give the following:

THEOREM 3. *Suppose the constant k in Theorem 2 is a zero of $P(s)$ of multiplicity m. Then the equation $Tz = e^{kt} A(t)$ has one, and only one, solution of the form $z = t^m C(t) e^{kt}$ where $C(t)$ is a polynomial. The degree of C is the same as that of A.*

If $k = a + ib$ is complex, the polynomial C in Theorem 3 is also, in general, complex. With $C = F + iG$ the solution given by Theorem 3 is

$$z = t^m e^{at}(F + iG)(\cos bt + i \sin bt).$$

Setting $z = x + iy$ and equating real and imaginary parts yields:

THEOREM 4. *Let $A(t)$ be a real polynomial, let T be real, and let $k = a+ib$ be a zero of $P(s)$ of multiplicity m. Then the equations*

$$Tx = A(t)e^{at} \cos bt, \quad Ty = A(t)e^{at} \sin bt$$

have solutions of the form

$$x = t^m e^{at}(F \cos bt - G \sin bt), \quad y = t^m e^{at}(G \cos bt + F \sin bt)$$

where $F + iG$ is a polynomial of the same degree as A.

Equations of the type considered in Theorems 1–4 have a host of applications in a branch of engineering known as systems science, which embraces the theory of mechanical and electrical circuits and certain aspects of the theory of feedback and control. Since these equations are the subject of a vast literature, it is of considerable interest that so much can be said about the form of their solutions. But the exponential shift theorem is usually preferable for calculation. As in the above proofs, it leads to the correct form automatically and reduces complicated cases to the case covered by Theorem 1.

EXAMPLE 1. What form do you expect for a particular solution?

$$(D-1)^3(D-2)y = 3t^2, \quad (D-8)^4 Dy = te^{7t}, \quad (D+1)(D^2+4)^2 y = \sin t$$

Since 0 is not a root of the first characteristic equation, nor 7 a root of the second, nor i a root of the third, the form is similar to that of the right-hand side. Namely, Theorems 1, 2, and 4 give the respective trial solutions

$$a + bt + ct^2, \quad (a+bt)e^{7t}, \quad a\sin t + b\cos t.$$

EXAMPLE 2. What form do you expect for a particular solution now?

$$D^3(D-2)y = 3t^2, \quad (D-7)^4 y = te^{7t}, \quad (D+1)(D^2+1)^2 y = \sin t$$

Here the characteristic equations have a root 0 of multiplicity 3, a root 7 of multiplicity 4, and a root i of multiplicity 2, respectively. By Theorems 3 and 4 the trial solutions of Example 1 must be multiplied respectively by t^3, t^4, and t^2.

(b) Asymptotic stability

In practical investigations one is often interested in the long-time behavior, that is, in the behavior of solutions as $t \to \infty$. The long-time behavior is also referred to as the *steady state*. The following informal discussion may give some idea of what the subject is about and why it is important.

For some equations the behavior as $t \to \infty$ depends strongly on the initial conditions. For example, if the equation is

$$y' - y = 5$$

the particular solution $y = -5$ is overwhelmed by the term ce^t in the complementary function. To be sure, the initial conditions could by accident make $c = 0$, but in any realistic case $c \neq 0$ is a virtual certainty. For this equation there is no well-defined steady state.

For the equation $y' + y = 5$, however, the particular solution $y = 5$ is almost unaltered, for large t, by the addition of the complementary function ce^{-t}. Every solution tends to 5 as a limit and, indeed, with exponential rapidity. Since every solution tends to 5, it is said that the solution $y = 5$ is *asymptotically stable*.

In problems of circuit analysis the steady state is ordinarily attained in a small fraction of a second. In other areas—ecology, economics, biology—the effect of the initial conditions generally persists for a long time, but still the concepts of "asymptotic stability" and "steady-state solution" retain their usefulness. Here we give a precise formulation and elaboration of these ideas, starting with the following:

DEFINITION. A particular solution u of the equation $Ty = f$ is asymptotically stable if every solution v of this equation satisfies

$$\lim_{t \to \infty} |u(t) - v(t)| = 0.$$

EXAMPLE 3.   To get a particular solution u of the equation

$$(D^4 + 4D^3 + 6D^2 + 5D + 2)y = 260 \sin 2t$$

replace the right-hand member by $260e^{2it}$ and try $u = Ae^{2it}$. This leads to

$$(16 - 32i - 24 + 10i + 2)A = 260 \quad \text{or} \quad (-6 - 22i)A = 260.$$

By a short calculation $A = -3 + 11i$ and hence

$$u = \text{Im} \ (-3 + 11i)(\cos 2t + i \sin 2t) = 11 \cos 2t - 3 \sin 2t.$$

If v is any other solution, the principle of the complementary function gives

(4) $v(t) = u(t) + c_1 e^{a_1 t} + c_2 e^{a_2 t} + c_3 e^{a_3 t} + c_4 e^{a_4 t}$

where the a_j are the roots of the characteristic equation $P(s) = 0$. In Problem 4 it is found that each a_j satisfies

$$\text{Re} \ a_j < 0, \quad \text{hence} \quad \lim_{t \to \infty} e^{a_j t} = 0.$$

This shows that $v(t) = u(t) + \varepsilon(t)$ where $\lim_{t \to \infty} \varepsilon(t) = 0$. Thus $v(t) - u(t) \to 0$ as $t \to \infty$ so that u is asymptotically stable.

A moment's thought shows that if *any* solution of $Ty = f$ is asymptotically stable then *every* solution is asymptotically stable and every solution is equally valid as a description of the steady state. Since we can often find a particular solution of $Ty = f$ without solving the characteristic equation, the notion of asymptotic stability leads to great simplification in practice.

The following theorem shows that the question of asymptotic stability depends only on the homogeneous equation, not on f:

PRINCIPLE OF ASYMPTOTIC STABILITY. Let u be a solution of $Tu = f$. Then u is asymptotically stable if, and only if, every solution of the homogeneous equation $Ty = 0$ tends to 0 as $t \to \infty$.

For proof, let v be any solution of $Tv = f$ and let $w = u - v$. By linearity $Tw = Tu - Tv = f - f = 0$. If *every* solution of the homogeneous equation tends to 0, then w does, and this shows that u is asymptotically stable. On the other hand if some solution w of the homogeneous equation does not tend to 0 we could pick $v = u + w$. Then $Tv = f$, but $u - v$ does not tend to 0. This shows that u was not asymptotically stable and completes the proof.

It is not difficult to show that

$$\lim_{t \to \infty} t^m e^{st} = 0, \qquad \text{Re } s < 0, \ m = 0, 1, 2, \cdots$$

and conversely, Re s must be negative if the limit is 0. Using this together with (3), we get the following:

COROLLARY. The solutions of $Ty = f$ are asymptotically stable if, and only if, all roots of the characteristic equation have negative real parts.

Under the hypothesis of the corollary any particular solution of $Tu = f$ can be used to describe the steady state, and the complementary function v represents a *transient* that depends on the initial conditions. The transient decays exponentially with time, and that is the reason why initial conditions are not emphasized in this chapter. Separation of a solution $y = u + v$ into a steady-state part u and a transient v will be seen again in connection with the Laplace transform.

PROBLEMS

1. What form do you expect for particular solutions of the following equations?

(a) $(D^4 + D^2)y = t^2$ (b) $(D^2 - 1)^3 y = te^t$ (c) $(D^2 + 1)^4 y = t \sin t$

2. For an equation of the form $Ty = Ae^{kt}$, where A and k are constant, a useful strategy is to try $y = Be^{kt}$ with constant B. If $P(k) \neq 0$ the process gives an immediate solution. If $P(k) = 0$ you have not wasted your time, because you now know that $D - k$ is a factor of $P(D)$. As an illustration, consider the following equation where $C = 5$ or $C = 6$:

$$(D^4 + 2D^3 + 3D + C)y = 7e^{-2t}$$

(a) Try $y = Be^{-2t}$ and show that this gives an immediate solution if $C = 5$, and it guarantees that $P(D)$ is divisible by $D + 2$ if $C = 6$.

(b) By division get the other factor when $C = 6$ and solve by the shift theorem. Or use the shift theorem directly, without factoring $P(D)$.

3. Obtain a particular solution (a) when $C = 2$ and (b) when $C = 1$:

$$(D^4 + 2D^3 + 2D^2 + 2D + C)y = 8e^{-t}$$

4. Check that the characteristic polynomial in Example 3 has the following factorization, and hence that the zeros a_j satisfy Re $a_j < 0$:

$$P(s) = (s^2 + 3s + 2)(s^2 + s + 1)$$

5. The solutions of $Tx = A \cos \omega t$ and of $Ty = B \sin \omega t$ are said to be *resonant* if $i\omega$ is a root of the characteristic equation. Show that the equation on the left has the resonant solution on the right:

$$(D + 3)(D^2 + 1)^5 x = 640 \cos t, \qquad x = \frac{t^5}{5!}(-2 \cos t + 6 \sin t)$$

The point is that this solution grows like t^5 while the complementary function has the order of t^4 at most. Thus any particular solution gives the dominant behavior for large t, even though the solutions are not asymptotically stable. Hint: Consider $(D+3)(D+i)^5(D-i)^5 z = 640e^{it}$ and use the shift theorem.

6. Show that the equation on the left has the resonant solution on the right:

$$(D + 2)(D^2 + 1)^3 y = 240 \sin t, \qquad t^3(2 \cos t + \sin t)$$

7. Let $P(s) = s^2 + p_1 s + p_0 = (s - r_1)(s - r_2)$ be a quadratic with real coefficients p_j and real or complex zeros r_j. Show that Re $r_1 < 0$ and Re $r_2 < 0$ both hold if, and only if, $p_1 > 0$ and $p_0 > 0$ both hold. Results such as this make it possible to decide about asymptotic stability without solving the characteristic equation.

8. This problem requires knowledge of higher-order determinants. The *Routh-Hurwitz criterion* gives a necessary and sufficient condition for all the zeros of a real polynomial

$$b_0 s^n + b_1 s^{n-1} + b_2 s^{n-2} + \cdots + b_{n-1} s + b_n, \qquad b_0 > 0$$

to have negative real parts. The criterion is as follows. Set $b_j = 0$ for $j > n$ and form the determinant

$$D_n = \begin{vmatrix} b_1 & b_0 & 0 & 0 & 0 & 0 & \cdots & 0 \\ b_3 & b_2 & b_1 & b_0 & 0 & 0 & \cdots & 0 \\ b_5 & b_4 & b_3 & b_2 & b_1 & b_0 & \cdots & 0 \\ \vdots & \vdots & \vdots & \vdots & \vdots & \vdots & \ddots & \vdots \\ b_{2n-1} & b_{2n-2} & b_{2n-3} & b_{2n-4} & b_{2n-5} & b_{2n-6} & \cdots & b_n \end{vmatrix}$$

Then all of the roots have negative real parts if, and only if, D_n and all its principle minors

$$b_1 \quad \begin{vmatrix} b_1 & b_0 \\ b_3 & b_2 \end{vmatrix}, \quad \begin{vmatrix} b_1 & b_0 & 0 \\ b_3 & b_2 & b_1 \\ b_5 & b_4 & b_3 \end{vmatrix}, \quad \begin{vmatrix} b_1 & b_0 & 0 & 0 \\ b_3 & b_2 & b_1 & b_0 \\ b_5 & b_4 & b_3 & b_2 \\ b_7 & b_6 & b_5 & b_4 \end{vmatrix}, \quad \cdots$$

are positive. Check that the determinant for the polynomial in Example 3 is the following determinant D_4 and show that it satisfies the Routh-Hurwitz criterion:

$$D_4 = \begin{vmatrix} 4 & 1 & 0 & 0 \\ 5 & 6 & 4 & 1 \\ 0 & 2 & 5 & 6 \\ 0 & 0 & 0 & 2 \end{vmatrix}$$

9. This problem pertains to the double integral

$$y = \int_0^t \int_0^r se^{as} \sin bs \, ds \, dr$$

where a and b are real nonzero constants.

(a) By considering $D^2 z = te^{(a+ib)t}$ reduce the problem to a simpler form.

(b) Solve completely when $a = b = 1$.

10. If the a_j are unequal constants, real or complex, show that any n functions $e^{a_j t}$ are *linearly independent* on any given interval I. That is, if

$$c_1 e^{a_1 t} + c_2 e^{a_2 t} + \cdots + c_n e^{a_n t} = 0, \qquad t \in I$$

the constants c_j are all 0.

Hint: Divide by $e^{a_n t}$ and differentiate. This reduces it to the case of $n - 1$ functions.

———————————————ANSWERS———————————————

1. (a) $t^2(at^2 + bt + c)$ (b) $t^3(at + b)e^t$ (c) $t^4((a + bt)\sin t + (c + dt)\cos t)$
2. (a) $y = -7e^{-2t}$ $P(D) = (D^3 + 3)(D + 2)$ (b) $5y = -7te^{-2t}$
3. (a) $8e^{-t}$ (b) $2t^2 e^{-t}$ 6. (b) $2y = e^t(\cos t + \sin t - t \cos t) - 1 - t$
9. (b) $2y = e^t(\cos t + \sin t - t \cos t) - 1 - t$

An elegantly executed proof is a poem in all but the form
in which it is written.

                                         —Morris Kline

Chapter 9

APPLICATIONS

MANY PROBLEMS lead to a system of two or more differential equations in two or more unknowns. Such systems can often be reduced to a single differential equation of higher order and solved as in the previous chapter. Using this technique we analyze the transfer of energy between coupled pendulums, the behavior of a radioactive chain in which the decay products are themselves radioactive, and the diffusion of a solute between compartments separated by a membrane. Similar methods apply to boundary-value problems in which the side conditions are specified not just at one point but at two. As examples, we consider the buckling of a supporting column in a building and the stability of a rotating propeller shaft.

1. Applications of systems

Here we show how linear systems arise in the study of mechanics. Applications to other disciplines are given in the accompanying problems.

(a) A coupled mass-spring system

Let masses m_1 and m_2 be suspended in tandem from springs of stiffness coefficients k_1 and k_2 as shown in Figure 1. If x and y denote the displacements of the masses m_1 and m_2 from their equilibrium positions, a procedure similar to that in Section 7.1 gives the two equations

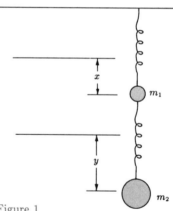

Figure 1

$$m_1 x'' = k_2(y - x) - k_1 x, \quad m_2 y'' = -k_2(y - x)$$

when friction is negligible. Upon division by m_1 and m_2 these equations become

$$x'' = c(y - x) - ax, \quad y'' = -b(y - x)$$

or also

$$x'' + (a + c)x = cy, \quad y'' + by = bx$$

where $a = k_1/m_1$, $b = k_2/m_2$, and $c = k_2/m_1$. If we define operators S and T by

$$S = D^2 + a + c, \quad T = D^2 + b$$

the equations take the form

$$Sx = cy, \quad Ty = bx.$$

This is a system of two simultaneous differential equations for the displacements x and y.

To find y let us operate on the second equation with S and substitute for Sx the value given by the first equation. The result is $STy = bSx = bcy$ or, written in full,

$$(D^2 + a + c)(D^2 + b)y = bcy.$$

Multiplying out yields

(1) $$(D^4 + (a + b + c)D^2 + ab)y = 0.$$

After we have solved this fourth-order equation for y the equation

(2) $$bx = y'' + by$$

gives x. There will be four arbitrary constants, since the equation for y is of order 4 and no new constants are introduced in the calculation of x. By Section 8.2, Theorem 1, the constants enable us to satisfy the initial conditions

$$x(0) = x_0, \quad x'(0) = x_1, \quad y(0) = y_0, \quad y'(0) = y_1.$$

These conditions are physically reasonable, since they specify the initial position and velocity of each mass.

(b) Natural frequencies

The characteristic equation

$$s^4 + (a + b + c)s^2 + ab = 0$$

associated with (1) is a quadratic in s^2. The discriminant is positive because

$$(a + b + c)^2 > (a + b)^2 \geq 4ab.$$

(The latter inequality follows from $(a - b)^2 \geq 0$.) Since the discriminant is positive the roots s^2 must be real and distinct, and they are negative because the coefficients are positive. Thus

$$s = \pm i\omega_1, \quad s = \pm i\omega_2$$

where ω_i are real and distinct. The corresponding solution is

$$y = c_1 \cos \omega_1 t + c_2 \sin \omega_1 t + c_3 \cos \omega_2 t + c_4 \sin \omega_2 t$$

and the equation (2) gives a similar expression for x.

The analysis indicates that the motion of each mass is a combination of two simple harmonic motions of angular frequencies ω_1 and ω_2. These are the *natural frequencies* of the system.

(c) Coupled pendulums

Figure 2 shows an oscillating pendulum of length ℓ and mass m. It is assumed that the bob can be regarded as a mass-point and that the rod joining it to the point of support is of negligible weight. Air resistance and friction at the pivot are also neglected.

Let θ be the angle in radians between the pendulum and the vertical as shown in the figure. The corresponding arc is $s = \ell\theta$ and the tangential acceleration is $s'' = \ell\theta''$. The force due to gravity on the pendulum bob is mg and the tangential component of this force is $mg \sin \theta$ as shown by the small triangle in the figure. Hence the equation of motion is

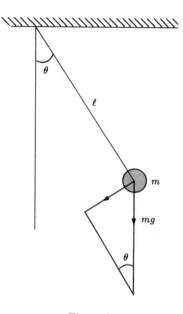

$$m\ell\theta'' = -mg \sin \theta \quad \text{or} \quad \ell\theta'' = -g \sin \theta.$$

Figure 2

The minus sign is needed because the force of gravity is directed downwards and tends to move the pendulum back to the equilibrium position. The normal component of the force and the centrifugal force are canceled by the reaction of the supporting rod.

Since l'Hospital's rule gives

$$\lim_{\theta \to 0} \frac{\sin \theta}{\theta} = \lim_{\theta \to 0} \frac{\cos \theta}{1} = 1$$

we have $\sin \theta \approx \theta$ when $|\theta|$ is small. Hence the equation for small oscillations is

$$\ell\theta'' = -g\theta.$$

The motion is simple harmonic motion with angular frequency

$$\omega = \sqrt{\frac{g}{\ell}}.$$

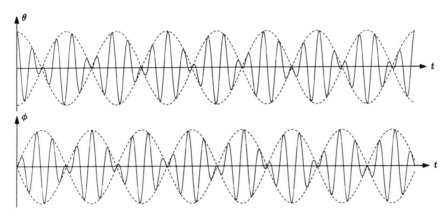

Figure 5

PROBLEMS——————————————————————————————————

1. A two-compartment problem. Figure 6 shows a tank with two compartments separated by a permeable mem-
brane. The volumes of the compartments are V_1 and V_2, independent of
time. At time t a quantity Q_1 of dis-
solved substance is in the first compart-
ment and a quantity Q_2 in the second.
The concentration is $C_i = Q_i/V_i$. The
rate of diffusion through the membrane
is proportional to the difference in con-
centrations on the two sides, and the direction is from higher to lower concentration. Thus

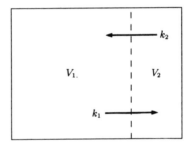

Figure 6

$$\frac{dQ_1}{dt} = k_1(C_2 - C_1), \qquad \frac{dQ_2}{dt} = k_2(C_1 - C_2)$$

where the constants k_i measure the diffusivity of the membrane in the two
directions.

(a) Divide the first equation by k_1, the second by k_2, and add to deduce

$$C_1(t)\frac{V_1}{k_1} + C_2(t)\frac{V_2}{k_2} = C_1(0)\frac{V_1}{k_1} + C_2(0)\frac{V_2}{k_2}$$

(b) With $Q_i = C_i V_i$, divide the first equation by V_1, the second by V_2, and
subtract to get an equation for $C_1 - C_2$. Thus deduce

$$C_1(t) - C_2(t) = (C_1(0) - C_2(0))e^{-kt} \quad \text{where} \quad k = \frac{k_1}{V_1} + \frac{k_2}{V_2}$$

2. A radioactive chain. Let P, Q, R denote the amounts of three radioactive substances at time t. Suppose $P \to Q \to R$ in the sense that the decay product of P is Q and the decay product of Q is R. This means

$$\frac{dP}{dt} = -aP, \qquad \frac{dQ}{dt} = aP - bQ, \qquad \frac{dR}{dt} = bQ - cR$$

where a, b, c are positive constants related to the half-lives of P, Q, R respectively. The process starts with an amount I of the first substance, so that $P(0) = I$, and none of the others has yet been produced. If a, b, c are all unequal, show that at time t

$$\frac{R}{I} = \frac{ab}{(a-b)(a-c)} e^{-at} + \frac{ab}{(b-c)(b-a)} e^{-bt} + \frac{ab}{(c-a)(c-b)} e^{-ct}$$

Hint: Find first P, then Q, then R.

3. Isolated coupled masses. Figure 7 shows two freight cars joined by a spring and constrained to move on a straight frictionless track. Since the track cancels the force due to gravity there is no external force on the system as a whole, and in this sense the system is isolated. The position and mass of the cars are x_1, m_1 for the first and x_2, m_2 for the second, with $x_1 \le x_2$. We assume Hooke's law for the spring whether it is in tension or compression, so that the stretching beyond the equilibrium length ℓ_0 can be either positive or negative. This stretching is not equal to the *length* of the spring $x_2 - x_1$, however, but to the *elongation* $x_2 - x_1 - \ell_0$. We take the positive direction to the right. Thus the effect of the spring on the first car is to pull it to the right, and on the second car, to pull it to the left. These considerations lead to the following system where k is the spring constant:

$$m_1 \frac{d^2 x_1}{dt^2} = k(x_2 - x_1 - \ell_0), \qquad m_2 \frac{d^2 x_2}{dt^2} = -k(x_2 - x_1 - \ell_0)$$

(a) By adding the equations show that the center of mass

$$\overline{x} = \frac{m_1 x_1 + m_2 x_2}{m_1 + m_2}$$

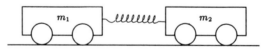

moves with constant velocity. Similar behavior is found for more general mass systems if the total external force on the system is zero.

Figure 7

(b) The quantity $u = x_2 - x_1 - \ell_0$ represents the elongation of the spring at time t. Check that

$$m_1 x_1'' = ku, \qquad m_2 x_2'' = -ku, \qquad u'' = x_2'' - x_1''$$

(c) Deduce from Part (b) that u executes simple harmonic motion according to the equation

$$mu'' + ku = 0 \quad \text{where} \quad \frac{1}{m} = \frac{1}{m_1} + \frac{1}{m_2}$$

4. Coupled electrical circuits. The currents i_1 and i_2 in the two coupled circuits shown in Figure 8 satisfy the following differential equations:

$$M\frac{d^2 i_2}{dt^2} + L_1\frac{d^2 i_1}{dt^2} + R_1\frac{di_1}{dt} + \frac{i_1}{C_1} = 0$$

$$M\frac{d^2 i_1}{dt^2} + L_2\frac{d^2 i_2}{dt^2} + R_2\frac{di_2}{dt} + \frac{i_2}{C_2} = 0$$

Here M is the *mutual inductance* between the inductors associated with L_1 and L_2. We assume $M > 0$, so the circuits are actually coupled.

Figure 8

(a) If operators T_i are defined by

$$T_i = L_i D^2 + R_i D + \frac{1}{C_i}$$

check that the above equations become

$$MD^2 i_2 = -T_1 i_1, \quad MD^2 i_1 = -T_2 i_2$$

(b) Operate on the second equation of Part (a) with MD^2 and use the first equation to show that

$$(M^2 D^4 - T_2 T_1)i_1 = 0$$

Since $T_1 T_2 = T_2 T_1$, the same result holds for i_2.

(c) Assuming $M^2 < L_1 L_2$, analyze the case $R_1 = R_2 = 0$. In particular, show that the characteristic equation reduces to a quadratic in s^2 in which all the coefficients have the same sign and the discriminant is

$$\left(\frac{L_1}{C_2} - \frac{L_2}{C_1}\right)^2 + \frac{4M^2}{C_1 C_2} > 0$$

Hence the response is purely oscillatory with two natural frequencies ω_1 and ω_2, as in Part (b) of the text.

2. Boundary-value problems

In all examples discussed above the independent variable has been the time t, and arbitrary constants in the solutions have been determined by prescribing the value and some of the derivatives at a single time, t_0. We now present two applications in which the independent variable is a space variable x and the side conditions are specified not just at one point but at two. Problems of this type are known as *boundary-value problems*. They have great importance in mathematics and in several branches of mathematical physics.

(a) The Bernoulli-Euler law

Consider a uniform horizontal beam under the action of vertical loads. We assume that all force-vectors acting on the beam lie in a vertical plane containing the central axis. Choose the x axis along the central axis of the beam in the undeformed state and the positive y axis downward as shown in Figure 9. Under the action of the external forces the beam is bent and its central axis is deformed. The deformed axis, shown by the dashed line in the figure, is called the *elastic curve* and it is an important problem in the theory of elasticity to determine its shape.

According to the *Bernoulli-Euler law*, a beam made of a material that obeys Hooke's law deforms in such a way that the curvature K of the elastic curve at a given point is proportional to the bending moment M at that point. More specifically,

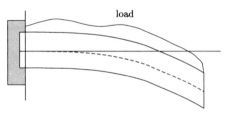

Figure 9

$$(1) \quad K = \frac{y''}{(1 + (y')^2)^{3/2}} = \frac{M}{EI}$$

where E is Young's modulus, I is the moment of inertia of the cross section of the beam, and y is the ordinate of the elastic curve. Both K and M are measured at x. The moment of inertia I is taken about a horizontal axis that lies in the plane of the cross section and passes through the center of mass of the cross section. The quantity EI, called the *flexural rigidity*, is assumed constant in the sequel.

When the beam is a structural element like a girder in a building, one would expect the deflection to be small, so that the slope y' is also small. If the term $(y')^2$ in (1) is neglected the result is the linear equation

$$(2) \qquad\qquad\qquad y'' = \frac{M}{EI}.$$

This is, by definition, the equation for small deflection of an elastic beam under loads as described above. Surprising as it may seem, the following analysis, based on (2), gives results that are mathematically exact when applied to (1).

(b) The Euler column

It is known from experiments that a long rectilinear rod subjected to the action of axial compressive forces is compressed and retains its initial shape as long as the compressive forces do not exceed a certain critical value. Upon gradual increase of the compressive load P a value $P = P_1$ is reached when the rod buckles suddenly and becomes curved. The deflection of a rod so compressed is extremely sensitive to minute changes of the load and increases rapidly as P increases. Since the rod is initially straight, analysis of the onset of this *instability* or *buckling* phenomenon can be based on the equation for small deflections.

Thus, consider a rod of uniform cross section and length ℓ compressed by the forces P applied to its ends. Initially the rod is straight but after the critical load P_1 is reached it becomes curved as shown in Figure 10. We denote the deflection of its central line at point x by y and we note that, as the process just begins, y is arbitrarily small. Hence (2) can be used. The bending moment in this case is $M = -Py$, since y is the distance from (x, y) to the line of action of the force. Thus

$$(3) \qquad y'' + k^2 y = 0, \qquad k^2 = \frac{P}{EI}$$

where the primes denote differentiation with respect to x. Equation (3) must be solved subject to the conditions

$$(4) \quad y(0) = 0, \quad y(\ell) = 0$$

since the ends of the rod remain on the axis.

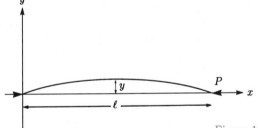

Figure 10

The *boundary-value problem* described by Equations (3) and (4) is quite different from the initial-value problems we have considered hitherto. Since the conditions (4) are given at two points, it is not obvious that there is any solution other than the trivial solution $y = 0$. We shall see, however, that for suitable choices of the parameter k, nontrivial solutions do exist.

The general solution of (3) is

$$y = c_1 \cos kx + c_2 \sin kx$$

and on imposing the two conditions (4) we get

$$c_1 = 0, \quad c_2 \sin k\ell = 0.$$

The choice $c_1 = c_2 = 0$ gives $y = 0$ corresponding to the rectilinear shape of the rod. If the rod does not remain straight we must have $c_2 \neq 0$, hence $\sin k\ell = 0$, and therefore

$$k = \frac{n\pi}{\ell}, \qquad n = 0,\ 1,\ 2,\ 3,\ \cdots.$$

The choice $n = 0$ gives $y = 0$. If $n = 1$ then $k = \pi/\ell$ and on recalling the definition of k from (3) we see that

$$P_1 = EI\frac{\pi^2}{\ell^2}.$$

This is the *critical*, or *Euler* load. The choice $n = 2,\ 3,\ \cdots$ gives other critical loads $P_2,\ P_3,\ \cdots$.

(c) Rotating shaft

If a long shaft such as a propeller shaft is allowed to rotate, its initially rectilinear shape is preserved only if the angular speed of rotation ω does not exceed a certain critical value ω_1. On approaching the speed ω_1 the shaft starts pulsating and its shape changes. On further increase of the speed another critical value ω_2 is reached when the shaft starts beating and its shape changes again, and so on. This phenomenon can be explained by calculations similar to those used in determining the Euler load.

Let us suppose that the shaft is supported by bearings at $x = 0$ and $x = \ell$ and is rotating with angular speed ω. An element of length Δs of the shaft experiences the centrifugal force

$$F\Delta s \doteq \rho\,\Delta s\,\omega^2 y$$

where ρ is the density per unit length of shaft and y is the deflection at point x. The inequality is only approximate because y is not constant. However in the limit we find that

$$(5) \qquad\qquad F = \rho\omega^2 y$$

is the force per unit length of shaft distributed along its length. For small deflections we need not distinguish the arc s from the coordinate x.

It is shown in books on the strength of materials that when forces F acting on a rod are normal to its axis then $F = M''$ where the bending moment M is given by the Bernoulli-Euler law $M = EIy''$. Combining these remarks we get

$$y^{(iv)} = \frac{F}{EI}.$$

The substitution of F from (5) gives the following equation for the rotating shaft:

$$(6) \qquad\qquad y^{(iv)} = k^4 y, \qquad k^4 = \frac{\rho \omega^2}{EI}.$$

Throughout the above analysis we have assumed small deflection, but this is permissible for the same reason as in the discussion of Part (b). Namely, we are interested in the deformation only when it has just barely started.

Since the characteristic equation for (6) is $s^4 = k^4$ its roots are

$$s = \pm k, \quad s = \pm ki$$

and the general solution is

$$(7) \qquad\qquad y = c_1 \cosh kx + c_2 \sinh kx + c_3 \cos kx + c_4 \sin kx.$$

If the bearings are short, that is, if they extend only a small distance along the axis of the shaft, the bending moment M at the bearings, as well as the deflection y is almost 0. Since M is proportional to y'' the initial conditions are then

$$(8) \qquad\qquad y(0) = 0, \quad y''(0) = 0, \qquad y(\ell) = 0, \quad y''(\ell) = 0.$$

Substitution of the first two equations (8) into (7) yields

$$c_1 + c_3 = 0, \quad c_1 - c_3 = 0.$$

Hence $c_1 = c_3 = 0$ and

$$y = c_2 \sinh kx + c_4 \sin kx.$$

The conditions (8) at ℓ now yield

$$c_2 \sinh k\ell + c_4 \sin k\ell = 0, \quad c_2 \sinh k\ell - c_4 \sin k\ell = 0.$$

This is a linear system for c_2 and c_4 with determinant $-2 \sinh k\ell \sin k\ell$. Since the determinant must be 0 for a nontrivial solution, we conclude that

$$k\ell = n\pi, \qquad n = 1, 2, 3, \cdots.$$

Upon recalling the value of k from (6) we see that the first critical speed is

$$\omega_1 = \frac{\pi^2}{\ell^2} \sqrt{\frac{EI}{\rho}}.$$

The critical speeds w_2, w_3, $\cdots$ are determined by taking $n = 2$, 3, $\cdots$.

PROBLEMS

1. If a wire is stretched between two fixed points, as in a piano or harp, analysis of its vibrations leads to the boundary-value problem

$$y'' + ky = 0, \qquad y(0) = y(\ell) = 0$$

where k is constant and ℓ is the length of the wire.

(a) Check that the general solution of the differential equation is

$$c_1 \cos ax + c_2 \sin ax, \qquad c_1 + c_2 x, \qquad c_1 \cosh ax + c_2 \sinh ax$$

when $k = a^2$ is positive, $k = 0$, or $k = -a^2$ is negative, respectively.

(b) Show that in two of these cases the sole solution is $y = 0$ but in the remaining case there is a nontrivial solution for certain values of k. What are these values?

2. In this problem the independent variable is the time t and primes denote differentiation with respect to t. An oscillating mass-spring system satisfies $my'' + ky = 0$ where m and k are positive constants. It is observed that $y = 0$ at $t = 0$ and at a later time T. What are the possible values of the ratio k/m?

3. Solve, if possible, subject to the boundary conditions $y(0) = y(\pi) = 0$:

(a) $y'' - y' = 0$ (e) $y'' + y = 0$ (i) $y'' + y = x^2 - \pi x + 2$
(b) $y'' - y' = 2x$ (f) $y'' + y = e^x$ (j) $y'' + y = x^2 - 3x + 2$
(c) $4y'' + y = 0$ (g) $y'' - 2y' + 2y = 0$ (k) $y'' - 4y' + 5y = 8 \sin x$
(d) $4y'' + y = 1$ (h) $y'' - 2y' + 2y = 2$ (l) $y'' - 4y' + 5y = 8e^x$

4. This problem and the next pertain to the situation illustrated in Figure 11. In the case of a uniformly loaded beam the moment $M(x)$ is a quadratic function of x and hence, for small deflection, $y^{(iv)} = $ const. By choice of units we can make the length of the beam 1, and we then multiply y by a constant so that the equation is $y^{(iv)} = 24$.

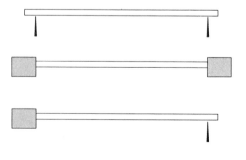

Figure 11

(a) If $y(0) = y(1) = 0$ show that
$$y = x^4 + ax^3 + bx^2 - (1 + a + b)x$$
where a and b are constant.

(b) If the ends are pivoted y satisfies $y''(0) = y''(1) = 0$ in addition to the boundary conditions of Part (a). Find y.

(c) If the ends are clamped y satisfies $y'(0) = y'(1) = 0$ in addition to the boundary conditions of Part (a). Find y.

(d) Show that the maximum deflection when the ends are clamped is $1/5$ as much as when they are pivoted. Since this information gives a check, no further answer is provided.

5. Suppose the beam in Problem 4 has the end $x = 0$ clamped and the end $x = 1$ pivoted. Find y and find the point x at which the deflection is maximum. What is the approximate value of the maximum deflection? The true value is given in the answers.

6.   If the bearings in the propeller problem keep the shaft horizontal as it enters the bearings, the boundary conditions are

$$y(0) = y'(0) = 0, \qquad y(\ell) = y'(\ell) = 0$$

(a) Show that the constant k giving the critical speeds now satisfies the transcendental equation $\cos k\ell \cosh k\ell = 1$.

(b) Plot the curves $h = \cos k\ell$, $h = \operatorname{sech} k\ell$ on a single diagram and deduce that the equation of Part (a) has infinitely many solutions k.

───────────────────────ANSWERS───────────────────────

1. $k = (n\pi/l)^2$, $n = 1, 2, 3, \cdots$ 2. $n^2\pi^2/T^2$, $n = 1, 2, 3, \cdots$

3. (a) $y = 0$ (g) $y = ce^x \sin x$

    (b) $y = \pi(\pi + 2)(e^x - 1)/(e^\pi - 1) - x^2 - 2x$ (h) no solution

    (c) $y = 0$ (i) $y = x(x - \pi) + c\sin x$

    (d) $y = 1 - \cos(x/2) - \sin(x/2)$ (j) no solution

    (e) $y = c\sin x$ (k) no solution

    (f) no solution   (l) no solution

5. $2y = 2x^4 - 5x^3 + 3x^2$, $2^4 x = 15 - \sqrt{33}$, $2^{13} y_{\max} = 117 + 165\sqrt{33}$

Chapter 10

LINEAR SECOND ORDER, VARIABLE

SECOND-ORDER linear equations with variable coefficients do not have elementary solutions, in general, and they provide an important means of defining new classes of functions. Confining attention to second-order equations, we begin with a study of the Wronskian and linear independence. A result known as Abel's identity yields the complete solution of the homogeneous equation from a single nonvanishing solution, and Lagrange's method of variation of parameters then gives the complete solution of the inhomogeneous equation. As an application, we use variation of parameters to construct Green's functions, which are an aid in the solution of both initial-value and boundary-value problems.

1. The Wronskian

The general linear equation has the form

(1) $p_n(x)y^{(n)} + p_{n-1}(x)y^{(n-1)} + \cdots + p_1(x)y' + p_0(x) = f(x).$

This differs from most of the linear equations discussed hitherto in that the coefficients need not be constant. We use x rather than t because equations with nonconstant coefficients are often encountered in boundary-value problems, in which the variable pertains to space rather than time.

If (1) is written in the form $Ty = f$, a discussion virtually identical to that for constant coefficients shows that T is linear. Hence the general principle of superposition, the principle of equating real parts, and all the other results established for linear operators in the foregoing discussion can be used for (1). When $n = 2$ the equation is of the second order. That case is emphasized here because it is typical of the general case and includes many equations of mathematical physics.

Let I be an open interval of the real axis. If $n = 2$ and $p_2(x) \neq 0$ for $x \in I$ we can divide by $p_2(x)$ and write the equation in the form

(2) $Ty = y'' + p(x)y' + q(x)y = f(x), \qquad x \in I.$

Until further notice, the operator T will be defined by (2) and p, q, f are assumed continuous on I. Under these conditions (2) has one, and only one, solution satisfying prescribed initial conditions

$$y(x_0) = y_0, \quad y'(x_0) = y_1$$

at any prescribed point $x_0 \in I$. For present purposes, the most important property is uniqueness, and this is established in Chapter A. A proof of existence for the nth-order linear equation is in DETA, page 341.

The *Wronskian* of two differentiable functions u and v is

$$W = \begin{vmatrix} u & v \\ u' & v' \end{vmatrix} = uv' - vu'$$

where the symbol in the center is a 2-by-2 determinant. Determinants of this form were introduced by the Polish mathematician Wronski in 1811, hence the name. We write $W(x)$ or $W(u,v)$ to emphasize dependence on x or on the functions u,v, as the case may be. For example

$$W(\cos x, \sin x) = \begin{vmatrix} \cos x & \sin x \\ -\sin x & \cos x \end{vmatrix} = \cos^2 x + \sin^2 x = 1.$$

Let $v \neq 0$ in I. According to the definition in Section 6.1, the functions u and v are linearly dependent on I if the ratio u/v is constant. In that case

$$\frac{d}{dx}\left(\frac{u}{v}\right) = \frac{vu' - uv'}{v^2} = 0$$

provided the derivatives exist. Comparing W with the numerator of the fraction in the center, we see that $W(u,v) = 0$. Conversely, if $W(u,v) = 0$ on an interval the above calculation shows that the derivative of u/v is 0 and hence u/v is constant.

The assumption that $v \neq 0$ is essential for the second part of this argument but not for the first. If $u = cv$ on an interval, the functions being differentiable, then

$$u = cv, \quad u' = cv'.$$

Here the second relation is obtained by differentiating the first. If we multiply the first by v', the second by v, and subtract, we get

$$uv' - u'v = cvv' - cv'v = 0.$$

This says that $W(u,v) = 0$. The same conclusion follows if $v = cu$, and hence it follows whenever u and v are linearly dependent. Thus we have established:

THEOREM 1. *Let u and v be differentiable on I.*

(a) *If u and v are linearly dependent, then $W(u,v) = 0$ on I.*

(b) *If $W(u,v) = 0$ and $v \neq 0$ on I, then u and v are linearly dependent.*

EXAMPLE 1. If a and b are unequal constants, real or complex,

$$W(e^{ax}, e^{bx}) = \begin{vmatrix} e^{ax} & e^{bx} \\ ae^{ax} & be^{bx} \end{vmatrix} = e^{(a+b)x}(b-a).$$

Since this is not 0 at any x, Theorem 1 indicates that the functions e^{ax} and e^{bx} are linearly independent on every interval.

It is seen in Problem 3 that the condition $W(u,v) = 0$ on an interval does not, in general, ensure that the functions u and v are linearly dependent. The following theorem shows, however, that a result of this kind does hold if u and v are solutions of $Ty = 0$:

THEOREM 2. *Let u and v satisfy the equation $Ty = 0$ on I, and suppose that their Wronskian satisfies $W(x_0) = 0$ at some point $x_0 \in I$. Then u and v are linearly dependent on I, and hence $W(x) = 0$ for all $x \in I$.*

The proof is simple. Since $W(x_0) = 0$ it is known from linear algebra that you can choose constants c_i, not both 0, such that

$$c_1 u(x_0) + c_2 v(x_0) = 0$$
$$c_1 u'(x_0) + c_2 v'(x_0) = 0.$$

(Note that $W(x_0)$ is the determinant of the system.) Hence the function

$$y = c_1 u(x) + c_2 v(x)$$

satisfies null initial conditions at x_0. Uniqueness now gives $y = 0$ on I, which shows that u and v are linearly dependent. The equation $W(x) = 0$ follows from Theorem 1.

An obvious consequence of Theorem 2 is:

THEOREM 3. *If u and v are solutions of the homogeneous equation $Ty = 0$ on I, either $W(x) = 0$ at every point of I or $W(x) = 0$ at no point of I.*

The Wronskian has an interesting application to initial-value problems. If the homogeneous equation $Ty = 0$ has solutions u and v on an interval I then $y = c_1 u + c_2 v$ is also a solution for any constants c_1 and c_2. To satisfy prescribed initial conditions

$$y(x_0) = y_0, \quad y'(x_0) = y_1$$

at $x_0 \in I$ we must determine the constants c_1 and c_2 in such a way that

(3)
$$c_1 u(x_0) + c_2 v(x_0) = y_0$$
$$c_1 u'(x_0) + c_2 v'(x_0) = y_1.$$

It is known from linear algebra that (3) can be solved for arbitrary choices of y_0 and y_1 if, and only if, the coefficient determinant is not zero. This determinant is $W(x_0)$. By Theorem 3, the condition $W(x_0) \neq 0$ for solvability of the initial-value problem does not depend on x_0.

These remarks give information about the possible form of a solution of $Ty = 0$. On a given interval I let u and v be particular solutions of $Ty = 0$ such that $W(u, v) \neq 0$ and let y be any solution whatever. If $x_0 \in I$ we can choose constants c_1 and c_2 such that (3) holds with $y_0 = y(x_0)$ and $y_1 = y'(x_0)$. This means that the given function y and the special function

$$\tilde{y} = c_1 u + c_2 v$$

both satisfy $Ty = 0$ with the same initial conditions at x_0. But the uniqueness theorem says that there can be *at most one* solution with prescribed initial conditions. Hence $y = \tilde{y}$. In view of Theorem 3, we summarize as follows:

THEOREM 4. *On I let $Ty = 0$ have two solutions u, v such that $W(u, v) \neq 0$ at some point $x \in I$. Then every solution of $Ty = 0$ can be written in the form $y = c_1 u + c_2 v$ and the constants c_j can be chosen so as to satisfy arbitrarily specified initial conditions $y(x_0) = y_0$, $y'(x_0) = y_1$ at any point $x_0 \in I$. In other words, the family $c_1 u + c_2 v$ is both the complete solution and the general solution of $Ty = 0$.*

The Wronskian satisfies an important identity that is discussed next. Starting with

$$Tv = v'' + pv' + qv, \quad Tu = u'' + pu' + qu$$

multiply the first equation by u, the second by v, and subtract. This gives

(4) $$uTv - vTu = (uv'' - vu'') + p(uv' - vu').$$

If $W = uv' - vu'$ we have

$$W' = (uv' - vu')' = uv'' + u'v' - vu'' - v'u' = uv'' - vu''$$

and hence (4) says that

(5) $$uTv - vTu = W' + pW.$$

As a first application, suppose $Tu = Tv = 0$. Then

$$W' + pW = 0$$

and, solving for W, we get *Abel's identity*

$$W(x) = ce^{-P(x)} \quad \text{where} \quad P(x) = \int p(x)\,dx.$$

Instead of $Tu = Tv = 0$ as above, let us now assume that $Tu = 0$ and $u \neq 0$ on I. Equation (5) shows that

$$Tv = 0 \iff W' + pW = 0$$

and hence, if v is chosen so as to satisfy

(6) $$uv' - vu' = ce^{-P},$$

then v is a solution of $Tv = 0$. Equation (6) is a first-order linear equation for v and admits the integrating factor $1/u^2$. In fact, if we divide by u^2, the left side of (6) becomes $(v/u)'$ and we get the following theorem:

THEOREM 5. *Suppose the equation $Tu = 0$ has a solution u that does not vanish anywhere on I. Then a second solution v of $Tv = 0$, independent of u, is given by*

$$\left(\frac{v}{u}\right)' = \frac{ce^{-P(x)}}{u^2}, \qquad c \neq 0.$$

The solutions are independent because the quotient v/u is not constant. Since $Tv = 0$ gives $T(cv) = 0$ for every constant c, the choice $c = 1$ involves no loss of generality. We include c nevertheless because a choice different from $c = 1$ is often convenient.

EXAMPLE 2. The trial solution $y = x^m$ shows that the equation

(7) $$x^2 y'' - 13xy' + 49y = 0, \qquad x > 0$$

has a solution $u = x^7$. To get a second solution, compute

$$p(x) = \frac{-13}{x}, \qquad P(x) = -13 \ln x, \qquad e^{-P(x)} = x^{13}.$$

Theorem 5 with $u = x^7$ and $c = 1$ gives

$$\frac{v}{u} = \int x^{13} x^{-14}\, dx = \ln x.$$

Hence $v = x^7 \ln x$. The constant of integration was ignored because only a single function v is wanted. The general solution of (7) is

$$x^7(c_1 + c_2 \ln x)$$

and it is here that we introduce the constants.

PROBLEMS————————————————————————————————

1. Using Theorem 1 or the definition of linear dependence, whichever you prefer, test the following pairs of functions for linear dependence:

(a) e^t, e^{2t}

(b) e^{t+1}, e^t

(c) $(t+1)^2, t^2$

(d) te^t, e^t

(e) $t \sin t, t \cos t$

(f) $e^t \sin t, e^t \cos t$

2. Show that the following pairs of functions are linearly independent on every interval:

(a) $t^m \cos \omega t, \ t^m \sin \omega t, \quad \omega \neq 0$ (b) $t^m e^{at}, t^m e^{bt}, \quad a \neq b$.

3. By considering $u(x) = x^3$, $v(x) = |x|^3$ show that the hypothesis $v \neq 0$ cannot be dropped in Theorem 1(b).

4. Obtain two linearly independent solutions of $y'' - 10y' + 74y = 0$ and verify that their Wronskian satisfies $W = ce^{10t}$.

5. If $u = 1$ the equation of Theorem 5 is $v' = ce^{-P}$. Apply this to get linearly independent solutions $(1, v)$ of:

(a) $xy'' + 2y' = 0$ (b) $xy'' + y' = 0$ (c) $y'' \sin x = y' \cos x$

6. With α and β constant, this problem pertains to the operator $T = x^{-2}S$ where for $x > 0$

$$Sy = x^2 y'' + (1 - \alpha - \beta)xy' + \alpha\beta y, \qquad e^P = x^{1-\alpha-\beta}$$

(a) Verify the above equation for e^P.

(b) For what values of the constant m, if any, does the trial solution $y = x^m$ satisfy $Sy = 0$?

(c) Obtain two linearly independent solutions of $Ty = 0$ when $\alpha = \beta$.

7. Here are some expressions Sy, containing a constant parameter m, that are frequently used in applied mathematics. The operator is denoted by S instead of T because the leading term in Sy is not y''. We give first the name of the mathematician with whom the equation $Sy = 0$ is identified, then Sy, then the most common notation for certain solutions of $Sy = 0$, and finally e^P for the associated operator T. You are asked to verify the given value of e^P and to state Abel's equation for the Wronskian $W(u, v)$ when $Su = Sv = 0$. Since division is needed to go from $Sy = 0$ to the standard form $Ty = 0$, assume $x > 0$ in (ad) and $-1 < x < 1$ in (bc).

(a) Laguerre $xy'' + (1 - x)y' + my$ L_m $e^P = xe^{-x}$

(b) Legendre $(1 - x^2)y'' - 2xy' + m(m+1)y$ P_m $e^P = 1 - x^2$

(c) Chebychev $(1 - x^2)y'' - xy' + m^2 y$ T_m $e^P = \sqrt{1 - x^2}$

(d) Bessel $x^2 y'' + xy' + (x^2 - m^2)y$ $J_{\pm m}$ $e^P = x$

8. This is difficult. When $m = 1$ each equation $Su = 0$ in Problem 7 (abc) has a solution of the form $u = x + c$ where c is constant. Find c and show that a second solution v is

$$\text{(a) } (x - 1) \int \frac{e^x}{x(x-1)^2}\, dx \qquad \text{(b) } 1 - x\tanh^{-1} x \qquad \text{(c) } \sqrt{1 - x^2}$$

9. Deduce Theorem 3 from Abel's identity.

————————————————————ANSWERS————————————————————

1. The pair (b) is linearly dependent on every interval, all others are linearly independent on every interval.
5. (a) $1/x$ for $x \neq 0$ (b) $\ln|x|$ for $x \neq 0$ (c) $\cos x$
6. (b) $m = \alpha$ or β (c) $x^\alpha, \quad x^\alpha \ln x$ 7. $W = c/e^P$

2. Variation of parameters

We continue our study of the equation

$$Ty = y'' + p(x)y' + q(x)y = f(x), \qquad x \in I$$

introduced in Section 1. It was discovered by Lagrange that if two linearly independent solutions of the homogeneous equation $Ty = 0$ are known, the inhomogeneous equation can be solved by a single integration. Lagrange's procedure extends to equations of order n and represents a major advance in the theory of differential equations.

The basis of the method is easily described. Let u and v be a pair of linearly independent solutions of $Ty = 0$ and form the expression $y = au + bv$. If a and b are constant parameters, this represents the general solution of $Ty = 0$. Lagrange solves $Ty = f$ by choosing a trial solution of this same form, but with a and b functions of x rather than constants. Since the formerly constant parameters are now variables, the method is called the method of *variation of parameters*.

A similar idea has already been used several times in this book, starting with Section 2.4, Problem 9. In studying equations with constant coefficients, we replaced the expression ce^{kt}, which occurs in connection with the homogeneous equation, by ue^{kt}. Still another example is given in Theorem 1 of the following section.

(a) The Lagrange procedure

Suppose then that $Tu = 0$, $Tv = 0$, and

$$(1) \qquad\qquad y = au + bv$$

where a and b are differentiable functions of x. Differentiation gives

$$y' = (au' + bv') + (a'u + b'v).$$

Now comes an important point. The calculation of y'' will be materially simplified if a and b are so chosen that the second expression in parentheses vanishes. Accordingly we set

$$(2) \qquad\qquad a'u + b'v = 0$$

and get $y' = au' + bv'$. Hence

$$y'' = au'' + bv'' + a'u' + b'v'$$

$$p: \qquad y' = au' + bv'$$

$$q: \qquad y = au + bv$$

where the second and third equations have been repeated for convenience. The symbols p and q at the left mean that you are to multiply the second equation by p, the third by q, and add the three equations. Since

$$Ty = y'' + py' + qy$$

and similarly for Tu and Tv, the result of addition is

$$Ty = aTu + bTv + a'u' + b'v' = a'u' + b'v'.$$

The second equality holds because $Tu = Tv = 0$ by hypothesis.

Thus (2) and the desired equation $Ty = f$ are

$$a'u + b'v = 0$$

$$a'u' + b'v' = f.$$

This is a linear system in the unknowns (a', b'). By Cramer's rule

$$\begin{vmatrix} u & v \\ u' & v' \end{vmatrix} a' = \begin{vmatrix} 0 & v \\ f & v' \end{vmatrix}, \qquad \begin{vmatrix} u & v \\ u' & v' \end{vmatrix} b' = \begin{vmatrix} u & 0 \\ u' & f \end{vmatrix}.$$

The coefficient determinant is the Wronskian $W(u,v)$, so that

$$a' = \frac{-fv}{W(u,v)}, \quad b' = \frac{fu}{W(u,v)}.$$

An integration gives a and b and (1) then gives the solution,

$$y = u(x)\int \frac{-fv}{W(u,v)}\,dx + v(x)\int \frac{fu}{W(u,v)}\,dx. \quad leqno(3)$$

Since the constants of integration produce an added term $c_1u + c_2v$ the result is actually the *general* solution of $Ty = f$.

EXAMPLE 1. As an illustration let us solve

$$x^2y'' - 2xy' + 2y = x^2 f(x), \quad x \neq 0.$$

The side condition $x \neq 0$ means that the equation is considered on an interval I that does not contain the origin. It is easily checked that a pair of linearly independent solutions and their Wronskian is

$$u = x, \quad v = x^2, \quad W(u,v) = x^2.$$

To reduce the equation to standard form we must divide by x^2. The factor x^2 has been inserted on the right so that the notation after division will be consistent with that in the general theory. Lagrange's formula (3) gives

$$y = -x\int f(x)\,dx + x^2 \int \frac{f(x)}{x}\,dx.$$

It is left for you to verify that the choice $f(x) = x^m$ with m constant yields the solutions

$$-x\ln x, \quad x^2\ln x, \quad \frac{x^{m+2}}{m(m+1)}$$

for $m = -1$, for $m = 0$, and for other values of m, respectively. The general solution when $f(x) = x^m$ is obtained by adding $c_1x + c_2x^2$ to these particular solutions.

(b) Green's function: initial values

The basic formula (3) has several applications. Let x_0 be any point on the interval I where the differential equation is to hold and let the indefinite integrals defining a and b be replaced by definite integrals. Thus

$$(4) \qquad a(x) = \int_{x_0}^x \frac{-f(t)v(t)}{W(t)}\,dt, \quad b(x) = \int_{x_0}^x \frac{f(t)u(t)}{W(t)}\,dt$$

where $W(t) = W(u(t), v(t))$. Since the coefficients $u(x)$ and $v(x)$ in (3) are constant with respect to the integration variable t, they can be moved under the integral sign. The result is the formula

$$(5) \qquad y(x) = \int_{x_0}^{x} \frac{u(t)v(x) - u(x)v(t)}{u(t)v'(t) - u'(t)v(t)} f(t)\, dt$$

when the Wronskian $W(t)$ is replaced by its definition.

    To interpret this formula let us recall the equations

$$(6) \qquad y(x) = a(x)u(x) + b(x)v(x), \quad y'(x) = a(x)u'(x) + b(x)v'(x)$$

with which our discussion began. Since (4) gives $a(x_0) = b(x_0) = 0$, it follows that $y(x_0) = 0$ and $y'(x_0) = 0$. Thus, (5) gives the rest solution of $Ty = f$ as referred to the initial point x_0.

EXAMPLE 2. For the equation considered in Example 1, namely

$$x^2 y'' - 2xy' + 2y = x^2 f(x), \qquad x \neq 0$$

substitution of $u = t, v = t^2$ into (5) yields the rest solution

$$y(x) = x \int_{x_0}^{x} (x - t) \frac{f(t)}{t}\, dt.$$

For example, if $f(t) = t \sin t$ partial integration gives

$$\int_{x_0}^{x} (x - t) \sin t\, dt = (x - t)(- \cos t) \Big|_{t=x_0}^{x} - \int_{x_0}^{x} \cos t\, dt.$$

Evaluating this expression and multiplying by x we get

$$y(x) = x(x - x_0) \cos x_0 - x(\sin x - \sin x_0).$$

    The result (5) can be written

$$(7) \quad y(x) = \int_{x_0}^{x} G_0(x,t) f(t)\, dt \quad \text{where} \quad G_0(x,t) = \frac{u(t)v(x) - u(x)v(t)}{u(t)v'(t) - u'(t)v(t)}.$$

The subscript 0 suggests that the formula is concerned with initial values at x_0 and serves to distinguish G_0 from another function, G, that occurs in connection with boundary-value problems.

(c) Green's function: boundary values

We denote the basic interval I by $[x_0, x_1]$ where $x_0 < x_1$. In the previous pages we gave several examples of boundary-value problems in which the conditions are specified not just at one point but at two. Three typical boundary conditions are

$$y(x_0) = y(x_1) = 0, \quad y'(x_0) = y'(x_1) = 0, \quad y(x_0) = y'(x_1) = 0.$$

All three of these are included in

$$\alpha_0 y(x_0) + \beta_0 y'(x_0) = 0, \quad \alpha_1 y(x_1) + \beta_1 y'(x_1) = 0$$

where α_j and β_j are constant.

It is a remarkable fact that if u satisfies the boundary conditions at x_0 and v satisfies them at x_1, we can arrange matters so that y satisfies them at *both* x_0 and x_1. The only restriction is the basic condition $W(u, v) \neq 0$ which is always needed.

To see how this is done recall the conditions (6), which we repeat for convenience:

$$(8) \qquad y(x) = a(x)u(x) + b(x)v(x), \quad y'(x) = a(x)u'(x) + b(x)v'(x).$$

We want to choose a and b so that y behaves like $u(x)$ near $x = x_0$ and like $v(x)$ near $x = x_1$. To this end let

$$a(x_1) = 0, \quad b(x_0) = 0.$$

Recalling the earlier formulas for a' and b' we set

$$a(x) = \int_x^{x_1} \frac{v(t)}{W(t)} f(t)\, dt, \quad b(x) = \int_{x_0}^{x} \frac{u(t)}{W(t)} f(t)\, dt.$$

Then a' and b' have the correct values and the result is the formula

$$(9) \qquad y(x) = \int_{x_0}^{x} \frac{u(t)v(x)}{W(t)} f(t)\, dt + \int_x^{x_1} \frac{u(x)v(t)}{W(t)} f(t)\, dt.$$

To verify the boundary conditions use (8) and get

$$y(x_0) = a(x_0)u(x_0), \quad y'(x_0) = a(x_0)u'(x_0).$$

If u satisfies the conditions at x_0 then

$$\alpha_0 y(x_0) + \beta_0 y'(x_0) = a(x_0)(\alpha_0 u(x_0) + \beta_0 u'(x_0)) = 0$$

and hence y satisfies them too. The behavior of v at x_0 does not matter because $b(x_0) = 0$. In a like manner, if v satisfies the boundary conditions at x_1 then y does also, since $a(x_1) = 0$.

If we define

$$(10) \qquad G(x,t) = \frac{u(t)v(x)}{W(t)}, \qquad G(x,t) = \frac{u(x)v(t)}{W(t)}$$

for $x_0 \leq t \leq x$ and $x \leq t \leq x_1$ respectively, Equation (9) becomes

$$(11) \qquad y(x) = \int_{x_0}^{x_1} G(x,t)f(t)\, dt.$$

The functions G_0 and G are called *Green's functions* after George Green, the self-taught son of an English miller. In a long-neglected paper on electricity and magnetism (1828) Green gave formulas of this general type for partial differential equations.

EXAMPLE 3. Let us solve the boundary-value problem

$$y'' + y = f(x), \qquad y'(0) = 0, \quad y(\pi) = 0.$$

Independent solutions of the homogeneous equation satisfying the first and second boundary conditions, respectively, are

$$u(x) = \cos x, \quad v(x) = \sin x.$$

The Wronskian is $W(u,v) = 1$ and by (10) we have

$$G(x,t) = \cos t \sin x, \quad G(x,t) = \cos x \sin t$$

according as $t \leq x$ or $t \geq x$. The solution is given by (11), but for numerical evaluation it is better to use (9). For example, when $f(t) = 1$ we get

$$y(x) = \int_0^x \cos t \sin x\, dt + \int_x^\pi \cos x \sin t\, dt = 1 + \cos x.$$

PROBLEMS————————————————————————————

1. The sole purpose of this problem is to help you understand the text. Do not use the formulas (3), (5), or (9). Instead, use the trial function below the equation and solve by the procedure that led to (3). In the last case $x > 0$:

(a) $y'' + y = 1$ (b) $y'' - y' = 1$ (c) $xy'' - y' = 1$

$a(x)\sin x + b(x)\cos x$ $a(x) + b(x)e^x$ $a(x) + x^2 b(x)$

2P. The Lagrange procedure is one of those rare procedures that is easier to carry out in general than in special cases. The reason is that if you have specific functions like $x \ln x$ or $e^x \sin x$ for u and v, you will be led to compute u', u'', v', u'' when you follow the pattern of Problem 1. It is a waste of time to find these derivatives, however, because they drop out of the calculation.

For this reason you are not asked to derive the basic formulas anew in the following problems. Instead, you are now asked to study the derivation of the equations

$$a'u + b'v = 0, \qquad a'u' + b'v' = f$$

leading to (3) so thoroughly that you can write it without the aid of the text.

3. In this problem there are four things to do. By trying $y = x^m$ obtain linearly independent solutions $(u, v) = (x, x^4)$ for the equation on the left below. Then solve the equation on the right by (3), evaluate when $f(x) = 1$, and verify that the resulting function satisfies the differential equation:

$$x^2 y'' - 4xy' + 4y = 0, \qquad x^2 y'' - 4xy' + 4y = x^2 f(x), \qquad x > 0.$$

4. In this problem there are three things to do. By Example 2 of Section 1, the equation on the left below has the solutions $(u, v) = (x^7, x^7 \ln x)$. Using these, proceed as in Problem 3:

$$x^2 y'' - 13xy' + 49y = 0, \qquad x^2 y'' - 13xy' + 49y = x^2 f(x)$$

5. If α and β are constant parameters, solve (a) when $\alpha \neq \beta$ and (b) when $\alpha = \beta$. Take $(u, v) = (x^\alpha, x^\beta)$ in the first case and $(x^\alpha, x^\alpha \ln x)$ in the second. These values were found in Section 1, Problem 6:

$$x^2 y'' + (1 - \alpha - \beta)xy' + \alpha\beta y = x^2 f(x)$$

6. This problem depends on the formulas (5) and (9). Using the functions $(1, \cos x)$ from Section 1, Problem 5(c), find (a) the solution satisfying $y(\pi/2) = y'(\pi/2) = 0$ and (b) the solution satisfying $y(\pi/2) = y(\pi) = 0$. Evaluate when $f(x) = \sin x$ and check that the result satisfies both the differential equation and the initial or boundary conditions:

$$y'' \sin x - y' \cos x = f(x) \sin x$$

Hint for (b): By consideration of $c_1 + c_2 \cos x$ construct u satisfying the first boundary condition and v satisfying the second.

7. For the equation $x^2 y'' - 4xy' + 6y = x^2 f(x)$ find the solution satisfying

$$(a) \; y(1) = y'(1) = 0 \qquad (b) \; y(1) = 0, \; y(2) = y'(2)$$

Evaluate when $f(x) = x^2$ and verify that the result satisfies both the differential equation and the initial or boundary conditions.

Hint: Solve the homogeneous equation by trying $y = x^m$.

8. If ω is a positive constant, obtain the formula on the right for the solution of the initial-value problem on the left:

$$y'' + \omega^2 y = f(x), \quad y(x_0) = y'(x_0) = 0; \quad y = \frac{1}{\omega} \int_{x_0}^{x} \sin\omega(x - t) f(t) \, dt$$

9. Show that the boundary conditions $y(0) = y(\pi) = 0$ can be imposed in Problem 8 if, and only if, $\omega \neq$ integer.

Hint: For the boundary condition $u(0) = 0$ it will be found that $u = \sin\omega x$ is, essentially, the only choice. You need a function $v = c_1 \cos\omega x + c_2 \sin\omega x$, vanishing at π, which is not a multiple of u.

—————————————————ANSWERS———————————————

1. (a) $c_1 \cos x + c_2 \sin x + 1$ (b) $c_1 + c_2 e^x - x$ (c) $c_1 + c_2 x^2 - x$
3. 4. See answer to Problem 5.
5. (a) $(\alpha - \beta)y = x^\alpha \int x^{1-\alpha} f(x) \, dx - x^\beta \int x^{1-\beta} f(x) \, dx$
   (b) $y = x^\alpha \ln x \int x^{1-\alpha} f(x) \, dx - x^\alpha \int x^{1-\alpha} f(x) \ln x \, dx$
6. (a) $y = \int_{\pi/2}^{x} (\cos t - \cos x)(f(t)/\sin t) \, dt$
      $= \sin x - 1 - (x - \pi/2) \cos x$ when $f(x) = \sin x$
   (b) $y = \int_{\pi/2}^{x} \cos t (1 + \cos x)(f(t)/\sin t) \, dt + \int_{x}^{\pi} \cos x (1 + \cos t)(f(t)/\sin t) \, dt$
      $= \sin x + (\pi - x - 1) \cos x - 1$ when $f(x) = \sin x$
7. (a) $\int_{1}^{x} (x/t)^2 (x - t) f(t) \, dt = x^2 (x - 1)^2/2$ when $f(x) = x^2$
   (b) $\int_{1}^{x} (x/t)^2 (1 - t) f(t) \, dt + \int_{x}^{2} (x/t)^2 (1 - x) f(t) \, dt$
      $= x^2 (1 - x)(3 - x)/2$ when $f(x) = x^2$
   We took $u = x^2 - x^3$ and $v = x^2$ to satisfy the boundary conditions.

3. Supplementary topics

As throughout this chapter, we define

$$Ty = y'' + p(x)y' + q(x)y, \qquad x \in I.$$

A useful identity is obtained by setting $y = uz$ where u and z are twice-differentiable functions on I. Since

$$y = uz, \quad y' = uz' + zu', \quad y'' = uz'' + 2u'z' + zu''$$

the result of substitution is

$$T(uz) = (uz'' + 2u'z' + zu'') + p(uz' + zu') + quz.$$

The terms with z as a factor together give zTu, so that

(1) $$T(uz) = uz'' + (2u' + pu)z' + zTu.$$

(a) The inhomogeneous equation

By means of (1) we can solve the inhomogeneous equation $Tw = f$ when a single nonvanishing solution u of the homogeneous equation is known. Let $Tu = 0$ and let a solution of the general equation $Tw = f$ be sought in the form $w = uz$. Since $Tu = 0$, the term zTu in (1) drops out, and the condition $Tw = f$ becomes

$$uz'' + (2u' + pu)z' = f.$$

This is a first-order linear equation in the unknown z'. After constructing an integrating factor as in Section 2.4, we get the following theorem:

THEOREM 1. *Suppose the equation $Tu = 0$ has a solution u that does not vanish on I. Then a solution of the inhomogeneous equation $Tw = f$ is given by $w = uz$ where*

(2) $$(e^P u^2 z')' = e^P uf.$$

If $f = 0$, Theorem 1 gives

$$(e^P u^2 z')' = 0, \quad \text{hence} \quad e^P u^2 z' = c.$$

This yields the same linearly independent solution v of the homogeneous equation as is given by Section 1, Theorem 5. The general, and also the complete, solution of $Ty = f$ is

$$y = c_1 u + c_2 v + w.$$

EXAMPLE 1. Taking $u = x^7$ express a solution of the following equation by means of integration:

$$x^2 y'' - 13xy' + 49y = x^2 f(x), \qquad x > 0.$$

Theorem 1 assumes that the leading term in Ty is y''. With this in mind, the factor x^2 was introduced on the right to give $f(x)$ after the equation is divided by x^2. That $Tu = 0$ was verified in Section 1, Example 2 where it was also seen that

$$e^{P(x)} = e^{-13 \ln x} = x^{-13}.$$

Hence the equation (2) for z becomes

$$(x^{-13} x^{14} z')' = x^{-13} x^7 f(x).$$

After simplification we integrate and divide by x to get

$$z' = \frac{1}{x} \int \frac{f(x)}{x^6} \, dx.$$

A second integration gives z, and $w = uz$ is the solution sought. As an illustration, it is left for you to verify that the choice $f(x) = x^m$ gives, aside from constants of integration,

$$w = \frac{x^{m+2}}{(m-5)^2}, \quad m \neq 5; \qquad w = \frac{1}{2} x^7 (\ln x)^2, \quad m = 5.$$

The first of these could have been found by the trial solution $w = Ax^{m+2}$ but the second is less obvious.

(b) Elimination of a term

Let us again use (1), but instead of $Tu = 0$ we now set

(3) $$2u' + pu = 0.$$

In this case (1) gives $T(uz) = uz'' + zTu$ and, if $w = uz$, the equation $Tw = f$ reduces to

(4) $$uz'' + zTu = f.$$

The advantage is that (4) lacks the term z'. Since $P' = p$, Equation (3) gives

$$u = e^{-P/2}, \quad u' = e^{-P/2} \left(-\frac{p}{2} \right), \quad u'' = e^{-P/2} \left(-\frac{p'}{2} + \frac{p^2}{4} \right).$$

(We do not introduce a constant of integration for u because only a single solution is wanted.) By a short calculation

$$Tu = e^{-P/2}\left(q - \frac{p'}{2} - \frac{p^2}{4}\right).$$

Substitution of $u = e^{-P/2}$ and of Tu into (4) yields the following theorem:

THEOREM 2. *Let p be differentiable and let*

$$P = \int p(x)\,dx, \quad Q = q - \frac{p'}{2} - \frac{p^2}{4}, \quad z = e^{P/2}w.$$

Then the following equations are equivalent. That is, if either holds, both do:

(5) $$z'' + Qz = e^{P/2}f, \quad w'' + pw' + qw = f.$$

EXAMPLE 2. The *Hermite equation* is

$$Ty = y'' - 2xy' + 2my = 0$$

where m is a constant parameter. To remove the second term from this equation we take $p = -2x$, $q = 2m$ and hence

$$e^{P(x)} = e^{-x^2}, \quad Q(x) = 2m + 1 - x^2.$$

By Theorem 2 the equation for z, which lacks the second term, is

(6) $$z'' + (2m + 1 - x^2)z = 0.$$

If $H_m(x)$ satisfies the Hermite equation then the function

$$z = e^{-x^2/2}H_m(x)$$

satisfies (6) and conversely.

(c) Interlacing of zeros

The functions $\sin x$ and $\cos x$ are linearly independent solutions of $y'' + y = 0$. They have the remarkable property that their zeros interlace. That is, if one of them has consecutive zeros x_1 and x_2 then the other is not zero at x_1 or x_2 but has exactly one zero between x_1 and x_2. The following theorem generalizes this result:

THEOREM 3. *Let u and v be differentiable functions on an interval I and suppose the Wronskian $W(x) = W(u, v)$ does not vanish at any point of I. Then the zeros of u and v interlace.*

For proof, let x_1 and x_2 be consecutive zeros of one of the functions, which without loss of generality we take to be u. If $v(x_1) = 0$ then the first row of the Wronskian is 0 at x_1, and this makes $W(x_1) = 0$ contrary to hypothesis. Similarly for x_2. Hence v does not vanish at x_1 or x_2.

If v does not vanish anywhere on the interval (x_1, x_2) then the function $y = u/v$ is differentiable for $x_1 \leq x \leq x_2$ and vanishes at the ends of this interval. By Rolle's theorem $y'(c) = 0$ at some value c. This gives $W(c) = 0$, which again contradicts the hypothesis. Hence v must vanish at some point of (x_1, x_2).

Finally, v cannot have two or more zeros between x_1 and x_2. If it did, the function u would have to vanish somewhere between them, contradicting the fact that x_1 and x_2 are consecutive zeros of u. This completes the proof.

EXAMPLE 3. The *Bessel equation* is

$$x^2 y'' + xy' + (x^2 - m^2)y = 0, \qquad x > 0.$$

If m is not an integer it is known that the solutions $J_m(x)$ and $J_{-m}(x)$ are linearly independent and that both have infinitely many zeros. Linear independence gives the hypothesis $W \neq 0$ of Theorem 3. Hence the zeros of either of these functions separate those of the other.

PROBLEMS

1. If $Ty = y'' + qy$, so that the second term is missing, check that Abel's equation for the Wronskian, the procedure for getting a second solution v from u, and the method for solving $Ty = f$ take the following simple form:

$$W(u, v) = c, \qquad \left(\frac{v}{u}\right)' = \frac{c}{u^2}, \qquad (u^2 z')' = uf$$

2. Let $Tu = 0$ and $u \neq 0$ where T is the Hermite operator in Example 2.

(a) Obtain the two equations

$$\left(\frac{v}{u}\right)' = \frac{e^{x^2}}{u^2}, \qquad (e^{-x^2} u^2 z')' = e^{-x^2} uf$$

The first is the recipe of Theorem 1 for a solution of $Tv = 0$ and the second is the recipe for a solution $w = uz$ of $Tw = f$.

(b) In Part (a) obtain the explicit formulas

$$v(x) = u(x) \int \frac{e^{x^2}}{u(x)^2}\, dx, \qquad z'(x) = \frac{e^{x^2}}{u(x)^2} \int e^{-x^2} u(x) f(x)\, dx$$

3. If the expression Q of Theorem 2 is a positive constant ω^2, Theorem 2 reduces the inhomogeneous equation to

$$z'' + \omega^2 z = g(x), \qquad g(x) = e^{P(x)/2} f(x)$$

Using this show that the general solution of $y'' + py' + qy = 0$ is

$$y = (c_1 \cos \omega x + c_2 \sin \omega x)e^{-P(x)/2}$$

4. Show that the Bessel equation $Ty = 0$ satisfies the condition $Q = \text{const}$ of the preceding problem if, and only if, $m = \pm 1/2$. Thus show that the general solution when $m = \pm 1/2$ is

$$x^{-1/2}(c_1 \cos x + c_2 \sin x)$$

If T is the Bessel operator with $m = 1/2$ the methods of this section yield an elementary solution for $Ty = f$, that is, a solution by quadrature. It can be shown that the result of this problem also gives

$$J_{1/2}(x) = \sqrt{\frac{2}{\pi x}} \sin x, \qquad J_{-1/2}(x) = \sqrt{\frac{2}{\pi x}} \cos x$$

5. Verify that $(\sin x)/x$, e^x, x satisfy the following equations, respectively, and thus obtain a general solution. Part (c) leads to a tricky integral:

(a) $\qquad xy'' + 2y' + xy = 0$ $\hspace{4cm}$ $(x > 0)$
(b) $\qquad (2x - x^2)y'' + (x^2 - 2)y' + (2 - 2x)y = 0$ $\hspace{1cm}$ $0 < x < 2$
(c) $\qquad (2x - 1)y'' - 4xy' + 4y = 0$ $\hspace{2.6cm}$ $2x > 1$

6. Explain why $y'' + py' + qy = f$ can be solved by quadrature, in general, on any interval on which one of the following holds:

(a) $p(x) + xq(x) = 0$ $\qquad$ (b) $1 + q(x) = p(x)$ $\qquad$ (c) $2q + qx^2 = px$

7. (a) Taking $u = e^h$ show that $Ty = f$ can be solved by quadrature if

$$h'' + (h')^2 + ph' + q = 0$$

(b) Check that the following equation has a solution of the form $y = e^{ax^2}$, find a, and obtain the formula on the right for a second solution:

$$y'' + (x + x^2)y' + (x^3 + 1)y = 0, \qquad y = e^{-x^2/2} \int e^{x^2/2 - x^3/3} \, dx$$

8. This pertains to $Ty = (1 - x)y'' + xy' - y = g(x)$. Given that x satisfies $Ty = 0$, (a) find a second solution of $Ty = 0$, (b) solve the inhomogeneous equation by the method of Lagrange and (c) evaluate when $g(x) = (1 - x)^2$. Part (a) leads to a tricky integral.

9. Remove the second term from Bessel's equation, Example 3, and thus show that the functions on the left below satisfy the equation on the right:

$$\sqrt{x}J_m(x), \qquad \sqrt{x}J_{-m}(x); \qquad x^2 z'' + (x^2 - m^2 + \frac{1}{4})z = 0$$

10. (a) Check that Green's function for $y'' = f(x)$, $y(0) = y(1) = 0$ is

$$G(x,t) = t(x-1), \qquad t < x; \qquad G(x,t) = x(t-1), \qquad x < t$$

(b) Using $G(x,t)$ compute y when $f(x) = x^m$, $m \geq 0$ and verify that the expected properties hold. No answer is given.

11. For the problem $y'' - 2y' + y = f(x)$:

(a) Find the solution satisfying $y(x_0) = y'(x_0) = 0$.

(b) Find Green's function for the boundary conditions $y(0) = y'(1) = 0$.

─────────────────────────ANSWERS─────────────────────────

5. (a) $(c_1 \cos x + c_2 \sin x)/x$ (b) $c_1 e^x + c_2 x^2$ (c) $c_1 x + c_2 e^{2x}$
6. Partial answer: Try $u = x^m$ and $u = e^{ax}$.
8. (c) $y = c_1 x + c_2 e^x + x^2 + 1$
11. (a) $y = \int_{x_0}^{x} (x - t)e^{x-t} f(t)\, dt$
     (b) $2G(x,t) = (x-2)te^{x-t}$ for $0 < t < x$;
        $2G(x,t) = (t-2)xe^{x-t}$ if $x < t < 1$.

───

> Mathematics is a vast adventure in ideas; its history reflects some of the noblest thoughts of countless generations.
>
> —Dirk Struik

> Mathematics may be defined as the subject in which we never know what we are talking about, nor whether what we say is true.
>
> —Bertrand Russell

Chapter 11

LINEAR HIGHER ORDER, VARIABLE

THE CONCEPTS of uniqueness, linear dependence, general solution, and complete solution are developed for linear equations of arbitrary order with variable coefficients. The homogeneous case is emphasized because a general solution of the homogeneous equation yields a general solution of the inhomogeneous equation by variation of parameters. Much of our work up to this point can be regarded as a preview of topics that are presented here from a comprehensive point of view.

A review of determinants is in Chapter E.

1. Mathematical foundations

Throughout this chapter the letters c_j denote constants, p_j, u_j, y, and f are real- or complex-valued functions on a given interval I, and the operator T is defined by

$$Ty = y^{(n)} + p_{n-1}y^{(n-1)} + \cdots + p_1 y' + p_0 y.$$

The results apply to complex functions because the algebra of linear equations and determinants is the same for complex numbers as for real.

In the preceding pages the theory of linear differential equations was developed for nth-order equations with constant coefficients and for 2nd-order equations with variable coefficients. Now we consider the general case, nth-order equations with variable coefficients. The basic vocabulary was already introduced in the earlier discussion, where we defined such notions as general solution, linear dependence, and initial values. For clarity, however, these terms are defined again here.

(a) Uniqueness

A *solution* of the equation $Ty = f$ is a function $y = y(x)$ that satisfies the equation on I. Associated with the concept of solution are the *initial conditions*

(1) $$y(x_0) = y_0, \quad y'(x_0) = y_1, \quad \cdots, \quad y^{(n-1)}(x_0) = y_{n-1}$$

where x_0 is a given point of I. The following uniqueness theorem is one of the main results of the theory:

THEOREM 1. *If the coefficients p_k are bounded, the equation $Ty = f$ has at most one solution satisfying the initial conditions (1).*

For proof, let z be another solution, so that $Tz = f$. With $u = y - z$ linearity gives

$$Tu = T(y - z) = Ty - Tz = f - f = 0$$

which is equivalent to

$$u^{(n)} = -p_{n-1}u^{(n-1)} - \cdots - p_1 u' - p_0 u.$$

If $|p_k| \leq K$ for $k = 0, 1, 2, \cdots, n - 1$ taking absolute values yields

$$|u^{(n)}| \leq K(|u^{(n-1)}| + \cdots + |u'| + |u|).$$

Since y and z satisfy the same initial conditions, the initial conditions for u are 0. Chapter A, Theorem 2 now gives $u = 0$, hence $y = z$.

(b) The Wronskian

If $Tu_j = 0$ for $j = 1, 2, \cdots, n$ the function

(2)
$$y = c_1 u_1 + c_2 u_2 + \cdots + c_n u_n$$

satisfies $Ty = 0$ by the superposition principle. A family of solutions such as this is a *general solution* of $Ty = 0$ if the constants c_j in it can be chosen so as to satisfy arbitrarily specified initial conditions at the arbitrarily specified point $x_0 \in I$. If the family (2) contains every solution of the equation $Ty = 0$, it is the *complete solution*.

Substitution of (2) into the initial-value equations (1) yields

(3)
$$c_1 u_1^{(j)}(x_0) + c_2 u_2^{(j)}(x_0) + \cdots + c_n u_n^{(j)}(x_0) = y_j$$

for $j = 0, 1, 2, \cdots, n - 1$. This is a system of n linear equations in the n unknowns c_j. The coefficient determinant is the *Wronskian* $W(x_0)$ where, by definition,

$$W = \begin{vmatrix} u_1 & u_2 & \cdots & u_n \\ u_1' & u_2' & \cdots & u_n' \\ \vdots & \vdots & \vdots & \vdots \\ u_1^{(n-1)} & u_2^{(n-1)} & \cdots & u_n^{(n-1)} \end{vmatrix}$$

For example, when $n = 3$ the linear system is

$$c_1 u_1(x_0) + c_2 u_2(x_0) + c_3 u_3(x_0) = y_0$$

$$c_1 u_1'(x_0) + c_2 u_2'(x_0) + c_3 u_3'(x_0) = y_1$$

$$c_1 u_1''(x_0) + c_2 u_2''(x_0) + c_3 u_3''(x_0) = y_2$$

and its determinant is

$$W(x_0) = \begin{vmatrix} u_1(x_0) & u_2(x_0) & u_3(x_0) \\ u_1'(x_0) & u_2'(x_0) & u_3'(x_0) \\ u_1''(x_0) & u_2''(x_0) & u_3''(x_0) \end{vmatrix}.$$

By definition, the functions u_j are *linearly dependent on I* if there are constants c_j, not all 0, such that

(4) $$c_1 u_1(x) + c_2 u_2(x) + \cdots + c_n u_n(x) = 0, \qquad x \in I.$$

If no such constants exist, the functions are *linearly independent on I*. The qualifying phrase "on I" is understood whether mentioned or not, and may be omitted in the following discussion.

Differentiation of (4) yields the equations

(5) $$c_1 u_1^{(j)}(x) + c_2 u_2^{(j)}(x) + \cdots + c_n u_n^{(j)}(x) = 0$$

for $j = 0, 1, 2, \cdots, n - 1$. This is a homogeneous linear system in the un-knowns c_j. Both systems (3) and (5) have essentially the same determinant, $W(x_0)$ in the first case and $W(x)$ in the second. For example, when $n = 3$ the system (5) is the 3-by-3 system displayed above with x_0 replaced by x and with $y_j = 0$.

(c) The Fredholm alternative

The main fact we need to know about linear systems pertains to a system

$$a_{i1} c_1 + a_{i2} c_2 + \cdots + a_{in} c_n = y_i, \qquad i = 0, 1, \cdots, n - 1$$

of n equations in n unknowns c_j. It states that *exactly one* of the following alternatives holds:

(i) the system has a unique solution for all y_j, or

(ii) the homogenous system has a nontrivial solution.

Furthermore, (i) happens if the coefficient determinant is not zero, and (ii) if it is zero. We refer to these statements collectively as the *Fredholm alterna-tive*, after the Swedish mathematician Ivar Fredholm who formulated similar properties in a more general setting.

THEOREM 2.   Let $Tu_j = 0$ for $j = 1, 2, \cdots, n$ and let $W(x)$ be the Wronskian of the functions u_j. Then:

(a) The expression $c_1u_1 + c_2u_2 + \cdots + c_nu_n$ is a general solution of $Ty = 0$ if, and only if, $W(x) \neq 0$ holds at every point of I.

(b) The functions u_j are linearly independent if $W(x) \neq 0$ holds at a single point of I.

The initial-value problem is solvable at x_0 if the c_j can be determined so as to satisfy (3) for arbitrary choices of the y_j. According to the Fredholm alternative this can be done if, and only if, $W(x_0) \neq 0$. If x_0 as well as y_j is arbitrary we conclude that $W(x) \neq 0$ for every $x \in I$. In other words the functions u_j produce a general solution by way of (2) if, and only if, $W(x) \neq 0$ holds at every point of I. This gives Part (a) of Theorem 2.

On the other hand if the functions are linearly dependent then the system (4) has a nontrivial solution c_j at every value $x \in I$. By the Fredholm alternative we must have $W(x) = 0$. Thus, the functions u_j can be linearly dependent on I only if $W(x) = 0$ at every point of I. This gives Part (b) of the theorem.

For most functions u_j it is to be expected that $W(x) = 0$ at some points of I and $W(x) \neq 0$ at others. Solutions of $Ty = 0$, however, exhibit the following surprising behavior:

THEOREM 3.   On a given interval I let u_j be solutions of $Tu_j = 0$ for $j = 1, 2, \cdots, n$ where T has bounded coefficients p_j. Suppose that $W(x_0) = 0$ at some point $x_0 \in I$. Then the functions u_j are linearly dependent and $W(x) = 0$ at every $x \in I$.

It follows, in particular, that either $W(x) = 0$ at every $x \in I$, or $W(x) = 0$ at no $x \in I$.

By the Fredholm alternative we can choose constants c_j, not all 0, in such a way that the function

(6) $$y = c_1u_1 + c_2u_2 + \cdots + c_nu_n$$

satisfies null initial conditions at the point x_0. The solution given by the zero function 0 satisfies these same null conditions, hence by uniqueness $y = 0, x \in I$. This shows that the u_j are linearly dependent, and $W(x) = 0$ follows from the remarks above.

Another consequence of uniqueness together with the Fredholm alternative is:

THEOREM 4.   Let u_j satisfy $Tu_j = 0$ on I for $j = 1, 2, \cdots, n$, where T has bounded coefficients p_j. Suppose further that $W(x_0) \neq 0$ at some point $x_0 \in I$. Then the expression (6) is the complete solution of $Ty = 0$.

The concluding statement means that every solution of $Ty = 0$ can be written in the form (6).

For proof, note that the constants c_j can be chosen so that y in (6) satisfies arbitrary initial conditions at x_0. This follows from the Fredholm alternative, since $W(x_0) \neq 0$. If z is any solution of $Tz = 0$, let us determine the c_j in such a way that the initial conditions for y at x_0 agree with those for z. Since $Ty = 0$ by the superposition principle and $Tz = 0$ by hypothesis, uniqueness gives $y = z$. Hence the arbitrary solution z is in the family (6), which shows that (6) is the complete solution.

PROBLEMS

Assume all functions sufficiently differentiable to justify the calculations.

1. If $u_1 = x^2 + 2x$, $u_2 = x^3 + x$, $u_3 = 2x^3 - x^2$ find constants c_1 and c_2 such that $u_3 = c_1 u_1 + c_2 u_2$. What does this indicate about the Wronskian? Verify your answer by direct computation.

2. Check for linear dependence by any method:

(a) $(x+1)^2$, $(x-1)^2$, $3x$ (d) e^{ix}, $\cos x$, $\sin x$ (g) 1, x, x^2, x^3

(b) $x^2 - 2x$, $3x - 1$, $\sin x$ (e) x, x^2, x^3 (h) e^{-x}, e^x, 1, $\sinh x$

(c) 1, $\sin x$, $\cos x$ (f) e^x, xe^x, $x^2 e^x$ (i) e^x, e^{2x}, e^{3x}, e^{4x}

3. Reduction of order. If u is a solution of $Tu = 0$ that is nowhere zero on a given interval I, the substitution $y = uv$ reduces the equation $Ty = f$ on I to an equation of lower order in the unknown v'. The reduction depends on formulas for differentiating a product such as

$$(uv)''' = u'''v + 3u''v' + 3u'v'' + uv'''$$

By this method reduce the equation $y''' + py'' + qy' + ry = f$ to

$$uv''' + (3u' + pu)v'' + (3u'' + 2pu' + qu)v' = f$$

4. Reduce the following equation to one of second order in v' by the choice (a) $u = x$ or (b) $u = x^2$ or (c) $u = x^3$ in Problem 3. Only the answer (a) is provided:

$$x^3 y''' - 3x^2 y'' + 6xy' - 6y = g(x), \qquad x > 0$$

--------------------------------ANSWERS--------------------------------

1. $W = 0$ 2. (adh) are dependent, the rest independent.

4. (a) $x^4 v''' = g(x)$

2. Basis

Let the functions u_1, u_2, $\cdots$ u_n be solutions of the homogeneous equation $Ty = 0$. These functions are said to form a *basis* if every solution of $Ty = 0$ can be written in the form

$$(1) \qquad\qquad y = c_1 u_1 + c_2 u_2 + \cdots + c_n u_n$$

and this representation is unique. We shall establish:

THEOREM 1. *Let u_j for $j = 1, 2, \cdots, n$ be n linearly independent solutions of $Tu_j = 0$ where T has bounded coefficients. Then the u_j form a basis for the solutions of $Ty = 0$.*

Since the functions u_j are linearly independent, Theorem 3 of the previous section shows that $W(x_0) \neq 0$ at each $x_0 \in I$. Theorem 4 of that section then shows that every solution can be written in the form (1). To see why the representation is unique, suppose there is another such representation with constants d_j instead of c_j. By subtraction

$$(c_1 - d_1)u_1 + (c_2 - d_2)u_2 + \cdots + (c_n - d_n)u_n = 0.$$

Since the u_j are linearly independent each coefficient $c_j - d_j = 0$. Hence $d_j = c_j$, which shows that the new representation is the same as the old. This completes the proof.

(a) Constant coefficients

Let us show that there is a basis consisting only of elementary functions when the coefficients p_j are constant. We know from Chapter 8 that the equation $Ty = 0$ has n distinct solutions of the form

$$x^j e^{q_j x}$$

where the j are nonnegative integers and the q_j are real or complex constants. We shall show that these n solutions are linearly independent on every interval I and therefore form a basis.

To this end, suppose a linear combination of these functions is 0 on I. Let the *distinct* exponents q_j be

$$r_1, r_2, \cdots, r_k$$

and group terms with the same exponent. The result is an equation

$$(2) \qquad c_1(x)e^{r_1 x} + c_2(x)e^{r_2 x} + \cdots + c_k(x)e^{r_k x} = 0, \qquad x \in I,$$

where the c_j are polynomials. The total number of constants is n. Thus n is the sum of the numbers $d_j + 1$ where d_j is the degree of c_j.

It will be established by induction on n that all $c_j = 0$. This is obvious if $n = 1$, since there is only a single term ce^{rx}. Suppose then that the result is known for $n - 1$. To get it for n, multiply (2) by $e^{-r_1 x}$. The result is

$$(3) \qquad c_1(x) + c_2(x)e^{s_2 x} + \cdots + c_k(x)e^{s_k x} = 0, \qquad x \in I,$$

where $s_j = r_j - r_1$. The s_j, like the r_j, are all unequal, and none of them equals the first exponent, 0, associated with c_1. If we differentiate (3) with respect to x we get an equation with $n - 1$ constants to which the induction hypothesis applies. Hence all the coefficients are 0 and in particular $c_1'(x) = 0$. This shows that c_1 is constant. A similar argument applies to each c_j and shows that each c_j is constant. However, the case of constant coefficients c_j was already settled in Section 8.3, Problem 10. This completes the proof.

(b) The Euler-Cauchy equation

There are only a few interesting classes of nth-order linear equations for which a basis can be expressed by means of elementary functions. One of these is the class of equations with constant coefficients, as we have seen. Another is the class of equations of the form

$$x^n y^{(n)} + p_{n-1} x^{n-1} y^{(n-1)} + \cdots + p_1 x y' + p_0 y = 0, \qquad x > 0$$

where the p_j are constant. An equation of this form is often called *Cauchy's equation* though such equations were studied earlier by Euler. The coefficients x^j exactly compensate for the lowering of the exponent by differentiation, hence the substitution $y = x^\rho$ leads to a polynomial equation for ρ. If this polynomial equation has n distinct roots ρ_j the functions x^{ρ_j} so obtained form a basis. Thus every solution of the equation is uniquely representable in the form

$$y = c_1 x^{\rho_1} + c_2 x^{\rho_2} + \cdots + c_n x^{\rho_n}.$$

For complex roots $\rho = \sigma + i\tau$ we write

$$x^\rho = x^{\sigma + i\tau} = x^\sigma x^{i\tau} = x^\sigma e^{i\tau \ln x}.$$

If the coefficients p_j are real, taking real and imaginary parts gives the two solutions

$$x^\sigma \cos(\tau \ln x), \quad x^\sigma \sin(\tau \ln x).$$

These correspond to the complex root $\sigma + i\tau$.

If ρ is a repeated root of multiplicity m the corresponding family of solutions is

$$x^\rho, \quad x^\rho \ln x, \quad x^\rho (\ln x)^2, \quad \cdots, \quad x^\rho (\ln x)^{m-1}.$$

The family of n distinct functions obtained by this process again forms a basis. These results are easy to prove because the Euler-Cauchy equation can be reduced to an equation with constant coefficients as in the following problems.

PROBLEMS

1. The Euler-Cauchy equation. If $Ty = x^3 y''' + p_2 x^2 y'' + p_1 xy' + p_0 y$ for $x > 0$ and ρ is constant, check that

$$Tx^\rho = (\rho(\rho - 1)(\rho - 2) + p_2 \rho(\rho - 1) + p_1 \rho + p_0)x^\rho$$

Hence $Tx^\rho = 0$ holds if ρ satisfies a certain cubic equation. Also $Ty = Ax^\rho$ has a solution of the form $y = Bx^\rho$ if ρ is not a root of that cubic equation.

2. (a) If $x = e^t$, $D_x = d/dx$, $D_t = d/dt$ show that sufficiently differentiable functions defined by $y(x) = z(t)$ satisfy $D_x y = e^{-t} D_t z$,

$$D_x^2 y = e^{-2t} D_t (D_t - 1)z, \quad D_x^3 y = e^{-3t} D_t (D_t - 1)(D_t - 2)z,$$

and so on. Hint: $dy/dx = (dy/dt)/(dx/dt)$, hence $D_x = e^{-t} D_t$.

(b) With $D = D_t$ in Part (a) show that the Euler-Cauchy equation for $n = 3$ can be reduced to the following; note the resemblance to the expression obtained in Problem 1:

$$(D(D - 1)(D - 2) + p_2 D(D - 1) + p_1 D + p_0)z = f(e^t)$$

3. By the result of Problems 1 or 2 obtain two linearly independent solutions:

(a) $x^2 y'' + 4xy' = 54y$ (c) $x^2 y'' + 5y = 3xy'$ (e) $x^2 y'' + 4y = 3xy'$

(b) $x^2 y'' + y = xy'$ (d) $x^2 y'' + 6y = 4xy'$ (f) $x^2 y'' + 50y = xy'$

4. Reduce to an equation with constant coefficients by setting $x = e^t$ and thus obtain a particular solution:

(a) $x^2 y'' + 4xy' + 2y = \cos(\ln x)$ (b) $x^2 y'' + xy' + y = 5x \sin(\ln x)$

     (c) $x^2 y'' + 5xy' + 7y = 17x \cos(\ln x) + 5x \sin(\ln x)$

5. Obtain three linearly independent solutions of $x^3 y''' + xy' = y$ by the method of Problem 2.

6. Show that for arbitrary n the results of Problems 1 and 2 are respectively

$$(\rho(\rho - 1) \cdots (\rho - n + 1) + \cdots + p_2 \rho(\rho - 1) + p_1 \rho + p_0)x^\rho = Tx^\rho$$

$$(D(D - 1) \cdots (D - n + 1) + \cdots + p_2 D(D - 1) + p_1 D + p_0)z = f(e^t)$$

────────────────────────ANSWERS────────────────────────

3. (a) x^{-9}, x^6 (c) $x^2 \cos(\ln x)$, $x^2 \sin(\ln x)$ (e) x^2, $x^2 \ln x$
  (b) x, $x \ln x$ (d) x^3, x^2 (f) $x \cos(7 \ln x)$, $x \sin(7 \ln x)$
4. (a) $.1 \cos(\ln x) + .3 \sin(\ln x)$ (b) $x \sin(\ln x) - 2x \cos(\ln x)$
              (c) $x \cos(\ln x) + x \sin(\ln x)$
5. $c_1 x + c_2 x \ln x + c_3 x (\ln x)^2$

---

3. Variation of parameters

In Chapter 10 we presented the method of Lagrange for solving the second-order equation $Ty = f$ by using two linearly independent solutions of the homogeneous equation $Ty = 0$. Lagrange extended his method to equations of order n, as described now. Let

$$(1) \qquad\qquad c_1 u_1 + c_2 u_2 + \cdots + c_n u_n$$

be a general solution of $Ty = 0$. Since it is a general solution, results of Section 1 show that the Wronskian does not vanish anywhere on I. Replacing the constants c_j by functions v_j, we seek a solution of $Ty = f$ in the form

$$(2) \qquad\qquad y = u_1 v_1 + u_2 v_2 + \cdots + u_n v_n.$$

To determine the v_j, differentiate (2) and set the sum of the terms involving the differentiated functions v_i' equal to 0. Differentiate again, using the fact that the first step leads to simplification, and again set the sum of terms involving the functions v_i' equal to 0. Continue until you have $n-1$ equations. The nth equation is obtained by inserting the simplified derivatives in the equation $Ty = f$. Using Cramer's rule, you get the formula

$$y(x) = \sum_{i=1}^{n} u_i(x) \int \frac{W_i(u_1, u_2, \cdots, u_n)}{W(u_1, u_2, \cdots, u_n)} f(x) \, dx$$

provided f is continuous. Here W is the Wronskian and W_i is the determinant obtained from W when the ith column is replaced by a column whose entries, from top to bottom, are $(0, 0, \cdots, 0, 1)$. For $n = 2$ the formula reduces, as it should, to the result obtained in Section 10.2.

The method of Problem 5 leads to *Abel's identity*, which states that

$$W(x) = W(x_0)e^{-P(x)} \quad \text{where} \quad P(x) = \int_{x_0}^{x} p_{n-1}(t) \, dt.$$

Here x_0 is any point of I and it is assumed that $p_{n-1}(x)$ is continuous so we can form the integral. This formula can often be used to get $W(x)$ without laborious computation.

EXAMPLE 1. Let us see how the Lagrange procedure works out for the third-order equation

$$(3) \qquad\qquad Ty = y''' + p_2 y'' + p_1 y' + p_0 y = f.$$

It is assumed that the coefficients p_j are bounded, that f is continuous, and that u_1, u_2, u_3 form a linearly independent set of solutions of the homogeneous equation.

Starting with

$$p_0: \qquad y = u_1 v_1 + u_2 v_2 + u_3 v_3$$

we differentiate and get

$$p_1: \qquad y' = u_1' v_1 + u_2' v_2 + u_3' v_3 + (u_1 v_1' + u_2 v_2' + u_3 v_3').$$

As a first condition on the v_j the expression in parentheses is set equal to zero. A second differentiation then gives

$$p_2: \qquad y'' = u_1'' v_1 + u_2'' v_2 + u_3'' v_3 + (u_1' v_1' + u_2' v_2' + u_3' v_3').$$

Again the expression in parenthesis is set equal to zero and a third differentiation gives

$$y''' = u_1''' v_1 + u_2''' v_2 + u_3''' v_3 + (u_1'' v_1' + u_2'' v_2' + u_3'' v_3').$$

The letters p_0, p_1, p_2 at the left mean that the first equation is to be multiplied by p_0, the second by p_1, and the third by p_2. These results are then added to the fourth equation. Using the definition of T as given in (3) we get

$$Ty = (Tu_1)v_1 + (Tu_2)v_2 + (Tu_3)v_3 + (u_1'' v_1' + u_2'' v_2' + u_3'' v_3').$$

By hypothesis $Tu_j = 0$ for $j = 1, 2, 3$. Hence the equations obtained by setting the above parentheses equal to zero, together with the equation $Ty = f$, yield the system

$$u_1 v_1' + u_2 v_2' + u_3 v_3' = 0$$

$$u_1' v_1' + u_2' v_2' + u_3' v_3' = 0$$

$$u_1'' v_1' + u_2'' v_2' + u_3'' v_3' = f.$$

This is a system of three equations in the three unknowns v_1' , v_2', v_3'. Since the coefficient determinant is the Wronskian, which is not zero, we can solve by Cramer's rule. Namely,

$$v_1' \begin{vmatrix} u_1 & u_2 & u_3 \\ u_1' & u_2' & u_3' \\ u_1'' & u_2'' & u_3'' \end{vmatrix} = \begin{vmatrix} 0 & u_2 & u_3 \\ 0 & u_2' & u_3' \\ f & u_2'' & u_3'' \end{vmatrix} = \begin{vmatrix} 0 & u_2 & u_3 \\ 0 & u_2' & u_3' \\ 1 & u_2'' & u_3'' \end{vmatrix} f$$

where the third expression shows the connection with the general formula stated above. This equation and those for v_2' and v_3' are

$$W v_1' = \begin{vmatrix} u_2 & u_3 \\ u_2' & u_3' \end{vmatrix} f, \qquad W v_2' = \begin{vmatrix} u_3 & u_1 \\ u_3' & u_1' \end{vmatrix} f, \qquad W v_3' = \begin{vmatrix} u_1 & u_2 \\ u_1' & u_2' \end{vmatrix} f.$$

Division by the Wronskian W gives v_j' and an integration gives v_j.

PROBLEMS

1. Here and below, f is continuous. Consider

$$x^3 y''' - 3x^2 y'' + 6xy' - 6y = x^3 f(x), \qquad x > 0$$

(a) By trying $y = x^m$ find the solutions x, x^2, x^3 for the homogeneous equation. Verify that the Wronskian of these three functions is $2x^3$ and that $W(x) = cx^3$ agrees with the integral formula in the text.

(b) Taking $u_1 = x$, $u_2 = x^2$, $u_3 = x^3$ in Example 1 check that the result of that example gives

$$2v_1' = xf(x), \qquad v_2' = -f(x), \qquad 2xv_3' = f(x)$$

and hence the inhomogeneous equation has the solution

$$y = \frac{x}{2} \int xf(x)\, dx - x^2 \int f(x)\, dx + \frac{x^3}{2} \int \frac{f(x)}{x}\, dx$$

(c) Let $x^3 f(x) = x^m$ assuming, for simplicity, that $m \neq 1, 2, 3$. Check that the general formula produces the same result as the trial solution $y = Ax^m$.

2. Discuss as in Problem 1. Since the process incorporates a check, no answer is provided:

(a) $x^3 y''' - x^2 y'' = x^3 f(x)$ (c) $x^3 y''' + 2x^2 y'' - 2xy' = x^3 f(x)$

(b) $x^3 y''' + 3x^2 y'' = x^3 f(x)$ (d) $x^3 y''' + x^2 y'' - 2xy' + 2y = x^3 f(x)$

3. This takes a good deal of time. Apply the method of Example 1 to

$$x^3 y''' + xy' - y = 24x \ln x, \qquad x > 0$$

with $u_1 = x$, $u_2 = x \ln x$, $u_3 = x(\ln x)^2$; by Section 2, Problem 5 these functions satisfy the homogeneous equation. It will be found that

$$v_1 = 3(\ln x)^4 \qquad v_2 = -8(\ln x)^3 \qquad v_3 = 6(\ln x)^2 \qquad y = x(\ln x)^4$$

4. Let W be the Wronskian of three functions u, v, w each of which is three times differentiable on an interval I. Show that dW/dx is obtained by differentiating the elements u'', v'', w'' of the bottom row and leaving the other rows unchanged.

5. If the three functions u, v, w satisfy the third-order equation

$$Ty = y''' + py'' + qy' + ry = 0$$

use first Problem 4 and then $Tu = Tv = Tw = 0$ to get

$$\frac{dW}{dx} = \begin{vmatrix} u & v & w \\ u' & v' & w' \\ -pu'' - qu' - ru & -pv'' - qv' - rv & -pw'' - qw' - rw \end{vmatrix}$$

Using row operations simplify the bottom row of the determinant and thus obtain Abel's equation, $dW/dx = -pW$. A similar method works in the general case.

What is provable should not be believed in science without proof.

                                     —J. W. R. Dedekind

Chapter 12

POWER-SERIES SOLUTIONS

EVEN WHEN they do not have elementary solutions, differential equations can often be solved by means of power series. The procedure is similar to the use of undetermined coefficients for polynomial solutions, except that there are infinitely many coefficients. Equating corresponding powers leads to a recurrence relation that expresses any given coefficient in terms of its predecessors. Thus they can be determined successively, one after another. In cases commonly encountered a formula for the general term can be surmised from the first few terms and proved by mathematical induction. The solution is then obtained as an infinite series.

A review of power series is given in Chapter B.

1. Fundamental principles

Unless the contrary is specifically stated, the word "power series" means "power series about the point $c = 0$." In inequalities of the form $|x| < r$ we agree that r denotes a positive number. This convention is easy to remember, because if $r \leq 0$ the inequality $|x| < r$ is satisfied by no value of x.

The following theorem shows why power series are useful in the solution of differential equations:

THEOREM 1. *Let y be a formal power-series solution of the differential equation*

$$(1) \qquad p_2 y'' + p_1 y' + p_0 y = 0$$

where the p_j are polynomials. Suppose further that the series y has a positive radius of convergence r. Then the function $y(x)$ defined by the sum of the series satisfies the differential equation for $|x| < r$.

The proof is simple. For $|x| < r$ the series on the left of (1) converges to

$$p_2(x)y''(x) + p_1(x)y'(x) + p_0(x)y(x)$$

where the primes denote the ordinary derivative as in calculus. Since y is a formal solution by hypothesis, each coefficient in the series on the left of (1) is 0; this is what is meant by a "formal solution." Hence

$$p_2(x)y''(x) + p_1(x)y'(x) + p_0(x) = \sum_{j=0}^{\infty} 0x^j = 0, \qquad |x| < r$$

which is what we wanted to show.

The argument can be generalized. Let y be a formal power-series solution of

$$p_n y^{(n)} + p_{n-1} y^{(n-1)} + \cdots + p_1 y' + p_0 = f$$

where all p_j and f have convergent power-series expansions for $|x| < r$. Suppose the series giving the formal solution y also converges for $|x| < r$. Then the function $y(x)$ defined by the sum of this series satisfies the differential equation for $|x| < r$. The proof is virtually identical to that given above.

EXAMPLE 1. To solve the equation

(2) $$y'' + y = xy'$$

we try the substitution

$$y = a_0 + a_1 x + a_2 x^2 + \cdots + a_j x^j + \cdots.$$

The general terms in the series for y, y', and y'' are respectively

$$a_j x^j, \quad j a_j x^{j-1}, \quad j(j-1) a_j x^{j-2}.$$

This yields the following table for the expressions that occur in the differential equation:

Expression:	y''	y	xy'
General term:	$j(j-1)a_j x^{j-2}$	$a_j x^j$	$j a_j x^j$
Coefficient of x^n:	$(n+2)(n+1)a_{n+2}$	a_n	$n a_n$

The second row gives the general term in the expansion of the entries in the first row and the third row is the coefficient of x^n in these expansions. It is obtained by the choices $j = n+2$, $j = n$, $j = n$ in the respective entries of the second row.

The equation $y + y'' = xy'$ holds in the sense of equality for formal power series if, and only if, the coefficients of x^n on both sides agree. Equating coefficients of x^n yields the *recurrence formula*

$$(n+2)(n+1)a_{n+2} + a_n = n a_n.$$

The formula allows us to determine all the a_j, starting from any prescribed values a_0 and a_1. To see how this is done, solve for a_{n+2}:

(3) $$a_{n+2} = \frac{n-1}{(n+2)(n+1)} a_n.$$

Setting $n = 0, 2, 4, \cdots$ we get, in succession,

$$a_2 = -\frac{1}{2}a_0, \qquad a_4 = \frac{1}{3\cdot 4}a_2 = -\frac{1}{4!}a_0, \qquad a_6 = \frac{3}{6\cdot 5}a_4 = -\frac{3}{6!}a_0,$$

and so on. Thus the even powers of x in the formal series give

$$(4) \qquad a_0\left(1 - \frac{1}{2!}x^2 - \frac{1}{4!}x^4 - \frac{1\cdot 3}{6!}x^6 - \frac{1\cdot 3\cdot 5}{8!}x^8 - \frac{1\cdot 3\cdot 5\cdot 7}{10!}x^{10} - \cdots\right).$$

The way the above series was constructed shows that it is a formal solution. We now investigate the convergence.

Since the series has only even powers of x, the ratio of two consecutive terms is

$$\left|\frac{a_{2(n+1)}x^{2(n+1)}}{a_{2n}x^{2n}}\right| = \left(\frac{|a_{2n+2}|}{|a_{2n}|}\right)|x^2|.$$

Using (3) with n replaced by $2n$ we get

$$\frac{a_{2n+2}}{a_{2n}} = \frac{2n-1}{(2n+2)(2n+1)}.$$

This tends to 0 as $n \to \infty$. Hence the above ratio tends to 0 and the series converges for all x. By Theorem 1, its sum satisfies the differential equation.

Note the way we avoided use of the general term of the series. It would be inefficient to work with the general term because the formula (3) already gives exactly what is needed. In fact, if you apply the ratio test directly to the series, you will find yourself going backward and getting (3) again.

We now use (3) when n is odd. Setting $n = 1$ in (3) yields $a_3 = 0$ and the formula then gives $a_5 = 0$, $a_7 = 0$, and so on. Thus the odd powers of x in the formal series reduce to the single term $a_1 x$, all other terms being 0. It is easily checked that $y = a_1 x$ satisfies the differential equation.

If $u(x)$ denotes the function multiplying a_0 in (4) and if $v(x) = x$, our solution is

$$(5) \qquad\qquad y = a_0 u(x) + a_1 v(x).$$

The Wronskian of u and v at 0 is 1 and hence these functions are linearly independent. It follows from Section 10.1 that (5) is both the general solution and the complete solution. The latter statement means that every solution can be represented in the form (5), and hence *every solution has a power-series expansion*. This fact was by no means obvious when we started our investigation.

For many purposes it is advantageous to define $a_j = 0$ for $j < 0$, while keeping the convention that $0x^j = 0$ for all x. Then

$$\sum_{j=0}^{\infty} a_j x^j = \sum_{j=-\infty}^{\infty} a_j x^j, \qquad a_{-1} = 0, \ a_{-2} = 0, \cdots.$$

As illustrated in the following example, this interpretation is to be used whenever you encounter a term a_j with negative j.

EXAMPLE 2. Airy's equation. The equation $y'' = xy$ was first analyzed by the English astronomer Sir George Airy and is named after him. We shall find solutions u and v with initial values

$$u(0) = 1, \quad u'(0) = 0; \qquad v(0) = 0, \quad v'(0) = 1.$$

By superposition, the solution satisfying arbitrary initial values is then

$$y = y_0 u(x) + y_1 v(x) \quad \text{where} \quad y_0 = y(0), \ y_1 = y'(0).$$

We begin with the table

Expression:	y	xy	y''
General term:	$a_j x^j$	$a_j x^{j+1}$	$j(j-1)a_j x^{j-2}$
Coefficient of x^n:	$--$	a_{n-1}	$(n+2)(n+1)a_{n+2}$

The equation $y'' = xy$ then yields the recurrence formula

(6) $$(n+2)(n+1)a_{n+2} = a_{n-1}.$$

If $y = u$ the initial values give $a_0 = 1$, $a_1 = 0$. Setting $n = 0$ in (6) we get

$$2a_2 = a_{-1} = 0$$

since $a_j = 0$ for negative j as explained above. The initial conditions give $a_1 = 0$. Each use of (6) increases the subscript on a by 3. Consequently

$$0 = a_1 = a_4 = a_7 = \cdots, \qquad 0 = a_2 = a_5 = a_8 = \cdots.$$

Thus the only nonzero a_j are those in which $j = 3n$, a multiple of 3. Setting $n + 2 = 3, \ 6, \ 9, \ \cdots$, we get $n - 1 = 0, \ 3, \ \cdots$ and hence

$$3 \cdot 2a_3 = a_0, \quad 6 \cdot 5a_6 = a_3, \quad 9 \cdot 8a_9 = a_6, \quad \cdots.$$

Successive determination of the a_j starting with $a_0 = 1$ yields

$$a_0 = 1, \quad a_3 = \frac{1}{3 \cdot 2}, \quad a_6 = \frac{1}{6 \cdot 5 \cdot 3 \cdot 2}, \quad \cdots.$$

Upon recalling that the other $a_j = 0$ we get the series

$$u(x) = 1 + \frac{x^3}{3 \cdot 2} + \frac{x^6}{6 \cdot 5 \cdot 3 \cdot 2} + \frac{x^9}{9 \cdot 8 \cdot 6 \cdot 5 \cdot 3 \cdot 2} + \cdots.$$

For the solution v we have $a_0 = 0$, $a_1 = 1$ and a similar process yields

$$v(x) = x + \frac{x^4}{4 \cdot 3} + \frac{x^7}{7 \cdot 6 \cdot 4 \cdot 3} + \frac{x^{10}}{10 \cdot 9 \cdot 7 \cdot 6 \cdot 4 \cdot 3} + \cdots.$$

It is easily checked by (6) that both series converge for all x; see Problem 5. Hence, both satisfy the differential equation.

The following example differs from the preceding in that the equation is of the first order and has an arbitrary constant m as one of its coefficients. That the procedure works for equations of arbitrary order was indicated in the remarks following Theorem 1.

EXAMPLE 3.   Binomial theorem.   We shall find a series solution for the initial-value problem

$$(1 + x)y' = my, \qquad y(0) = 1.$$

The table now is

Expression:	y	y'	xy'
General term:	$a_j x^j$	$j a_j x^{j-1}$	$j a_j x^j$
Coefficient of x^n:	a_n	$(n+1)a_{n+1}$	na_n

Since $y' = my - xy'$ we get the recurrence formula

(7) $\qquad (n+1)a_{n+1} = ma_n - na_n \quad \text{or} \quad a_{n+1} = \left(\frac{m-n}{n+1}\right)a_n.$

Successive use of this formula, starting with $a_0 = 1$, yields

$$a_0 = 1, \quad a_1 = m, \quad a_2 = \frac{m(m-1)}{2!}, \quad a_3 = \frac{m(m-1)(m-2)}{3!},$$

and so on, and the resulting series is the binomial series. By (7) the ratio of two successive terms satisfies

$$\left|\frac{a_{n+1}x^{n+1}}{a_n x^n}\right| = |x|\left|\frac{a_{n+1}}{a_n}\right| = |x|\left|\frac{m-n}{n+1}\right|.$$

As $n \to \infty$ the limit is $|x|$. Hence the series converges for $|x| < 1$, diverges for $|x| > 1$, and the radius of convergence is 1.

By separation of variables the solution is

$$y = (1 + x)^m.$$

Equating this to the series solution found above we get the *binomial theorem*,

$$(1+x)^m = 1 + mx + \frac{m(m-1)}{2!}x^2 + \frac{m(m-1)(m-2)}{3!}x^3 + \cdots, \qquad |x| < 1.$$

The interest of the method is that it yields the binomial theorem for all real exponents m.

If m is a positive integer in Example 3 the series terminates at x^m; that is, all terms $a_j x^j$ for $j > m$ are 0. Hence, although the series solution converges only for $|x| < 1$ in general, when m is a positive integer it reduces to a polynomial and converges for all x. In the next section we discuss a second-order equation that shows similar behavior.

Having presented a number of examples, we now return to the general case. If $p_2(0) \neq 0$ in Theorem 1, one can divide the equation by $p_2(x)$ for small $|x|$ and reduce the equation to the standard form in which the coefficient of y'' is 1. The following theorem holds:

THEOREM 2. (a) *Suppose the equation $y'' + p(x)y' + q(x)y = 0$ has two power-series solutions $u(x)$ and $v(x)$ for $|x| < r_1$. Suppose further that the Wronskian W of u and v satisfies $W(0) \neq 0$. Then p and q have power-series expansions on some interval $|x| < r$.*

(b) *Conversely, if p and q have power-series expansions for $|x| < r$, there exist solutions u and v with $W(0) \neq 0$ that have power-series expansions at least for $|x| < r$.*

Whenever one says that a function "has a power-series expansion," one means that the series converges on some interval and that the function is given by its sum. For example the hypothesis of (a) means that $u(x)$ and $v(x)$ are the sums of convergent power series at least for $|x| < r_1$ and that they both satisfy the equation on this interval. The conclusion says that $p(x)$ and $q(x)$ are sums of convergent power series on some possibly smaller interval $|x| < r$.

For proof of Part (a), write the equations satisfied by u and v in the form

$$pu' + qu = -u'', \quad pv' + qv = -v''.$$

This can be regarded as a linear system in the two unknowns p, q. Solving by Cramer's rule we get

$$p = \frac{u''v - v''u}{W}, \quad q = \frac{u'v'' - u''v'}{W}$$

where W is the Wronskian. Since $W(0) \neq 0$ these expressions have convergent power-series expansions on some interval $|x| < r$, where $r > 0$. This completes the proof.

A similar result holds for the linear equation of order n. Given n solutions u_i for such an equation, we can write

$$p_0 u_i + p_1 u_i' + \cdots + p_{n-1} u_i^{(n-1)} = -u_i^{(n)}, \qquad i = 1, 2, \cdots, n$$

and an analog of Part (a) follows at once.

Proof of Part (b) is given in DETA, pages 282–284. That this part also extends to equations of order n.

PROBLEMS

1P. In this and subsequent problems assume a power series for y with coefficients a_j as in the text. Here we are going to solve the equation $y'' = xy'$ by power series.

(a) Make a table showing the general term for y', y'', xy'. Thus get the coefficient of x^n in the series for y'' and xy' and from this deduce the recursion formula

$$a_{n+2} = \frac{n}{(n+2)(n+1)} a_n$$

(b) Taking first $n = 0, 2, 4, \cdots$ and then $n = 1, 3, 5, \cdots$ obtain

$$y = a_0 + a_1 \left(x + \frac{x^3}{3!} + \frac{3 \cdot 1}{5!} x^5 + \frac{5 \cdot 3 \cdot 1}{7!} x^7 + \cdots \right)$$

The solution is valid for all x. Why?

2. Solve by series and express your answer in terms of elementary functions:

(a) $y' = y$ (b) $y' = -2y$ (c) $y'' = y$ (d) $y'' = -y$ (e) $y' = 2xy$

3. If a_j are the coefficients of a power-series solution, obtain the recursion formula shown at the right of the equation:

(a) $x(1-x)y' = y$ $(n-1)a_n = (n-1)a_{n-1}$

(b) $(x^2+1)y'' = 2y$ $(n+2)a_{n+2} = (2-n)a_n$

(c) $(x^2-1)y'' + 4xy' + 2y = 0$ $a_{n+2} = a_n$

(d) $y'' = xy' + y$ $(n+2)a_{n+2} = a_n$

(e) $(1+x)y' = 3y$ $(n+1)a_{n+1} = (3-n)a_n$

(f) $y'' + y = xy'$ $(n+2)(n+1)a_{n+2} = (n-1)a_n$

(g) $(x^2+x)y' = (2x+1)y$ $(n-1)a_n = (3-n)a_{n-1}$

(h) $xy' = (2x^2+1)y$ $(n-1)a_n = 2a_{n-2}$

(i) $y'' = x^2 y$ $(n+2)(n+1)a_{n+2} = a_{n-2}$

4. (abcdefghi) Obtain series solutions for the equations in Problem 3.

5. Let m be a positive integer. An *m-step recursion formula* gives a_{j+m} in terms of a_j, starting from some definite value a_k. Usually one starts at a_0 or a_1.

(a) Write the following recursion relations in terms of $j = n-1$, n, $n-2$, as the case may be, and thus convince yourself that these are 1-step, 2-step and 4-step formulas respectively:

$$na_n = a_{n-1} \qquad (n^2+1)a_{n+2} = (2n)^2 a_n \qquad 16na_{n+2} = (n+1)a_{n-2}$$

(b) Starting with a_k, the m-step formula gives a_{k+m}, a_{k+2m}, and so on. Hence it determines the coefficients in the series

$$a_k x^k + a_{k+m} x^{k+m} + a_{k+2m} x^{k+2m} + \cdots + a_{k+nm} x^{k+nm} + \cdots$$

Show that the limit relation below yields the formula on the right for the radius of convergence of this series:

$$(j = k + nm), \qquad \lim_{j \to \infty} \left| \frac{a_{j+m}}{a_j} \right| = L, \qquad r = \frac{1}{\sqrt[m]{L}}$$

The equation $(j = k+nm)$ means that j has the form $k+nm$ as $n \to \infty$. This avoids coefficients a_j that are not in the above series and may be zero.

(c) How should you interpret the cases $L = 0$, $L = \infty$?

(d) Check that the formula of Part (b) yields $r = \infty$, $1/2$, 2 in the three examples of Part (a) unless the series terminates (in which case the limit condition of Part (b) is meaningless and $r = \infty$).

6. (abcdefghi) Assuming that the starting value is chosen so that the series does not terminate, use the formula of Problem 5 to find the radius of convergence of the power-series solutions in Problem 3. Also verify agreement with the result of Theorem 2(b) in those cases in which this theorem applies.

7. Problems (abcd) are special cases of the equations of Legendre, Hermite, Chebychev, and Laguerre respectively. Give a recursion formula for the coefficients of a series solution. Taking $(a_0, a_1) = (1,0)$ or $(0,1)$, as the case may be, use your formula to obtain a polynomial solution, and check the latter by substitution into the differential equation (three things to do):

    (a) $(1 - x^2)y'' + 12y = 2xy'$ (c) $(1 - x^2)y'' + 9y = xy'$

    (b) $y'' + 8y = 2xy'$ (d) $xy'' + 2y = (x - 1)y'$

8. (a) Solve the equation $x^2 y'' - 2xy' + 2y = 0$ by trying $y = x^s$ where s is constant and thus show that the hypothesis "$W(0) \neq 0$" in Theorem 2(a) cannot be replaced by "u and v are linearly independent for $|x| < r_1$."

(b) Under the hypothesis of Theorem 2(b) does every solution y necessarily have a power-series expansion for $|x| < r$? Why or why not?

9. The recursion formula $(n+1)a_{n+1} = (m-n)a_n$ with the initial condition $a_0 = 1$ was obtained for the binomial coefficients a_n in Example 3. Prove by mathematical induction that

$$a_n = \frac{m(m-1)(m-2)\cdots(m-n+1)}{n!}, \qquad n \geq 1$$

————————————————ANSWERS————————————————

1. By Theorem 1 and the ratio test.
2. Partial answer: (a) $a_0 e^x$ (b) $a_0 e^{-2x}$ (c) $a_0 \cosh x + a_1 \sinh x$
   (d) $a_0 \cos x + a_1 \sin x$ (e) $a_0 e^{x^2}$
4. (a) $a_1 x/(1-x)$

   (b) $a_0(1+x^2) + a_1 \left(x + \dfrac{1}{3}x^3 - \dfrac{1}{3\cdot 5}x^5 + \dfrac{1}{5\cdot 7}x^7 - \dfrac{1}{7\cdot 9}x^9 + \cdots \right)$

   (c) $(a_0 + a_1 x)/(1-x^2)$

   (d) $a_0 e^{x^2/2} + a_1 \left(x + \dfrac{x^3}{3} + \dfrac{x^5}{3\cdot 5} + \dfrac{x^7}{3\cdot 5\cdot 7} + \cdots \right)$ (e) $a_0(1+x)^3$

   (f) $a_0 \left(1 - \dfrac{x^2}{2!} - \dfrac{1}{4!}x^4 - \dfrac{3\cdot 1}{6!}x^6 - \dfrac{5\cdot 3\cdot 1}{8!}x^8 - \cdots \right) + a_1 x$

   (g) $a_1(x + x^2)$ (h) $a_1 x e^{x^2}$

   (i) $a_0 \left(1 + \dfrac{x^4}{3\cdot 4} + \dfrac{x^8}{3\cdot 4\cdot 7\cdot 8} + \cdots \right) + a_1 \left(x + \dfrac{x^5}{4\cdot 5} + \dfrac{x^9}{4\cdot 5\cdot 8\cdot 9} + \cdots \right)$

5. (c) If $L = 0$ it converges for all x, if $L = \infty$ it converges only for $x = 0$.
6. (a) 1 (b) 1 (c) 1 (d) ∞ (e) terminates (f) ∞ (g) terminates (h) ∞ (i) ∞
7. (a) $(n+2)(n+1)a_{n+2} = a_n(n-3)(n+4)$, $x - (5/3)x^3$
   (b) $(n+2)(n+1)a_{n+2} = 2(n-4)a_n$, $1 - 4x^2 + (4/3)x^4$
   (c) $(n+2)(n+1)a_{n+2} = (n+3)(n-3)a_n$, $x - (4/3)x^3$
   (d) $(n+1)^2 a_{n+1} = (n-2)a_n$, $1 - 2x + (1/2)x^2$
8. (a) Partial answer: x, x^2.
   (b) Yes, because every solution can be written in the form $y = c_1 u + c_2 v$.

2. Supplementary topics

Legendre's equation

$$(1 - x^2)y'' - 2xy' + m(m + 1)y = 0$$

was introduced in Section 10.1; it has many applications in mathematics and mathematical physics. To solve this equation by power series we make the following table:

y''	$-x^2y''$	$-2xy'$	$m(m+1)y$
$j(j-1)a_jx^{j-2}$	$-j(j-1)a_jx^j$	$-2ja_jx^j$	$m(m+1)a_jx^j$
$(n+2)(n+1)a_{n+2}$	$-n(n-1)a_n$	$-2na_n$	$m(m+1)a_n$

The differential equation says that the sum of the expressions in the first row is 0. Hence the sum of those in the third row must also be 0. That is,

$$(n+2)(n+1)a_{n+2} - n(n-1)a_n - 2na_n + m(m+1)a_n = 0.$$

Simplifying and moving all terms in a_n to the right yields

$$(n+2)(n+1)a_{n+2} = -a_n(m^2 + m - n^2 - n)$$

or equivalently,

(1) $$(n+2)(n+1)a_{n+2} = -a_n(m-n)(m+n+1).$$

The coefficients for even n are determined by a_0 as follows:

$$a_2 = -a_0\frac{m(m+1)}{2 \cdot 1}, \quad a_4 = -a_2\frac{(m-2)(m+3)}{4 \cdot 3}, \quad a_6 = -a_4\frac{(m-4)(m+5)}{6 \cdot 5}$$

and so on. It is left for you to determine the a_{2j} from this and to carry out a similar calculation for odd n starting with $n = 1$. The result is

$$y = a_0\left(1 - \frac{m(m+1)}{2!}x^2 + \frac{m(m-2)(m+1)(m+3)}{4!}x^4 - \cdots\right)$$

$$+ a_1\left(x - \frac{(m-1)(m+2)}{3!}x^3 + \frac{(m-1)(m-3)(m+2)(m+4)}{5!}x^5 + \cdots\right).$$

Since (1) gives

$$\lim_{n \to \infty} \left| \frac{a_{n+2}}{a_n} \right| = \lim_{n \to \infty} \left| \frac{(m-n)(m+n+1)}{(n+2)(n+1)} \right| = 1$$

the radius of convergence, in general, is 1; see Problem 5 of the last section. However, if m is a positive integer or 0, the choice $n = m$ in (1) gives $a_{m+2} = 0$, and the same holds for subscripts $m + 4$, $m + 6$, and so on. Thus the series terminates. Hence the coefficient of a_0 or a_1 reduces to a polynomial of degree m according as m is even or odd.

Choosing a_0 or a_1 for $m = 0, 1, 2, \cdots$ in such a way that the value of the polynomial is 1 when $x = 1$, we get a sequence of polynomials P_m as follows:

$$1, \quad x, \quad \frac{3}{2}x^2 - \frac{1}{2}, \quad \frac{5}{2}x^3 - \frac{3}{2}x, \quad \frac{35}{8}x^4 - \frac{15}{4}x^2 + \frac{3}{8}, \quad \cdots.$$

These are the *Legendre polynomials*. For any integer $m \geq 0$ the Legendre polynomial $P_m(x)$ is the unique polynomial of degree m that satisfies Legendre's equation with parameter m and also satisfies $P_m(1) = 1$. In the following problems we discuss three other equations that also lead to important classes of polynomials.

Concluding this discussion of power-series solutions, we show how the radius of convergence can be determined, in many cases, by inspection of the differential equation. The most important step is to replace the real variable x by a complex variable z.

A complex-valued function f of the complex variable z is *analytic* at a point z_0 if it has a convergent power-series expansion about z_0. It is analytic in a disk $|z - z_0| < r$ if it is analytic at every point of the disk. The following theorem, which we accept without proof, is established in a branch of mathematics known as the theory of functions of a complex variable:

THEOREM A. *The function $f(z)$ has a series expansion about $z = c$ with radius of convergence at least r if, and only if, $f(z)$ is analytic at every point of the disk $|z - c| < r$.*

To get at the meaning of this theorem, let us define a *singularity* of $f(z)$ to be a point z_0 at which f fails to be analytic. For example, the function

$$\frac{z^2 + 4}{(z + 2)(z^2 + 1)}$$

has singularities at -2, i, $-i$ where the denominator vanishes. These are the only singularities, since results stated in Chapter B indicate that this function has a power-series expansion about every other point. Theorem A can now be reworded as follows:

THEOREM 1. *The radius of convergence of the power series for $f(z)$ about $z = c$ is the distance from c to the nearest singularity of f.*

If c is real, a moment's thought shows that the disk of convergence $|z - c| < r$ in the complex plane leads to the same value r as that associated with the interval of convergence $|x - c| < r$ on the real axis. When combined with Section 1, Theorem 2(b), this fact gives information about the interval of convergence for power-series solutions $y(x)$ when we know the location of the singularities of the coefficients in the complex plane.

EXAMPLE 1. Solutions of the following differential equation are expressed as power series about the point $x = 0$. What radius of convergence is to be expected?

$$(x + 2)(x^2 + 1)(x^2 - 4x + 5)y'' + (x^2 + 4)y = 0.$$

If this is written in standard form $y'' + q(x)y = 0$ the function q has, as sole singularities, the points

$$-2, \quad i, \quad -i, \quad 2 + i, \quad 2 - i$$

where the denominator vanishes. As seen in Figure 1, the distance from 0 to the nearest singularity is 1 and hence the expected radius of convergence for the solutions is 1. We say "expected" because at least this radius is guaranteed by Theorem 1, but the actual radius is sometimes larger. See Problem 8.

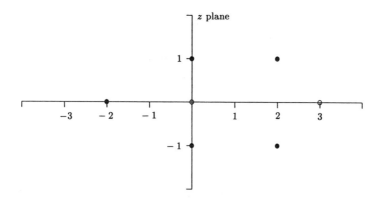

Figure 1

PROBLEMS

1. This pertains to *Hermite's equation $y'' - 2xy' + 2my = 0$.*

(a) Obtain a recursion formula for a power-series solution, find the corresponding radius of convergence, and verify agreement with Theorem 2(b) of Section 1. Taking $y(0) = 1$, $y'(0) = 0$, or $y(0) = 0$, $y'(0) = 1$, get

(b) $y = 1 - \dfrac{2m}{2!}x^2 + \dfrac{2^2 m(m-2)}{4!}x^4 - \dfrac{2^3 m(m-2)(m-4)}{6!}x^6 + \cdots$

(c) $y = x - \dfrac{2(m-1)}{3!}x^3 + \dfrac{2^2(m-1)(m-3)}{5!}x^5 - \cdots$

2. When m is a nonnegative integer the series in Problem 1(b) or 1(c) terminates. The resulting polynomials, multiplied by a constant to give 2^m as the leading coefficient, are the *Hermite polynomials* $H_m(x)$. Verify that the first five Hermite polynomials are

$$1, \qquad 2x, \qquad 4x^2 - 2, \qquad 8x^3 - 12x, \qquad 16x^4 - 48x^2 + 12$$

3. This pertains to *Chebychev's equation $(1 - x^2)y'' - xy' + m^2 y = 0$.*

Part (a) is the same as in Problem 1. For Parts (bc) obtain the solutions

(b) $y = 1 - \dfrac{m^2}{2!}x^2 + \dfrac{m^2(m^2 - 2^2)}{4!}x^4 - \dfrac{m^2(m^2 - 2^2)(m^2 - 4^2)}{6!}x^6 + \cdots$

(c) $y = x - \dfrac{m^2 - 1^2}{3!}x^3 + \dfrac{(m^2 - 1^2)(m^2 - 3^2)}{5!}x^5 - \cdots$

4. When m is a nonnegative integer the series in Problem 3(b) or 3(c) terminates. The resulting polynomials, multiplied by a constant to give 2^{m-1} as the leading coefficient, are the *Chebychev polynomials* $T_m(x)$ for $m \geq 1$, and by convention $T_m(x) = 1$ when $m = 0$. Verify that the first five are

$$1, \qquad x, \qquad 2x^2 - 1, \qquad 4x^3 - 3x, \qquad 8x^4 - 8x^2 + 1$$

5. This pertains to *Laguerre's equation $xy'' + (1 - x)y' + my = 0$.*

(a) Obtain a recursion formula for a series solution, determine the corresponding radius of convergence, and discuss the applicability of Section 1, Theorem 2.

(b) Obtain the following solution:

$$y = 1 - mx - \dfrac{m(1 - m)}{(2!)^2}x^2 - \dfrac{m(1 - m)(2 - m)}{(3!)^2}x^3 - \cdots$$

6. If m is a nonnegative integer in Problem 5 the series terminates. The resulting polynomials, multiplied by a constant to give $y(0) = m!$, are the *Laguerre polynomials* $L_m(x)$. Verify that the first four are

$$1, \qquad 1 - x, \qquad 2 - 4x + x^2, \qquad 6 - 18x + 9x^2 - x^3$$

7. (a) Check that the equation $z'' + (2m + 1 - x^2)z = 0$ leads to a three-term recursion formula for the coefficients of a power-series solution, and that the determination of z by this method presents a difficult task.

(b) Show however that the substitution $z = ye^{-x^2/2}$ leads to the Hermite equation for y, and from this z can be written by inspection.

8. Check that the equation $(2x - 1)y'' - 4xy' + 4y = 0$ has the linearly independent solutions x, e^{2x} although the expected radius of convergence for power-series solutions expanded about $x = 0$ is only $1/2$.

9. Solutions of the differential equation in Example 1 are expanded about the point $x = 3$. Show that the expected radius of convergence is $\sqrt{2}$.

Hint: By Figure 1 the singularities nearest to $x = 3$ are $2 + i$ and $2 - i$.

10. What is the expected radius of convergence for series solutions of the following equations if the expansion is about $x = 0$? $x = 4$? $x = -1$?

(a) $(x^2 + 1)y'' = y$ (c) $x(x^2 + 6x + 9)y'' + y' = xy$

(b) $(x^2 - 9)y'' = y'$ (d) $(x^2 + 2x + 2)(x^2 + 4)y'' = (xy)'$

---------------------------------ANSWERS---------------------------------

1. (a) $(n + 2)(n + 1)a_{n+2} = 2(n - m)a_n$, $r = \infty$
3. (a) $(n + 2)(n + 1)a_{n+2} = (n^2 - m^2)a_n$, $r = 1$
5. (a) $(n + 1)^2 a_{n+1} = (n - m)a_n$, $r = \infty$. Theorem 2(b) does not apply but Theorem 2(a) shows there cannot be a second power-series solution making $W(0) \neq 0$ for the pair.
10. (a) 1, $\sqrt{17}$, $\sqrt{2}$ (b) 3, 1, 2 (c) 0, 4, 1 (d) $\sqrt{2}$, $2\sqrt{5}$, 1

The function of mathematics in providing answers is often less important than its function in providing understanding.

—Ben Noble

Chapter 13

GENERALIZED POWER SERIES

$\mathbf{I}$NSTEAD of assuming that the coefficients p and q of the differential equation $y'' + py' + qy = 0$ have power-series expansions about $x = 0$, as in the previous chapter, we assume only that $xp(x)$ and $x^2 q(x)$ have such expansions. Under this weaker assumption the solutions are generally of the form $x^s A(x)$ where s is a constant and where $A(x)$ has a power-series expansion. It turns out that s must satisfy a certain quadratic equation, called the indicial equation, and the coefficients are then determined by recursion. Introduction of the factors x, x^2, x^s as described above seems to be only a modest change. Nevertheless it leads to a dramatic increase in the scope of the theory.

1. The Frobenius normal form

Throughout this chapter, the phrase "power-series expansion," unless further qualified, means "convergent power-series expansion on some interval $|x| < r$." Since $x - x_0$ can be taken as a new variable, the assumption that our expansions are about the point $x = 0$ involves no loss of generality.

Let us consider the general second-order linear equation

(1) $$p_2(x)y'' + p_1(x)y' + p_0(x)y = 0$$

where the coefficients have power-series expansions. We assume $p_2(x) \neq 0$ except possibly at $x = 0$. Multiplying the equation by $1/p_2(x)$ yields

$$y'' + p(x)y' + q(x)y = 0$$

where

$$p(x) = \frac{p_1(x)}{p_2(x)}, \quad q(x) = \frac{p_0(x)}{p_2(x)}.$$

If $p_2(0) = 0$ neither function is defined at $x = 0$ but we agree that

$$p(0) = \lim_{x \to 0} \frac{p_1(x)}{p_2(x)}, \quad q(0) = \lim_{x \to 0} \frac{p_0(x)}{p_2(x)}$$

provided the limits exist. This definition is motivated by a desire to make the functions p and q continuous at $x = 0$. When the continuous functions so obtained have power-series expansions, $x = 0$ is called an *ordinary point*. A sufficient, but by no means necessary, condition for 0 to be an ordinary point is that $p_2(0) \neq 0$.

Instead of multiplying by $1/p_2(x)$ we could multiply by $x^2/p_2(x)$. The result is

(2)
$$x^2 y'' + P(x)xy' + Q(x)y = 0$$

where

$$P(x) = \frac{xp_1(x)}{p_2(x)}, \quad Q(x) = \frac{x^2 p_0(x)}{p_2(x)}.$$

If $p_2(0) = 0$ neither function is defined for $x = 0$, but we agree that

$$P(0) = \lim_{x \to 0} \frac{xp_1(x)}{p_2(x)}, \quad Q(0) = \lim_{x \to 0} \frac{x^2 p_0(x)}{p_2(x)}$$

provided the limits exist. The factors x and x^2 greatly improve the behavior at the origin, and it is quite possible for P and Q to have power-series expansions when p and q do not. In that case the origin is said to be a *regular singular point*. The fact that (2) is an important and natural class of equations will become clear in the course of the following discussion. First we illustrate the definitions by an example.

EXAMPLE 1. Classify the singularity at the origin:

$$xy'' + (\sin x)y = 0, \quad x^2 y'' + (\sin x)y' + (\cos x)y = 0, \quad x^2 y'' + y' = 0.$$

If the left-hand equation is divided by x the coefficient of y becomes

$$q(x) = \frac{\sin x}{x} = 1 - \frac{x^2}{3!} + \frac{x^4}{5!} - \cdots$$

since, by our convention regarding continuity, $q(0) = 1$. The origin is an ordinary point.

The coefficient of y' in the second equation can be written

$$\sin x = xP(x) \quad \text{where} \quad P(x) = \frac{\sin x}{x}.$$

Taking $P(0) = 1$ gives a power series for $P(x)$ as seen above. Since $Q(x) = \cos x$ also has a series expansion, the origin is a regular singular point. It is not an ordinary point because neither $x^{-2} \sin x$ nor $x^{-2} \cos x$ has a power-series expansion.

In the third case the coefficient of y' is $(1/x)x$. The function $P(x) = 1/x$ does not have a series expansion and the origin is an example of an *irregular singular point*. This means that the equation has the form (1) with power-series coefficients as described above, but that the origin is neither an ordinary point nor a regular singular point.

The theory of (2) is due mainly to the German mathematicians Carl Friedrich Gauss, Bernhard Riemann, Georg Frobenius, and Lazarus Fuchs. Gauss and Riemann initiated the investigation by profound study of certain second-order equations (1812, 1857). The general equation of order n was dealt with by Fuchs (1866), and his methods were later simplified by Frobenius (1873). It was Frobenius who emphasized the convenience of introducing the factors x^2 and x, and (2) is sometimes said to be in *Frobenius normal form*. Without intending to disparage prior claims, we refer to (2) as "the Frobenius equation."

An example of an equation in Frobenius normal form is the Euler-Cauchy equation

$$x^2 y'' + P_0 xy' + Q_0 y = 0, \qquad x > 0$$

where P_0 and Q_0 are real constants. The Euler-Cauchy equation and its nth-order generalization were discussed in Chapter 11. All you need to know here is that the substitution

$$y = x^s, \quad y' = sx^{s-1}, \quad y'' = s(s-1)x^{s-2}$$

yields the quadratic equation

(3) $$s(s-1) + P_0 s + Q_0 = 0$$

for the constant s. This fact was illustrated in Chapter 2 and can be verified by inspection.

If (3) has two roots λ and μ the process gives the solutions

(4) $$u = x^\lambda, \quad v = x^\mu.$$

When a root $s = \sigma + i\tau$ is complex we define x^s for $x > 0$ by

$$x^s = (e^{\ln x})^{\sigma + i\tau} = e^{\sigma \ln x} e^{i\tau \ln x} = x^\sigma (\cos(\tau \ln x) + i\sin(\tau \ln x)).$$

The principle of equating real parts gives corresponding real solutions

$$x^\sigma \cos(\tau \ln x), \qquad x^\sigma \sin(\tau \ln x).$$

Conversely, if λ and μ are prescribed real or complex numbers we can define P_0 and Q_0 in such a way that

$$s(s-1) + P_0 s + Q_0 = (s - \lambda)(s - \mu).$$

Then (3) has roots λ, μ and the Euler-Cauchy equation has the solutions (4). Equating coefficients yields the formulas

$$P_0 = 1 - (\lambda + \mu), \quad Q_0 = \lambda\mu.$$

These equations show that P_0 and Q_0 are real if the prescribed roots λ and μ are real, and also if they are complex conjugates.

The point we want to make is that solutions of the Euler-Cauchy equation can have a great variety of behaviors near $x = 0$. The following pairs of linearly independent solutions (u, v) are obtained by suitable choices of P_0 and Q_0:

$$(x, x^2), \quad (x^{-1}, x^{-2}), \quad (x^{1/5}, x^{1/3}), \quad (x^{-1}, x^2).$$

The first shows that both solutions can have power-series expansions, the second shows that both solutions can be unbounded near 0, the third shows that both solutions can be continuous yet such that neither has a power-series expansion, and the fourth shows that one solution can be unbounded while the other has a power-series expansion.

Similar behavior is found for solutions of the general Frobenius equation, which is repeated for convenience:

$$(5) \qquad\qquad x^2 y'' + P(x)xy' + Q(x)y = 0.$$

When x is near 0, Equation (5) is approximately

$$(6) \qquad\qquad x^2 y'' + P(0)xy' + Q(0)y = 0.$$

This is the Euler-Cauchy equation with $P_0 = P(0)$ and $Q_0 = Q(0)$. The corresponding quadratic polynomial

$$(7) \qquad\qquad I(s) = s(s - 1) + sP(0) + Q(0)$$

is called the *indicial polynomial* of (5), the equation $I(s) = 0$ is called the *indicial equation*, and the roots λ, μ of the indicial equation are called the *exponents* at the regular singular point $x = 0$. Equation (6) retains its relevance as a means of approximation when P and Q are merely continuous at $x = 0$, without having power-series expansions. If P and Q do have power-series expansions, as assumed here, the quantities

$$P(0) = P_0, \quad Q(0) = Q_0$$

are the constant terms in these expansions.

The following theorem shows that solutions of (5) near the origin behave much like solutions of (6):

THEOREM 1. *Suppose P and Q have power-series expansions for $|x| < r$. Let (λ, μ) be the roots of the indicial equation, so labeled that $\lambda \geq \mu$ when both are real. Then the Frobenius equation has a solution*

$$(8) \qquad\qquad u(x) = x^s \sum_{j=0}^{\infty} a_j x^j, \qquad a_0 \neq 0$$

with $s = \lambda$. If $\lambda - \mu$ is not an integer there is a second solution $v(x)$ of this form with $s = \mu$. In both cases the series converges at least for $|x| < r$ and the given function satisfies the differential equation at least for $0 < x < r$.

For proof see DETA, pages 316–318. The following remarks give useful insight.

Even though we have assumed P and Q to be real, we must allow complex coefficients a_j. This is true because the coefficients a_j are almost always complex if the roots of the indicial equation are complex. The situation is similar to that for the equation

$$y'' + ay' + by = 0, \qquad a, b \text{ const.},$$

where the substitution $y = e^{st}$ may lead to complex solutions even if a and b are real.

Since the exponent s can be increased at will, the assumption $a_0 \neq 0$ involves no loss of generality. For example, if $a_0 = a_1 = 0$ but $a_2 \neq 0$ the series starts as follows:

$$a_2 x^{s+2} + a_3 x^{s+3} + \cdots.$$

This has the same form as the original series, with the role of a_0 taken by a_2 and with s replaced by $s+2$. When $a_0 \neq 0$, as assumed here, it will be found that s has to be a root of the indicial equation. That is why the indicial equation is important.

The differential equation is considered for $0 < x < r$ rather than for $|x| < r$ because the factors x^λ and x^μ can be infinite at $x = 0$ or, if finite, can be complex when $x < 0$ even if λ and μ are real. Examples are $1/x$ and $x^{5/2}$. If we replace x^λ and x^μ by

$$|x|^\lambda, \quad |x|^\mu$$

respectively, we get a solution valid for $-r < x < 0$ and hence, in some cases, a solution valid for $|x| < r$. See Problem 5.

If there are two solutions as described in Theorem 1, one for $s = \lambda$ and one for $s = \mu$, with $\lambda \neq \mu$, the two solutions are linearly independent on the interval $(0, r)$. Hence the expression

$$c_1 u(x) + c_2 v(x)$$

yields both the complete solution and the general solution on this interval. This follows from the fact that for x positive and close to zero the ratio $u(x)/v(x)$ is defined and not constant.

We now illustrate the procedure by an example.

EXAMPLE 2. For the equation

(9) $$x^2 y'' + (x^2 - 20)y = 0$$

we have $P(x) = 0$ and $Q(x) = x^2 - 20$. Hence the indicial polynomial (7) is

(10) $$I(s) = s(s-1) - 20 = (s-5)(s+4)$$

and the exponents are $\lambda = 5$, $\mu = -4$.

To solve Equation (9) we try a *generalized power series*

$$y = x^s \sum_{j=0}^{\infty} a_j x^j = \sum_{j=0}^{\infty} a_j x^{s+j}.$$

The series is easily differentiated and the general term is given by the following table:

y	y'	y''
$a_j x^{s+j}$	$(s+j)a_j x^{s+j-1}$	$(s+j)(s+j-1)a_j x^{s+j-2}$

This yields another table for the expressions in the differential equation:

$x^2 y''$	$x^2 y$	$-20y$
$(s+j)(s+j-1)a_j x^{s+j}$	$a_j x^{s+j+2}$	$-20a_j x^{s+j}$
$(s+n)(s+n-1)a_n$	a_{n-2}	$-20a_n$

The second row gives the general term in the expansion of the entries of the first row, and the third row is the coefficient of x^{s+n} in these expansions. It is obtained by the choices $j = n$, $j = n - 2$, $j = n$ in the respective entries of the second row.

Setting the coefficient of x^{s+n} equal to 0 in (9), we get the recursion formula

$$((s+n)(s+n-1) - 20)\, a_n + a_{n-2} = 0.$$

Since $s(s-1) - 20 = I(s)$ by (10), this can be written

$$I(s+n)a_n = -a_{n-2}.$$

When $n = 0$ we get $I(s)a_0 = -a_{-2} = 0$ and hence $I(s) = 0$. The two roots $\lambda = 5$ and $\mu = -4$ were already found above. If $s = 5$ the recursion formula becomes

$$n(n+9)a_n = -a_{n-2}.$$

It is easily checked that $a_n = 0$ when n is odd and that the a_n with even n yield the solution

$$a_0 x^5 \left(1 - \frac{x^2}{2 \cdot 11} + \frac{x^4}{(4 \cdot 2)(13 \cdot 11)} - \frac{x^6}{(6 \cdot 4 \cdot 2)(15 \cdot 13 \cdot 11)} + \cdots\right).$$

The factor x^5 comes from x^s with $s = 5$.

If we set $s = -4$, which is the other root μ, the recurrence formula becomes

$$n(n-9)a_n = -a_{n-2}.$$

By a short calculation

$$y = a_0 x^{-4}\left(1 + \frac{x^2}{2\cdot 7} + \frac{x^4}{(4\cdot 2)(5\cdot 7)} + \frac{x^6}{(6\cdot 4\cdot 2)(3\cdot 5\cdot 7)} + \cdots\right).$$

The pattern is clear from the recursion formula. Notice, however, that the signs alternate as soon as $n > 9$.

Since $\lambda - \mu = 9$ this example shows that the condition $\lambda - \mu \neq$ integer in Theorem 1 is not always necessary for the existence of two linearly independent solutions of the form (8).

PROBLEMS

1P. The notation in this and subsequent problems is the same as in the text. For the equation $x^2 y'' = (6 + 2x)y$:

(a) Find $P(0)$ and $Q(0)$ by inspection and check that the indicial polynomial is $I(s) = s^2 - s - 6 = (s-3)(s+2)$.

(b) Make a table showing the coefficient of x^{s+n} in $x^2 y''$, y, and xy and thus get the recurrence formula $I(s+n)a_n = 2a_{n-1}$.

(c) The choice $n = 0$ makes $I(s) = 0$, hence $s = 3$ or -2. If $s = 3$ show that $n(n+5)a_n = 2a_{n-1}$, which gives

$$y = a_0 x^3\left(1 + \frac{2x}{1!\cdot 6} + \frac{(2x)^2}{2!\cdot 6\cdot 7} + \frac{(2x)^3}{3!\cdot 6\cdot 7\cdot 8} + \cdots\right)$$

(d) If $s = -2$ show that $n(n-5)a_n = 2a_{n-1}$. This equation for $n = 1, 2, 3, 4, 5$ makes $a_0 = 0$, contrary to our basic assumption. Why did the method fail to give a second solution?

2P. Multiply $(x + x^2)y'' + (2x+1)y' + (5x+4)y = 0$ by $x^2/(x + x^2)$ to get the Frobenius normal form

$$x^2 y'' + \frac{2x+1}{x+1}xy' + \frac{5x^2+4x}{x+1}y = 0$$

Thus show that the indicial polynomial is $I(s) = s^2$. Now that you know $I(s)$, which is better for finding a series solution: the original form of the equation, or the normal form?

3. There are four things to do in each part: (i) Find the indicial polynomial $I(s)$. (ii) Derive the recurrence formula given at the right of the equation. (iii) Determine s by setting $n = 0$ and specialize the recurrence relation accordingly. (iv) Get one or two solutions when $a_0 = 1$. For two problems, which you will recognize when you come to them, the recurrence formula has been written with $n - 1$ replacing n.

(a) $x^2 y'' = xy$ $\qquad\qquad$ $I(s+n)a_n = a_{n-1}$

(b) $xy'' + y' = xy$ $\qquad\qquad$ $I(s+n)a_n = a_{n-2}$

(c) $xy'' = xy' + 2y$ $\qquad\qquad$ $I(s+n)a_n = (s+n+1)a_{n-1}$

(d) $x^2 y'' = (6 + 2x^2)y$ $\qquad\qquad$ $I(s+n)a_n = 2a_{n-2}$

(e) $x^2 y'' + 4xy' + 2y = 0$ $\qquad\qquad$ $I(s+n)a_n = 0$

(f) $4x^2 y'' = (4x - 1)y$ $\qquad\qquad$ $I(s+n)a_n = a_{n-1}$

(g) $x^2 y'' + 6xy' = (x^2 - 6)y$ $\qquad\qquad$ $I(s+n)a_n = a_{n-2}$

(h) $x^2 y'' + xy' = (1 + x^2)y$ $\qquad\qquad$ $I(s+n)a_n = a_{n-2}$

(i) $2x(1 - x)y'' + (1 - 8x)y' = 4y$ $\quad$ $I(s+n)a_n = (s+n)(s+n+1)a_{n-1}$

(j) $4x^2 y'' = (4x^2 - 1)y$ $\qquad\qquad$ $I(s+n)a_n = a_{n-2}$

(k) $2xy'' + y' = 2xy' + y$ $\qquad\qquad$ $I(s+n)a_n = (s+n-\frac{1}{2})a_{n-1}$

(l) $2x^2 y'' + y = xy'$ $\qquad\qquad$ $I(s+n)a_n = 0$

(m) $9x^2 y'' = (3x - 2)y$ $\qquad\qquad$ $I(s+n)a_n = \frac{1}{3}a_{n-1}$

(n) $xy'' = (6 + 5x)y' + 5y$ $\qquad\qquad$ $I(s+n)a_n = 5(s+n)a_{n-1}$

(o) $16x^2 y'' = (8x^2 - 3)y$ $\qquad\qquad$ $I(s+n)a_n = \frac{1}{2}a_{n-2}$

(p) $3x(1 - x)y'' + (1 - 7x)y' = y$ $\quad$ $I(s+n)a_n = (s+n)(s+n-\frac{2}{3})a_{n-1}$

(q) $4x^2 y'' + 4xy' = (4x^2 + 1)y$ $\qquad\qquad$ $I(s+n)a_n = a_{n-2}$

4. Since $I(s + n)$ is essentially n^2 for large n the radius of convergence corresponding to the recursion relations in Problem 3 can be determined more or less by inspection. In all but two cases the radius is infinite. What are the two exceptions and what is the radius of convergence in those cases?

5. In this problem $t = -x$, $y(x) = z(t)$, and λ, μ are the exponents associated with the Euler-Cauchy equation. Check that the Euler-Cauchy equation for y as a function of x yields the same equation for z as a function of t. Since $t = -x = |x|$ for $x < 0$ this shows that $|x|^\lambda$, $|x|^\mu$ are solutions on $(-\infty, 0)$. If λ and μ are real, what conditions ensure that these solutions are valid on $(-\infty, \infty)$?

6. Change of variable. If P and Q have convergent power-series expansions about x_0, we say that x_0 is a regular singular point for the equation

$$(x - x_0)^2 y'' + P(x)(x - x_0)y' + Q(x)y = 0$$

Let $t = x - x_0$ and $y(x) = z(t)$ to obtain

$$t^2 z'' + P(x_0 + t)tz' + Q(x_0 + t)z = 0$$

where the primes on z denote differentiation with respect to t. Check that this is in the Frobenius normal form with $t = 0$ as a regular singular point, and that the series on the left below for z yields the series on the right for y:

$$z = t^\lambda \sum_{j=0}^{\infty} a_j t^j, \qquad y = (x - x_0)^\lambda \sum_{j=0}^{\infty} a_j (x - x_0)^j$$

--------------------------------ANSWERS--------------------------------

1. Because $\lambda - \mu = 5$, an integer.
2. The original form, because its coefficients are polynomials while those in the normal form involve infinite series. In theory, nonterminating series do no harm; but in practice they complicate the recurrence formulas.
3. (a) general term $nx^n/(n!)^2$ when $s = 1$, no second solution
   (b) general term $x^{2n}/(2^n n!)^2$ when $s = 0$, no second solution
   (c) $(x + x^2/2)e^x$ when $s = 1$, no second solution.
   (d) The following when $s = 3$ and $s = -2$ respectively:

   $$x^3 \left(1 + \frac{x^2}{1! \, 7} + \frac{x^4}{2! \, 7 \cdot 9} + \frac{x^6}{3! \, 7 \cdot 9 \cdot 11} + \cdots \right)$$

   $$x^{-2} \left(1 - \frac{x^2}{3 \cdot 1!} + \frac{x^4}{3 \cdot 2!} + \frac{x^6}{3 \cdot 3! \cdot 1} + \frac{x^{10}}{3 \cdot 5! \cdot 1 \cdot 3 \cdot 5} + \cdots \right)$$

   (e) $1/x$, $1/x^2$
   (f) general term $x^{n+1/2}/(n!)^2$ when $s = 1/2$, no second solution
   (g) $(\cosh x)/x^3$, $(\sinh x)/x^3$
   (h) general term $x(x/2)^{2n}/((n+1)! \, n!)$ when $s = 1$, no second solution
   (i) the following when $s = 0$ and $s = 1/2$, respectively:

   $$1 + \frac{2!}{1}(2x) + \frac{3!}{1 \cdot 3}(2x)^2 + \frac{4!}{1 \cdot 3 \cdot 5}(2x)^3 + \cdots,$$

   $$x^{1/2} \left(1 + \frac{5}{2}x + \frac{7 \cdot 5}{4 \cdot 2}x^2 + \frac{9 \cdot 7 \cdot 5}{6 \cdot 4 \cdot 2}x^3 + \cdots \right)$$

   (j) general term $x^{2n+1/2}/(2^n n!)^2$, no second solution   (k) e^x and

   $$x^{1/2} \left(1 + \frac{2x}{3} + \frac{(2x)^2}{3 \cdot 5} + \frac{(2x)^3}{3 \cdot 5 \cdot 7} + \cdots \right)$$

(l) x, $x^{1/2}$ (m) The following for $s = 1/3$ and $s = 2/3$, respectively:

$$x^{1/3} \left(1 + \frac{x}{1!\,2} + \frac{x^2}{2!\,2\cdot 5} + \frac{x^3}{3!\,2\cdot 5\cdot 8} + \cdots \right)$$

$$x^{2/3} \left(1 + \frac{x}{1!\,4} + \frac{x^2}{2!\,4\cdot 7} + \frac{x^3}{3!\,4\cdot 7\cdot 10} + \cdots \right)$$

(n) $x^7 e^{5x}$, no second solution

(o) The following for $s = 1/4$ and $3/4$ respectively:

$$x^{1/4} \left(1 + \frac{x^2}{(2)(3)} + \frac{x^4}{(2\cdot 4)(3\cdot 7)} + \frac{x^6}{(2\cdot 4\cdot 6)(3\cdot 7\cdot 11)} + \cdots \right)$$

$$x^{3/4} \left(1 + \frac{x^2}{(2)(5)} + \frac{x^4}{(2\cdot 4)(5\cdot 9)} + \frac{x^6}{(2\cdot 4\cdot 6)(5\cdot 9\cdot 13)} + \cdots \right)$$

(p) $1/(1-x)$, $x^{2/3}/(1-x)$ (q) $x^{-1/2}\sinh x$, $x^{-1/2}\cosh x$

4. $r = 1$ in (ip)   5. $\lambda \geq 2$, $\mu \geq 2$

---

2. Equations with parameters

In the examples and problems of the preceding section we considered only equations having specific numerical coefficients, such as 2 or -5. Now we consider equations in which the coefficients depend on numerical parameters such as a, b, c, m. One equation of this kind is discussed in the text and others are taken up in the accompanying problems.

The *Bessel equation* is

(1) $$x^2 y'' + xy' + (x^2 - m^2)y = 0, \qquad m \geq 0$$

where m is a real constant. This is in the Frobenius normal form with

$$P(x) = 1, \quad Q(x) = x^2 - m^2.$$

Since $P(0) = 1$ and $Q(0) = -m^2$ the indicial polynomial is

$$I(s) = s(s-1) + P(0)s + Q(0) = s^2 - m^2,$$

the indicial equation is $s^2 = m^2$, and the exponents are $\lambda = m$, $\mu = -m$. The condition $m \geq 0$ ensures $\lambda \geq \mu$.

The general terms in the series for y, y', and y'' are, respectively,

$$a_j x^{s+j}, \quad (s+j)a_j x^{s+j-1}, \quad (s+j)(s+j-1)x^{s+j-2}a_j.$$

This yields the following table for the functions appearing in the Bessel equation:

$x^2 y''$	xy'	$x^2 y$	$-m^2 y$
$(s+j)(s+j-1)a_j x^{s+j}$	$(s+j)a_j x^{s+j}$	$a_j x^{s+j+2}$	$-m^2 a_j x^{s+j}$
$(s+n)(s+n-1)a_n$	$(s+n)a_n$	a_{n-2}	$-m^2 a_n$

The second row gives the general terms in the corresponding series and the third row gives the coefficients of x^{s+n}. These coefficients are obtained by the respective choices $j = n$, n, $n-2$, n in the second row. By (1) and the last line of the table we get the recurrence formula

$$(s+n)^2 a_n + a_{n-2} - m^2 a_n = 0.$$

Since $I(s) = s^2 - m^2$ the coefficient of a_n is $I(s+n)$ and hence

(2) $$\qquad\qquad I(s+n)a_n = -a_{n-2}.$$

The value $n = 0$ in (2) shows, as expected, that s must be one of the roots m or $-m$ of the indicial equation. The choice $s = m$ yields

$$I(s+n) = I(m+n) = (m+n)^2 - m^2 = n(2m+n)$$

and the recurrence formula (2) becomes

$$a_n = -\frac{a_{n-2}}{n(2m+n)}, \qquad n \geq 1.$$

Taking $n = 1, 3, 5, \cdots$ shows that $a_n = 0$ for all odd n. The values $n = 2, 4, 6, \cdots$ yield

$$a_2 = -\frac{a_0}{2(2m+2)}, \quad a_4 = -\frac{a_2}{4(2m+4)}, \quad a_6 = -\frac{a_4}{6(2m+6)},$$

and so on. Our solution is therefore

$$y = a_0 x^m \left(1 - \frac{x^2}{2(2m+2)} + \frac{x^4}{4 \cdot 2(2m+4)(2m+2)} - \cdots \right)$$

where the initial factor x^m is x^s with $s = m$. A solution for the exponent $s = -m$ is obtained by writing $-m$ in place of m provided the denominators so obtained do not vanish. This holds if, and only if, m is not a positive integer.

The terms can be simplified by factoring out 2 from each factor in the denominator and combining it with x in the numerator. The result is

$$(3) \qquad y = a_0 x^m \left(1 - \frac{(x/2)^2}{1!(m+1)} + \frac{(x/2)^4}{2!(m+2)(m+1)} - \cdots \right).$$

Assuming for the moment that m is a positive integer, we can write this in the form

$$y = a_0 x^m \left(\frac{m!}{m!} - \frac{m!(x/2)^2}{1!(m+1)!} + \frac{m!(x/2)^4}{2!(m+2)!} - \cdots \right).$$

If we now choose $a_0 = 1/(2^m m!)$ we get

$$y = \left(\frac{x}{2} \right)^m \left(\frac{1}{m!} - \frac{(x/2)^2}{1!(m+1)!} + \frac{(x/2)^4}{2!(m+2)!} - \cdots \right).$$

This function is denoted by $J_m(x)$ and is called *the Bessel function of order m*. Thus, by definition,

$$(4) \qquad J_m(x) = \left(\frac{x}{2} \right)^m \sum_{j=0}^{\infty} (-1)^j \frac{(x/2)^{2j}}{j!(m+j)!}.$$

The Bessel equation is meaningful even when m is not a positive integer, and so is the series (3). It is natural to inquire whether we can define $m!$ for arbitrary real m and thus give a meaning to the series (4). In Section 22.3 we see that such an extension is in fact possible. The extended definition of $m!$ agrees with the familiar definition when m is a positive integer and it satisfies the equation

$$(5) \qquad (m+1)\frac{1}{(m+1)!} = \frac{1}{m!}$$

for all m without exception. The universal validity of (5) depends on the following equation, which holds by definition:

$$(6) \qquad \frac{1}{m!} = 0 \quad \text{when} \quad m = -1, -2, -3, \cdots.$$

The definition is appropriate because $|m!|$ tends to ∞ when m tends to a negative integer.

Equation (5) shows that (4) remains a valid solution for all m, and hence $J_m(x)$ and $J_{-m}(x)$ both satisfy the Bessel equation. If m is not an integer it is easily checked that the two functions J_m and J_{-m} are linearly independent on $(0, \infty)$ and hence

$$y = c_0 J_m(x) + c_1 J_{-m}(x)$$

gives both the general solution and the complete solution of the Bessel equation on this interval. If $m = 0$, however, we obtain only the single function $J_0(x)$. Another exceptional case arises when m is a positive integer. In this case the series (3), with m replaced by $-m$, has a zero denominator and is meaningless. However the series (4) for the Bessel function remains well defined and satisfies

(7) $$J_{-m}(x) = (-1)^m J_m(x), \qquad m = 1, 2, 3, \cdots.$$

For proof, write $-m$ in place of m in (4). Since (6) shows that $1/(j-m)! = 0$ when $j < m$, the result is

$$J_{-m}(x) = \sum_{j=0}^{\infty} (-1)^j \frac{(x/2)^{2j-m}}{j!(j-m)!} = \sum_{j=m}^{\infty} (-1)^j \frac{(x/2)^{2j-m}}{j!(j-m)!}.$$

The truth of (7) becomes evident when this series and the one for $J_m(x)$ are both written in expanded form.

PROBLEMS

1. The Bessel equation with $m = 1/2$ is

$$x^2 y'' + xy' + \left(x^2 - \frac{1}{4} \right) y = 0$$

In this case $\lambda - \mu = m - (-m) = 2m = 1$, an integer. Using the method but not the result in the text, obtain series solutions that you will recognize as

$$a_0 x^{-1/2} \sin x \quad \text{and} \quad a_0 x^{-1/2} \cos x$$

for $s = 1/2$ and $s = -1/2$ respectively. It is known that $(-1/2)! = \sqrt{\pi}$, and $(1/2)!$ can then be found from $m! = m(m-1)!$. Thus verify that the choice $a_0 = 1/(2^m m!)$ used in the text yields

$$J_{1/2}(x) = \sqrt{\frac{2}{\pi x}} \sin x, \qquad J_{-1/2}(x) = \sqrt{\frac{2}{\pi x}} \cos x$$

2. Generalization of Section 1, Problem 3(bhq). A modification of the Bessel equation is

$$x^2 y'' + xy' = (x^2 + m^2)y, \qquad m \geq 0$$

where m is constant. Show that the theory parallels that for the Bessel function, except that the terms of the series do not alternate in sign.

3. Generalization of Section 1, Problem 3(ck). The *confluent hypergeometric equation* is $xy'' + ay' = xy' + by$ where a and b are constants.

(a) For $s = 0$ obtain the solution

$$y = 1 + \frac{b}{a} \frac{x}{1!} + \frac{b(b+1)}{a(a+1)} \frac{x^2}{2!} + \frac{b(b+1)(b+2)}{a(a+1)(a+2)} \frac{x^3}{3!} + \cdots$$

(b) Taking $s = 1 - a$ show that another solution is

$$y = x^{1-a} G(2 - a, b + 1 - a, x)$$

where $G(a, b, x)$ denotes the above series. It is assumed that a is not 0 or a negative integer in the first case, and that $2 - a$ is not 0 or a negative integer in the second.

4. Generalization of Section 1, Problem 3(ip). The *hypergeometric equation* is

$$x(1 - x)y'' + (c - (a + b + 1)x)y' - aby = 0, \qquad a, b, c \text{ const.}$$

Discuss (a) when $a = 1$ and (b) when $b = c$. No answer is provided but see Problem 5.

5. (a) For the hypergeometric equation of Problem 4 obtain the following solution corresponding to the exponent $s = 0$:

$$y = 1 + \frac{ab}{1!c}x + \frac{(a+1)a(b+1)b}{2!(c+1)c}x^2 + \frac{(a+2)(a+1)a(b+2)(b+1)b}{3!(c+2)(c+1)c}x^3 + \cdots$$

(b) The above series is called the *hypergeometric series* and is denoted by $F(a, b, c, x)$. Check that a solution with $s = 1 - c$ is given by

$$y = x^{1-c} F(a - c + 1, b - c + 1, 2 - c, x)$$

It is assumed that c is not 0 or a negative integer in Part (a) and that c is not a positive integer exceeding 1 in Part (b).

3. Supplementary topics

As in the preceding discussion $I(s)$ is the indicial polynomial and λ, μ are the corresponding exponents, with $\lambda \geq \mu$ when both are real. We denote by u a solution with exponent λ; namely, a solution of the form $x^\lambda A(x)$ where $A(x)$ has a power-series expansion and $A(0) \neq 0$. The phrase "second solution" means a solution $y = v$ that is valid on some interval $0 < x < r$ and is such that (u, v) are linearly independent.

If $\lambda = \mu$, there is only one exponent and the method of the preceding sections gives only one solution. There may also be difficulty if $\lambda - \mu$ is a positive integer. This difficulty comes from the recursion formula. In all examples and problems discussed hitherto the recursion formula had the form

$$I(n + s)a_n = c_n a_{n-1} \quad \text{or} \quad I(n + s)a_n = c_n a_{n-2}$$

where c_n were known constants. A similar result holds in the general case, except that the right-hand side can involve all a_j from $j = 0$ to $n - 1$. What we want to emphasize is that if $I(n + s) = 0$ for some positive integer n, the formula fails to give a_n. In fact it leads to a contradiction unless the right-hand side happens to be 0 for the corresponding value n.

Let us recall that the equation for $n = 0$ requires $s = \lambda$, where λ is a root of the indicial equation. If also $I(n+s) = 0$ we conclude that $n+s = \mu$, the other root. Hence $\mu - \lambda = n$. This equation cannot hold if λ and μ are complex, nor if $\lambda \geq \mu$. But it always holds for some n if $\lambda - \mu$ is a positive integer and we try to get a solution with exponent μ by interchanging the roles of λ and μ.

Regardless of the value of $\lambda - \mu$, we can obtain a second solution by the methods of Section 10.1. Define a linear operator T by

$$Ty = x^2 y'' + P(x)xy' + Q(x)y, \qquad x > 0$$

and then define a power series $R(x)$ by

$$R(x) = P_1 x + P_2 \frac{x^2}{2} + P_3 \frac{x^3}{3} + P_4 \frac{x^4}{4} + \cdots$$

where P_j are the coefficients of $P(x)$. The following theorem is proved in DETA, page 304:

THEOREM 1. *In the above notation, if $u(x)$ is a nonvanishing solution of $Ty = 0$ a second solution is*

$$v(x) = u(x) \int w(x)\, dx \quad \text{where} \quad w(x) = \frac{x^{\lambda+\mu-1}}{(u(x))^2} e^{-R(x)}.$$

If the first solution in Theorem 1 is $u(x) = x^\lambda A(x)$ where $A(0) \neq 0$, Theorem 1 gives information about a second solution. We take $A(0) = 1$ and we set $A(x) = 1 + h(x)$ to get

$$w(x) = x^{\mu - \lambda - 1}(1 + h)^{-2}e^R, \qquad h = h(x), \ R = R(x).$$

By Taylor series,

$$w(x) = x^{\mu - \lambda - 1}(1 - 2h + 3h^2 - 4h^3 + \cdots)(1 - R + \frac{R^2}{2!} - \frac{R^3}{3!} + \cdots).$$

This gives a means of approximation near the singular point. If a logarithmic term is needed it is produced automatically as in the following example.

EXAMPLE 1. In Problem 3(c) of Section 1 it was seen that the equation

$$xy'' = xy' + 2y$$

has the exponents $\lambda = 1$, $\mu = 0$ and that a solution corresponding to λ is

$$u(x) = x(1 + h) \text{ where } h = \frac{3}{2}x + x^2 + \cdots.$$

We call this a *partial series* since it has only the first few terms. In this case $\mu - \lambda - 1 = -2$ and $R(x) = x$. Keeping only terms up to x^2 we get

$$w(x) = x^{-2}(1 - 2(\frac{3}{2}x + x^2) + 3(\frac{3}{2}x)^2 + \cdots)(1 + x + \frac{1}{2}x^2 + \cdots).$$

When multiplied out this gives, to the same approximation,

$$w(x) = x^{-2}(1 - 2x + \frac{9}{4}x^2 + \cdots).$$

Integrating $w(x)$ and multiplying by $u(x)$ yields the desired second solution,

$$v(x) = u(x)(-x^{-1} - 2\ln x + \frac{9}{4}x + \cdots).$$

The logarithmic term is $-2u(x)\ln x$. If we keep this term, but use the partial series for u in the remaining terms, the final result is

$$v(x) = -1 - \frac{3}{2}x + \frac{5}{4}x^2 + \cdots - 2u(x)\ln x.$$

We have carried only enough terms to illustrate the principles. Calculations involving many terms should be done on a computer, not by hand.

In DETA, pages 305–306, the procedure illustrated in Example 1 is generalized to yield the following:

THEOREM 2. *Let $u(x) = x^\lambda A(x)$ be a solution for the exponent λ. Then there is a second solution of the form*

$$v(x) = x^\mu B(x) + cu(x) \ln x$$

where $B(x)$ has a convergent power-series expansion. The constant c is 0 if $\lambda - \mu$ is not an integer. But if $\lambda - \mu = n$, an integer, then c is the coefficient of x^n in the series for $C(x)$, where $CA^2 e^R = 1$.

It is easily checked that $c_0 \neq 0$ and hence, if $\lambda = \mu$, there is always a logarithmic term. But there may or may not be such a term when $\lambda - \mu$ is a positive integer. The following example illustrates a case, already familiar, in which the logarithmic term is absent.

EXAMPLE 2. For the Bessel equation $\lambda - \mu = 2m$ and $R(x) = 0$. It was seen in Section 2 that $J_m(x)$ has the form $x^m A(x)$ where $A(x)$ has only even powers of x in its series expansion. Hence $C(x) = 1/A(x)^2$ also has only even powers of x; in other words, if $n = 2m$, where $2m$ is odd, then $c_n = 0$. Theorem 2 now shows that the log term is missing. This agrees with Section 2, where it was found that the general solution of Bessel's equation is

$$c_1 J_m(x) + c_2 J_{-m}(x)$$

unless m itself (and not only $2m$) is an integer.

In an alternative approach due to Frobenius, the exponent s is not set equal to λ or μ but is left as a variable. A second solution is obtained by differentiating the first solution with respect to s, and the logarithmic term arises from the fact that

$$\frac{d}{ds} x^s = \frac{d}{ds} e^{s \ln x} = x^s \ln x, \qquad x > 0.$$

For details see DETA, Section 13.4, where it is shown that a second solution of the Bessel equation for $m = 0$ is

$$J_0(x) \ln x - \sum_{j=1}^{\infty} \frac{(-1)^j (x/2)^{2j}}{(j!)^2} h_j, \quad h_j = \sum_{i=1}^{j} \frac{1}{i}.$$

PROBLEMS

1P. If $xy'' + (3 - 2x)y' = 2y$ check that the exponents are $0, -2$ and obtain the recurrence formulas

$$(n+2)a_n = 2a_{n-1}, \qquad n \geq 1; \qquad n(n-2)a_n = 2(n-2)a_{n-1}, \qquad n \geq 0$$

for $s = 0$ and $s = -2$ respectively. Find series solutions with a_0 arbitrary in the first case, a_0 and a_2 arbitrary in the second.

2. Using Theorem 1 to get a second solution, solve:

(a) $x^2 y'' - 13xy' + 49y = 0$ (b) $x(1-x)y'' + (1-3x)y' = y$

3. Here are five equations from Section 1, Problem 3 for which we did not get a second solution. Using the partial series for u given at the right of the equation, get a partial series for a second solution as in Example 1.

(a) $xy'' = y$ $x(1 + \dfrac{x}{2} + \dfrac{x^2}{12} + \cdots)$

(b) $4x^2 y'' = (4x - 1)y$ $x^{1/2}(1 + x + \dfrac{x^2}{4} + \cdots)$

(c) $x^2 y'' + xy' = (1 + x^2)y$ $x(1 + \dfrac{x^2}{8} + \cdots)$

(d) $4x^2 y'' = (4x^2 - 1)y$ $x^{1/2}(1 + \dfrac{x^2}{4} + \cdots)$

(e) $xy'' = (6 + 5x)y' + 5y$ $x^7(1 + 5x + \dfrac{25x^2}{2} + \cdots)$

--------ANSWERS--------

1. $(a_0/2)x^{-2}(e^{2x} - 1 - 2x)$, $a_0 x^{-2}(1+2x) + (a_2/2)x^{-2}(e^{2x} - 2x - 1)$. Caution: Do not divide by $n - 2$ until $n > 3$.
2. (a) x^7, $x^7 \ln x$ (b) $1/(1-x)$, $(\ln x)/(1-x)$. This is the special case $a = b = c = 1$ of the hypergeometric equation.
3. For convenience in printing the following solutions have been multiplied by $-2, -4, -16, -4, -42$, respectively:
(a) $2 + x - x^2 + \cdots + 2u(x)\ln x$ (e) $6 - 5x + 5x^2 + \cdots$.
(b) $x^{1/2}(8x + 3x^2 + \cdots) - 4u(x)\ln x$ The coefficient of the log term is $c_7 = -5^7/7!$, which is beyond the scope of the calculation.
(c) $x^{-1}(8 + x^2 + \cdots) + 4u(x)\ln x$
(d) $x^{-1/2}(x^2 + \cdots) - 4u(x)\ln x$

Chapter 14

EXISTENCE THEORY

AFTER discussing differential equations that can be solved by explicit formulas, we present the method of Picard iteration for the first-order equation $y' = f(x, y)$. This yields both existence and uniqueness even when there is no elementary solution. Though details are not given here, Picard's method applies to higher-order equations and to systems, nonlinear as well as linear. Its development toward the end of the last century marked a major advance in the theory of ordinary differential equations, and since then it has had an impact on several other areas of mathematics.

1. Solution by quadrature

When the solution of a differential equation is expressed by a formula involving one or more integrations, the equation is said to be *solvable by quadrature*, and the formula is called a closed-form solution. The term "quadrature" has its historical origin in the connection of integration with area. In plane geometry, a problem of quadrature, such as the quadrature of the circle, is a problem about the area of a plane figure.

Not all differential equations can be solved by quadrature. For example the second-order equation

$$y'' + p(x)y' + q(x)y = f(x)$$

cannot be so solved, in general, even if $p = f = 0$. Except in special cases, the *Riccati equation*

$$y' = A(x) + B(x)y + C(x)y^2$$

also cannot be solved by quadrature. When we do have a closed-form solution, the problems of existence and uniqueness are greatly simplified. If the formula produces a solution, it provides an existence proof; and if every solution is given by the formula, that fact leads to uniqueness.

It is not customary to consider solution by quadrature as a part of existence theory. But historically, the first existence proofs were based on explicit formulas, and many equations solvable by quadrature actually fall outside the scope of general theorems. Thus, the results obtained in this section will not be superseded later. Since the basic techniques were already illustrated in the foregoing pages, we give only a brief indication of the proofs.

(a) The first-order linear equation

The following theorem can be established by construction of an integrating factor as in Section 2.4 or by variation of parameters as in DETA, page 643:

THEOREM 1. *If $p(x)$ and $f(x)$ are continuous on an open interval I containing the point x_0, the initial-value problem*

$$y' + p(x)y = f(x), \qquad y(x_0) = y_0$$

has one and only one solution, and it is given by the formula

$$y = e^{-P(x)} \left(\int_{x_0}^{x} e^{P(t)} f(t)\, dt + y_0 \right) \quad \text{where} \quad P(x) = \int_{x_0}^{x} p(t)\, dt.$$

Theorem 1 applies to nonlinear equations that can be linearized by a suitable substitution. For example, let us show that the following initial-value problem for the Bernoulli equation has a nontrivial solution:

$$y' + e^x y = y^{4/5} \cos x, \quad y^{(j)}(0) = 0, \qquad j = 0, 1, 2, 3, 4.$$

Setting $y = u^5$ as in Section 2.4 we get

$$5u^4 u' + e^x u^5 = u^4 \cos x, \qquad 5u' + e^x u = \cos x.$$

Theorem 1 gives a solution satisfying $u(0) = 0$, and the solution is nontrivial because $5u'(0) = 1$. A brief argument, which we omit, shows that $y = u^5$ satisfies the desired initial conditions. Uniqueness fails because the solution $y = 0$ was lost when we divided by u^4. If we impose the additional condition $y^{(5)}(0) = 0$, however, the sole solution is $y = u = 0$ and uniqueness holds.

(b) The separated equation

Let R denote the rectangular region defined by

(1) $$R = \{ (x, y) : a < x < b, \quad c < y < d \}$$

with constants a, b, c, d satisfying $a < b$, $c < d$. A function $f(x)$ given initially for $a < x < b$ can be extended in a natural way to be defined on R, and similarly for $g(y)$. Namely,

$$f(x, y) = f(x), \quad g(x, y) = g(y), \qquad (x, y) \in R.$$

The following theorem can be obtained by careful use of the techniques in Section 2.3. For details see DETA, pages 644–645:

THEOREM 2. *Suppose that $f(x)$ and $g(y)$ are continuous for $(x, y) \in R$ and suppose further that f and g are not simultaneously zero at any point of R. Then the equation*

$$f(x)\, dx = g(y)\, dy$$

has one, and only one, solution through each point $(x_0, y_0) \in R$. The solution is given implicitly by

$$\int_{x_0}^{x} f(\xi)\, d\xi = \int_{y_0}^{y} g(\eta)\, d\eta$$

and can be expressed on its entire interval of definition in the form $y = \phi(x)$ or in the form $x = \psi(y)$.

It is essential that f and g do not vanish simultaneously. For example, $x\,dx = -y\,dy$ has no solution through the origin and $x\,dx = y\,dy$ has two.

(c) The first-order homogeneous equation

A function $f(x, y)$ is *homogeneous of degree m* if the following holds whenever (x, y) is in its domain:

$$f(tx, ty) = t^m f(x, y), \qquad t > 0.$$

If P and Q are homogeneous functions of the same degree, and P/Q has a nonempty domain of definition, the calculation

$$\frac{P(tx, ty)}{Q(tx, ty)} = \frac{t^m P(x, y)}{t^m Q(x, y)} = \frac{P(x, y)}{Q(x, y)}$$

shows that $g = P/Q$ is homogeneous of degree 0. Setting $t = 1/x$ then gives

$$g(x, y) = g(tx, ty) = g\left(1, \frac{y}{x}\right) = f\left(\frac{y}{x}\right)$$

for $x > 0$, where $f(s) = g(1, s)$. Hence these equations are essentially equivalent:

$$P(x, y)\, dx = Q(x, y)\, dy, \qquad \frac{dy}{dx} = f\left(\frac{y}{x}\right).$$

Both are called *homogeneous*. Note, however, that this use of the term is different from its use in connection with a linear equation $Ty = 0$.

The second equation above can be solved by the substitution $u = y/x$. Setting $y = xu$ we get

$$\frac{dy}{dx} = x\frac{du}{dx} + u, \qquad x\frac{du}{dx} + u = f(u)$$

which yields the following:

THEOREM 3. *If $x \neq 0$ and $f(u) \neq u$ the substitution $u = y/x$ reduces the homogeneous equation to the separated form*

$$\text{(2)} \qquad \frac{du}{f(u) - u} = \frac{dx}{x}.$$

This falls within the scope of Theorem 2 and leads to a general existence and uniqueness theorem as soon as we specify suitable properties of the function f. The procedure fails at a point u_0 where $u_0 = f(u_0)$, but there is the obvious solution $y = u_0 x$ in that case.

EXAMPLE 1. If $xy \neq x^2$ the equation $y'(xy - x^2) = y^2$ can be written

$$\frac{dy}{dx} = \frac{y^2}{xy - x^2} = \frac{(y/x)^2}{(y/x) - 1}.$$

Thus $f(u) = u^2/(u - 1)$. The equation $u = f(u)$ has $u = 0$ as sole solution, and this gives $y = 0$. When $xu \neq 0$ Equation (2) yields

$$\frac{u - 1}{u} du = \frac{dx}{x}, \qquad u - \ln|u| = \ln|x| + c_1, \qquad u = \ln|cxu|$$

where $\ln|c| = c_1$. Choosing the sign of c to agree with the sign of xu, we drop the absolute value. Since $u = y/x$ the final result is

$$y = x \ln(cy) \quad \text{or} \quad cy = cx \ln(cy).$$

(d) Exact equations

We assume that functions P, Q, F are defined on the rectangle R of (1) and that

$$(x_0, y_0) \in R, \qquad (x, y) \in R.$$

Partial derivatives are denoted by subscripts; thus,

$$P_x = \frac{\partial P}{\partial x}, \quad Q_y = \frac{\partial Q}{\partial y}, \quad F_{xy} = \frac{\partial^2 F}{\partial y \partial x}.$$

The last equation agrees with the convention that $F_{xy} = (F_x)_y$.

An expression $\omega = P\, dx + Q\, dy$ is said to be *exact* in R if there exists a function F with continuous partial derivatives such that $\omega = dF$. In other words

$$\text{(3)} \qquad P(x, y)\, dx + Q(x, y)\, dy = F_x(x, y)\, dx + F_y(x, y)\, dy$$

for $(x, y) \in R$, where x, y, dx, and dy are independent variables. If ω is exact, the equation $\omega = 0$ is also said to be exact. Taking first $dx = 1$, $dy = 0$ and then $dx = 0$, $dy = 1$ we see that (3) holds if, and only if,

$$(4) \qquad P(x, y) = F_x(x, y), \quad Q(x, y) = F_y(x, y), \qquad (x, y) \in R.$$

These remarks lead to:

THEOREM 4. *Suppose that P and Q have continuous first partial derivatives in R. Then $P\,dx + Q\,dy$ is exact in R if, and only if, $P_y = Q_x$. In that case $P\,dx + Q\,dy = dF$ where*

$$(5a) \qquad F(x, y) = \int_{x_0}^{x} P(\xi, y)\, d\xi + \int_{y_0}^{y} Q(x_0, \eta)\, d\eta.$$

For proof, suppose $P\,dx + Q\,dy$ is exact. Then (4) gives

$$P_y = F_{xy}, \quad Q_x = F_{yx}$$

and these are equal. Conversely, if $P_y = Q_x$ let F be defined by (5a). Then the fundamental theorem of calculus gives $F_x = P$ and also

$$F_y(x, y) = \int_{x_0}^{x} P_y(\xi, y)\, d\xi + Q(x_0, y).$$

Since $P_y = Q_x$ this yields the desired relation $F_y = Q$, as follows:

$$F_y(x, y) = \int_{x_0}^{x} Q_x(\xi, y)\, d\xi + Q(x_0, y) = Q(x, y) - Q(x_0, y) + Q(x_0, y).$$

For more details, including a statement of auxiliary theorems upon which the argument depends, see DETA, page 649.

EXAMPLE 2. The expression $(2xy+1)\,dx + (x^2 + 4y)\,dy$ is exact since both P_y and Q_x are equal to $2x$. Equation (5a) with $x_0 = y_0 = 0$ gives

$$F(x, y) = \int_0^x (2\xi y + 1)\, d\xi + \int_0^y 4\eta\, d\eta = x^2 y + x + 2y^2.$$

EXAMPLE 3. Solve Example 2 by indefinite integrals. Since $F_x = P$ we try

$$F(x, y) = \int (2xy + 1)\, dx = x^2 y + x + C(y)$$

where the constant of integration C can depend on y. The equation $F_y = Q$ leads to $x^2 + C'(y) = x^2 + 4y$. Hence $C(y) = 2y^2 + c$, which gives

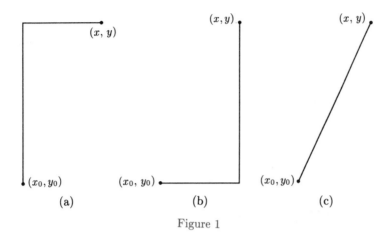

Figure 1

essentially the same result as before. This procedure has the advantage that it is easy to understand and easy to remember. But for theoretical discussion indefinite integrals are less convenient than definite integrals, and they are very unsuitable when the rectangle R is replaced by a more general region.

EXAMPLE 4. For the equation $f(x)\,dx = g(y)\,dy$ we have

$$P(x,y) = f(x), \quad Q(x,y) = -g(y), \quad P_y = 0, \quad Q_x = 0$$

and hence $P_y = Q_x$. Thus the separated equation is exact. Nevertheless, the general theory of exact equations does not yield Theorem 2, because in Theorem 2 neither f nor g is assumed differentiable.

(e) Line integrals

The formula (5a) is an example of a *line integral;* the path of integration is shown in Figure 1(a). The path in Figure 1(b) leads to

(5b) $$F(x,y) = \int_{x_0}^{x} P(\xi, y_0)\,d\xi + \int_{y_0}^{y} Q(x,\eta)\,d\eta$$

and if $(x_0, y_0) = (0,0)$ the path in Figure 1(c) gives

(5c) $$F(x,y) = x \int_{0}^{1} P(tx, ty)\,dt + y \int_{0}^{1} Q(tx, ty)\,dt.$$

Use of these equations does not presuppose familiarity with the theory of line integrals. However, for readers who have such familiarity, we mention that the general formula is

$$F(x,y) = \int_{(x_0,y_0)}^{(x,y)} (P\,dx + Q\,dy)$$

and instead of being a rectangle, R can be any simply connected region. An example in DETA, page 657 shows that simple connectivity is essential.

(f) Integrating factors

The gist of the result of Parts (d) and (e) is that if an exact equation

$$(6) \qquad\qquad P(x,y)\,dx + Q(x,y)\,dy = 0$$

has a general solution, the solution can be written $F(x,y) = c$. Under mild hypotheses a partial converse holds. Namely, if (6) has a general solution given implicitly by $F(x,y) = c$, then (6) becomes exact when multiplied by a suitable nonvanishing function r. Such a function is called an *integrating factor*. The condition $r \neq 0$ is required because otherwise an alleged "solution" might satisfy $r(P\,dx + Q\,dy) = 0$ without satisfying (6).

The following is established in DETA, pages 651–652:

THEOREM 5. *Let $F(x,y) = c$ give a general solution of* (6) *in R. Suppose that P, Q, F have continuous partial derivatives and that neither Q nor F_y is zero at any point of R. Then F_y/Q is an integrating factor for $P\,dx + Q\,dy$.*

Theorem 5 presupposes knowledge of the solution and does not give a practical method of finding the factor r. But r can often be found by use of the equation $(rP)_y = (rQ)_x$ from Theorem 4. Here is an example.

EXAMPLE 5. Let us see when $P\,dx + Q\,dy$ has an integrating factor of the form $r = x^a y^b$ for $x > 0$ and $y > 0$, where a and b are constant. Setting

$$(Px^a y^b)_y = (Qx^a y^b)_x$$

we get first

$$P\,x^a\,by^{b-1} + P_y\,x^a\,y^b = Q\,ax^{a-1}\,y^b + Q_x\,x^a\,y^b$$

and then

$$(7) \qquad\qquad xy(P_y - Q_x) = ayQ - bxP.$$

Assuming continuity of the relevant derivatives, (7) gives a condition for $P\,dx + Q\,dy = 0$ to be solvable by quadrature.

PROBLEMS

All functions are assumed continuous and sufficiently differentiable to justify the calculations. The letters a, b, c, sometimes with subscripts, denote constants.

1. (abc) Check that the following are exact and find F as in Example 3:

$$3x^2 y^2 \, dx + 2x^3 y \, dy \quad (xy^2 + x) \, dx + (yx^2 + y) \, dy \quad e^{xy}(dx + xy \, dx + x^2 \, dy)$$

(defgh) Assuming (x, y) suitably restricted, do the same here:

$$\frac{x \, dx + y \, dy}{\sqrt{x^2 + y^2}} \quad \frac{y \, dx - x \, dy}{y^2} \quad \frac{x^2 \, dx + y^2 \, dy}{x^3 + y^3} \quad \frac{x \, dy - y \, dx}{x\sqrt{x^2 - y^2}} \quad \frac{2xy \, dy - y^2 \, dx}{x^2 + y^4}$$

2. (abcdefgh) Explore the use of each equation (5) in Problem 1. Take $(x_0, y_0) = (0, 0)$ when possible, otherwise $(1, 0)$, otherwise $(0, 1)$.

3. If $dy/dx = f(ax + by)$ with $b \neq 0$ show that the substitution $u = ax + by$ leads to a separable equation on any interval on which $a + bf(u) \neq 0$. Show also that if $a + bf(u_0) = 0$ a solution is given by $ax + by = u_0$.

4. (a) Verify that each of these functions is homogeneous and find its degree:

$$x^2 + y^2 \quad (x + y)(x - y)^{-1} \quad (xy)^{-1/3} \quad (x + y)^e x^\pi \quad x^2 e^{y/x} \quad \ln x - \ln y$$

(b) If $f(x, y)$ is homogeneous of degree m and has continuous partial derivatives, derive Euler's equation $x f_x + y f_y = mf$.

Hint: Differentiate the defining equation with respect to t and set $t = 1$.

5. (abcd) Assuming (x, y) suitably restricted, solve:

$$xy' = x + y \quad (x^2 + y^2)y' + 2xy = 0 \quad (x - \sqrt{xy})y' = y \quad xy' - y = \sqrt{x^2 - y^2}$$

(efgh) Find the solution through $(1, 1)$:

$$(x + y)y' = x - y \quad (x^3 - y^3)y' = x^2 y \quad x^2 y' = xy - y^2 \quad x^3 y' = 2xy^2 - y^3$$

6. Change of origin. Let $a_1 x_0 + b_1 y_0 + c_1 = 0$ and $a_2 x_0 + b_2 y_0 + c_2 = 0$. Show that the substitution $u = x - x_0$, $v = y - y_0$ reduces the following equation, in general, to a homogeneous equation $dv/du = F(v/u)$:

$$\frac{dy}{dx} = f\left(\frac{a_1 x + b_1 y + c_1}{a_2 x + b_2 y + c_2}\right)$$

7. Let P and Q be homogeneous functions of the same degree m.

(a) If $P\,dx + Q\,dy$ is exact and $m > -1$ deduce from (5c) that

$$P\,dx + Q\,dy = dF \quad \text{where} \quad F(x,y) = \frac{xP(x,y) + yQ(x,y)}{m+1}$$

(b) Deduce the result (a) for all $m \neq -1$ by use of Problem 4(b).

8. (a) When is $(a_1 x^2 + b_1 xy + c_1 y^2)\,dx + (a_2 x^2 + b_2 xy + c_2 y^2)\,dy$ exact, and what is F in that case?

(b) The coefficients in Part (a) are homogeneous polynomials of degree 2. With this as a clue, state and solve a similar problem in which the coefficients are homogeneous polynomials of degree 3. No answer is provided.

9. This pertains to construction of an integrating factor r.

(a) Show that $r(x,y) = r(x)$ and $r(x,y) = r(y)$ lead respectively to

$$\frac{r'}{r} = \frac{P_y - Q_x}{Q}, \qquad \frac{r'}{r} = \frac{Q_x - P_y}{P}$$

If the right-hand side is a function of x alone in the first case, or of y alone in the second, integration gives $\ln r$ and exponentiation yields r.

(b) If the equation on the left holds, obtain r as on the right:

$$\frac{P_y - Q_x}{yQ - xP} = h(xy), \qquad H' = h, \qquad r = e^{H(xy)}$$

10. The d'Alembert-Lagrange equation. The equation $y = xf(p) + g(p)$ with $p = y'$ can be solved as follows. Differentiation with respect to x yields

$$p = xf'(p)\frac{dp}{dx} + f(p) + g'(p)\frac{dp}{dx}$$

Assuming $p \neq f(p)$, solve for dx/dp to get a linear equation. If the solution is $x = h(p)$ then a solution to the original equation is given parametrically, with the slope p as parameter, by

$$x = h(p), \qquad y = h(p)f(p) + g(p)$$

11. Clairaut's equation. For the equation $y = xy' + g(y')$ show that the result of Problem 10 reduces to $(dp/dx)(x + g'(p)) = 0$. Taking $p = c$ or $g'(p) = -x$ get $y = xc + g(c)$ or the parametric solution

$$x = -g'(p), \qquad y = g(p) - pg'(p)$$

12. This pertains to the Riccati equation $y' = A(x) + B(x)y + C(x)y^2$.

(a) If u is a given solution show that the substitution $y = u + v$ leads to a Bernoulli equation for v with exponent $n = 2$.

(b) Using the substitution $v = w^{1-n} = 1/w$ in Part (a) deduce that

$$y' = A + By + Cy^2 \iff w' + (B + 2Cu)w + C = 0, \qquad w \neq 0$$

Since w can be found by quadrature, you get a one-parameter family of solutions y (often the complete solution) from the single solution u.

(c) Show that the substitution $Cy = -z'/z$ with $Cz \neq 0$ yields

$$y' = A + By + Cy^2 \iff Cz'' - (BC + C')z' + AC^2 z = 0$$

Hence, if either equation can be solved by quadrature, so can the other.

(d) Let u and v satisfy the linear equation of Part (c) with the initial conditions $u(x_0) = 1$, $u'(x_0) = 0$, $v(x_0) = 0$, $v'(x_0) = -C(x_0)$. Show that the solution of the Riccati equation satisfying $y(x_0) = c$ is given, in general, by

$$y(x) = -\frac{1}{C(x)} \frac{u'(x) + cv'(x)}{u(x) + cv(x)}$$

————————————————————ANSWERS————————————————————

1. (abcd) $x^3 y^2$, $(x^2 + 1)(y^2 + 1)/2$, xe^{xy}, $\sqrt{x^2 + y^2}$

    (efgh) x/y, $(1/3)\ln(x^3 + y^3)$, $\sin^{-1}(y/x)$, $\tan^{-1}(y^2/x)$

2. Aside from an additive constant, the answers agree with those for Problem 1. In some cases, however, (5) may lead to a divergent integral.

4. (a)   2,   0, $-2/3$, $e + \pi$,   2,   0

5. (a) $y = x \ln cx$ (e) $y^2 + 2xy - x^2 = 2$

    (b) $3x^2 y + y^3 = c$ (f) $3y^3 \ln y + x^3 = y^3$

    (c) $y = 0$, $4x = y(\ln cy)^2$ (g) $y + y \ln x = x$

    (d) $y = x$, $y = -x$, $y = x \sin(\ln cx)$ (h) $y = x$

8. (a) $b_1 = 2a_2$, $b_2 = 2c_1$. Problem 7 gives F by inspection.

2. Picard iteration

If $y = \phi(t)$ satisfies the initial-value problem

$$(1) \qquad\qquad \frac{dy}{dt} = f(t, y), \qquad y(t_0) = y_0$$

then $f(t, y)$ gives the slope of the curve $y = \phi(t)$ at any point (t, y) on it. For broad classes of functions f, we can get an idea of the shape of this curve in the following way: Draw a line segment through the point (t_0, y_0) with slope $f(t_0, y_0)$ and proceed a short distance along this segment to a new point (t_1, y_1). Draw a line segment through (t_1, y_1) with slope $f(t_1, y_1)$, proceed a short distance along it to a new point (t_2, y_2), and so on. The result is a polygonal line, called the *Euler polygon*, which is built up of short segments. The general appearance of the Euler polygon in a typical case is shown in Figure 2.

Suppose now that $|f| \le M$ where M is constant. In this case the segments in the Euler polygon have slopes between $-M$ and M and the polygon lies between two lines of slopes $\pm M$ through the point (t_0, y_0) as shown in Figure 2. It is natural to surmise that the solution has the same property. This surmise can be proved, without using the Euler polygon and without assuming that the solution is unique. We shall establish the following:

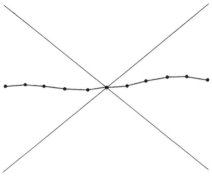

Figure 2

THEOREM 1. *Let I be an interval containing the point t_0, and let M be constant. Suppose $y = \phi(t)$ satisfies $|y'| \le M$ for $t \in I$ and $y(t_0) = y_0$. Then*

$$|y - y_0| \le M|t - t_0|, \qquad t \in I.$$

With $y = \phi(t)$ and $y_0 = \phi(t_0)$ the mean-value theorem gives

$$|y - y_0| = |(t - t_0)\phi'(c)| = |t - t_0||\phi'(c)|$$

where c is between t and t_0. The hypothesis $|y'| \le M$ yields $|\phi'(c)| \le M$ and Theorem 1 follows.

We used the Euler polygon as a point of departure for Theorem 1. The concept of iteration is introduced next, and the Euler polygon again provides a convenient point of departure. Let the successive points where the segments of the polygon join be

$$(2) \qquad (t_0, y_0), \qquad (t_1, y_1), \qquad (t_2, y_2), \qquad \cdots, \qquad (t_n, y_n), \qquad \cdots.$$

For definiteness, we go a constant horizontal distance h from one point to the next, where h is a small positive number. The Euler construction is then formalized by the equations

$$(3) \qquad t_{n+1} = t_n + h, \qquad y_{n+1} = y_n + hf(t_n, y_n).$$

Each point is determined from its predecessor by a uniform procedure, and the process as a whole is an example of an *iterative scheme*. Many methods in numerical analysis depend on iterative schemes such as this, with various refinements to improve the behavior of the errors $\phi(t_n) - y_n$.

In this chapter we use iteration in a somewhat different form. Instead of constructing a sequence of points (t_n, y_n) as in (2) we construct a sequence of functions $\{y_n(t)\}$ satisfying

$$y_1(t_0) = y_0, \qquad y_2(t_0) = y_0, \qquad \cdots, \qquad y_n(t_0) = y_0, \qquad \cdots$$

and

$$y_1' = f(t, y_0), \qquad y_2' = f(t, y_1), \qquad \cdots, \qquad y_{n+1}' = f(t, y_n), \qquad \cdots.$$

Under suitable hypotheses it will be shown that the sequence converges and yields a solution of the differential equation $y' = f(t, y)$. Note that each of the above equations can be solved by the methods of Chapter 2. Namely,

$$(4) \qquad y_{n+1}(t) = y_0 + \int_{t_0}^{t} f(s, y_n(s))\, ds.$$

If the integrand is continuous, as it will be in the sequel, the fundamental theorem of calculus gives

$$y_{n+1}'(t) = f(t, y_n(t))$$

and the condition $y_n(t_0) = y_0$ is also satisfied. Thus the y_j are determined one at a time, starting with the known value y_0. This process is called *Picard iteration* after the French mathematician Émile Picard, who used it in 1893 to develop a comprehensive existence theory for differential equations.

EXAMPLE 1. Let us apply Picard's method to the problem

$$y' = t^2 + y^2, \qquad y(0) = 0, \qquad 0 \le t < 1.$$

Setting $t_0 = y_0 = 0$, $f(t, y) = t^2 + y^2$ in (4), we get

$$y_{n+1} = \int_0^t (t^2 + y_n^2)\, dt, \qquad y_0 = 0.$$

The choice $n = 0$ gives

$$y_1 = \int_0^t t^2 \, dt = \frac{t^3}{3}.$$

Using this with $n = 1$ we get

$$y_2 = \int_0^t \left(t^2 + \frac{t^6}{9} \right) dt = \frac{t^3}{3} + \frac{t^7}{63}.$$

Using this with $n = 2$ we get y_3, and so on. In DETA, page 640, it is shown that

$$\left| y(t) - \frac{t^3}{3} \right| < 1.5 \cdot 10^{-4}, \qquad \left| y(t) - \frac{t^3}{3} - \frac{t^7}{63} \right| < 4 \cdot 10^{-13}$$

for $0 \le t \le .5$ and $0 \le t \le .1$, respectively. We mention this to point out that Picard's method is useful not only for the development of theory, but also for numerical analysis.

One of the main requirements for convergence is that the function $f(t, y)$ should not change too much when y is changed. In the Euler polygon, for example, it is desirable to have the successive values y_{n+1} and y_n in some sense close to each other. Since these values depend on y_n and y_{n-1} there is a cumulative effect. Rapid change of $f(t, y)$ can make y_n very different from y_{n-1}, and this difference may be further exaggerated by the effect of f at the next stage. Instead of converging to a definite curve $y = \phi(t)$ as we take smaller and smaller values of h in (3), the sequence of Euler polygons in such cases generally diverges. Similar remarks apply to the Picard method. An appropriate restriction on f is given by the following definition:

LIPSCHITZIAN FUNCTIONS. The function $f(t, y)$ is Lipschitzian with respect to y in a region R if there is a constant K such that

(5) $|f(t, y) - f(t, z)| \le K|y - z|, \qquad (t, y) \in R, \ (t, z) \in R.$

The region R can be any set of points, but no harm is done if you consider R to be a familiar region such as a rectangle or disk. What is really needed is that (5) shall hold for all values of the variables under consideration.

EXAMPLE 2.   The function $\sin y$ satisfies

$$\sin y - \sin z = (y - z) \cos c$$

for some c between y and z. This follows from the mean-value theorem of differential calculus. Hence $\sin y$ is Lipschitzian in the whole (t, y) plane and the Lipschitz constant K can be taken as 1. An extension of this reasoning yields the following theorem:

THEOREM 2.   Let $f(t, y)$ be such that $f_y = \partial f/\partial y$ is bounded in a rectangular region R. Then f is Lipschitzian with respect to y in R.

For proof, let $|f_y(t, y)| \le K$ in R where K is constant, and consider $f(t, y)$ to be a function of y alone. The mean-value theorem for functions of one variable yields

$$|f(t, y) - f(t, z)| = |(y - z)f_y(t, c)| \le |y - z|K$$

where c is between y and z. (The value of c depends on t, y, and z but that does not matter.) The inequality at the right follows from the hypothesis $|f_y| \le K$ and yields the conclusion; we assume, naturally, that (t, y) and (t, z) lie in R. A similar argument applies to more general regions.

PROBLEMS

1. If $p(t)$ is bounded for $a < t < b$, show that $p(t)y$ is Lipschitzian with respect to y in the strip $a < t < b$, $-\infty < y < \infty$. This is the reason why Lipschitzian functions are often met in the study of linear differential equations.

2. In a region in which $|y| \le M$, where M is constant, show that the function y^2 satisfies a Lipschitz condition with constant $K = 2M$. Show however that if y is unrestricted the function y^2 is not Lipschitzian. This failure of the Lipschitz condition is reflected in the behavior of the initial-value problem

$$y' = y^2, \qquad y(0) = y_0 > 0$$

Namely, the solution $y = y_0/(1 - ty_0)$ tends to ∞ as $t \to 1/y_0$.

3. Show that the function $y^{2/3}$ is not Lipschitzian in any strip $|y| < h$ containing the origin. This fact is reflected in the behavior of

$$y' = y^{2/3}, \quad y(0) = 0$$

Namely, there are two solutions, $y = 0$ and $y = (t/3)^3$.

4P. For the equation $y' = 2ty^2$, $y(0) = 1$, Picard's formula is

$$y_{n+1}(t) = 1 + \int_0^t 2sy_n(s)^2 \, ds, \qquad y_0 = 1.$$

(a) Find $y_2(t)$ and (b) compare with the exact solution $y(t)$.

5. Using Picard's method obtain the iterate y_j specified at the right of the problem:

(a) $y' = 1 + ty$ $y(0) = 0$ y_3 (d) $y' = y$ $y(0) = 1$ y_n

(b) $y' = 1 + ty$ $y(0) = 6$ y_2 (e) $y' = 1 + y^2$ $y(0) = 0$ y_3

(c) $y' = t + y$ $y(0) = -1$ y_{n-1} (f) $y' = 1 + ty^2$ $y(0) = 0$ y_3

6. In Example 1 show that

$$y_3(t) = \frac{t^3}{3} + \frac{t^7}{63} + \frac{2t^{11}}{2,079} + \frac{t^{15}}{59,535}$$

-----------------------------------ANSWERS-----------------------------------

4. (a) $y_2(t) = 1 + t^2 + t^4 + t^6/3$ (b) $y = 1/(1 - t^2) = 1 + t^2 + t^4 + \cdots.$

5. (a) $t + t^3/3 + t^5/15$ (d) $1 + t + t^2/2! + \cdots + t^n/n!$

   (b) $6 + t + 3t^2 + t^3/3 + 3t^4/4$ (e) $t + t^3/3 + 2t^5/15 + t^7/63$

   (c) $-1 - t + t^n/n!$ (f) $t + t^4/4 + t^7/14 + t^{10}/160$

3. A local existence theorem

      If $|f(t, y)| \leq M$, where M is constant, every solution of the initial-value problem

$$y' = f(t, y), \qquad y(t_0) = y_0$$

lies in the region defined by

(1) $|y - y_0| \leq M|t - t_0|.$

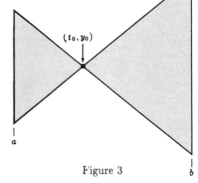

This follows from Section 2, Theorem 1. In general, the condition $|f(t, y)| \leq M$ does not hold for all t nor does the solution exist for all t. We therefore supplement the inequality (1) by $a < t < b$ where a and b are constants satisfying $a < t_0 < b$. The region defined by these two conditions is called a *standard region* and is denoted by

Figure 3

$$R(t_0, y_0, a, b, M) = \{(t, y) : a < t < b, \ |y - y_0| \leq M|t - t_0|\}.$$

A standard region is shown in Figure 3.

This section is devoted to a proof of the following:

LOCAL EXISTENCE THEOREM. In $R = R(t_0, y_0, a, b, M)$ let f be a continuous function that is Lipschitzian with respect to y and satisfies the inequality $|f(t, y)| \le M$. Then the initial-value problem

$$y' = f(t, y), \qquad y(t_0) = y_0$$

has a solution $y = \phi(t)$ for $a < t < b$ and the graph of y lies in R.

The theorem is *local* because in most cases a suitable standard region is available only in the immediate neighborhood of the point (t_0, y_0). A global theorem is proved in DETA, pages 332–333.

For proof, we use Picard iteration

$$y_1'(t) = f(t, y_0), \quad y_2'(t) = f(t, y_1(t)), \quad \cdots, \quad y_{n+1}'(t) = f(t, y_n(t)), \quad \cdots$$

with the initial value $y_n(t_0) = y_0$ in each case. Here t ranges over the interval $a < t < b$. The graph of the constant function $y = y_0$, $a < t < b$ clearly lies in R. If, for a specified n, the graph of y_n lies in R then

$$|f(t, y_n(t))| \le M, \qquad a < t < b.$$

Hence $|y_{n+1}'(t)| \le M$. Section 2, Theorem 1 yields

$$|y_{n+1}(t) - y_0| \le M|t - t_0|, \qquad a < t < b$$

which shows that the graph of y_{n+1} also lies in R. By mathematical induction the graph of every y_j lies in R, and this is the first step of the proof.

For the second step, we assume $t_0 \le t < b$; the case $a < t \le t_0$ is similar. Recall that Picard's procedure gives

(2) $$y_{n+1}(t) = y_0 + \int_{t_0}^t f(s, y_n(s)) \, ds.$$

If we choose a constant M_0 so that

$$|f(t, y_0)| \le M_0, \qquad a < t < b$$

Equation (2) with $n = 0$ yields

(3) $$|y_1(t) - y_0| \le \int_{t_0}^t M_0 ds = M_0(t - t_0).$$

This completes the second step of the proof. The constant M_0 could be replaced by M but M_0 gives a better estimate for the rate of convergence of y_n. It also has some theoretical advantages as seen in DETA, Chapter 14.

We now take $n \geq 1$. If (2) is written with $n-1$ instead of n and the results subtracted we get first

$$y_{n+1}(t) - y_n(t) = \int_{t_0}^t \left(f(s, y_n(s)) - f(s, y_{n-1}(s)) \right) ds$$

and then, with the Lipschitz constant K,

(4) $$|y_{n+1}(t) - y_n(t)| \leq \int_{t_0}^t K |y_n(s) - y_{n-1}(s)| \, ds.$$

This follows from the Lipschitz condition, which is available because the graphs of the y_j all lie in R. The inequality (3) with t replaced by s is

$$|y_1(s) - y_0| \leq M_0(s - t_0).$$

Substituting into (4) when $n = 1$ yields

$$|y_2(t) - y_1(t)| \leq K \int_{t_0}^t M_0(s - t_0) \, ds = K M_0 \frac{(t - t_0)^2}{2!}.$$

We will now prove that, more generally,

(5) $$|y_n(t) - y_{n-1}(t)| \leq M_0 K^{n-1} \frac{(t - t_0)^n}{n!}.$$

Indeed, if (5) holds for a specified value n, we write (5) with t replaced by s. Substitution into (4) yields

$$|y_{n+1}(t) - y_n(t)| \leq K \int_{t_0}^t M_0 K^{n-1} \frac{(s - t_0)^n}{n!} \, ds$$

or also

(6) $$|y_{n+1}(t) - y_n(t)| \leq M_0 K^n \frac{(t - t_0)^{n+1}}{(n + 1)!},$$

which is (5) with n replaced by $n + 1$. Hence (5) holds by mathematical induction. This is the third step.

After these preliminaries the proof is easily completed. Since

$$|y'_{n+1}(t) - y'_n(t)| = |f(t, y_n(t)) - f(t, y_{n-1}(t))|$$

the Lipschitz condition gives

$$|y'_{n+1}(t) - y'_n(t)| \leq K|y_n(t) - y_{n-1}(t)|$$

and hence, by (5),

(7)
$$|y'_{n+1}(t) - y'_n(t)| \leq M_0 K^n \frac{(t - t_0)^n}{n!}.$$

Recalling that $t_0 \leq t < b$, we replace $t - t_0$ by $b - t_0$ in the right-hand sides of (6) and (7). The resulting bounds are independent of t. By the theory of uniform convergence (DETA, Chapter F) there is a function $y = \phi(t)$ such that simultaneously

$$\lim_{n \to \infty} y_n(t) = \phi(t), \qquad \lim_{n \to \infty} y'_n(t) = \phi'(t).$$

The conditions $y(t_0) = y_0$, $|y - y_0| \leq M|t - t_0|$ are inherited from the corresponding conditions for y_n. The second of these shows that the graph of $y = \phi(t)$ is in R. The equation

$$\lim_{n \to \infty} f(t, y_n(t)) = f(t, \phi(t))$$

holds because f is continuous, or also because

$$|f(t, y_n(t)) - f(t, \phi(t))| \leq K|y_n(t) - \phi(t)|.$$

Hence letting $n \to \infty$ in $y'_{n+1}(t) = f(t, y_n(t))$ yields

$$\phi'(t) = f(t, \phi(t)), \qquad t_0 \leq t < b.$$

This completes the proof.

Mathematics possesses not only truth but supreme beauty, a beauty cold and austere, like that of sculpture, without appeal to any part of our weaker nature, and capable of a stern perfection such as only the greatest art can show.

—Bertrand Russell

Chapter 15
MATRICES AND LINEAR SYSTEMS

A LINEAR homogeneous system of n differential equations in n unknowns can be described by $dy/dt = Ay$ where A is an n-by-n matrix and y is an n-dimensional column vector. If $n=1000$ the system represents 1000 scalar equations in 1000 unknowns. Nevertheless, it behaves much like the single scalar equation obtained when $n = 1$. The general theory is surprisingly easy, and it is only when we turn to specific examples that the underlying complexity becomes manifest. Many numerical examples are presented in Chapters 16 and 17. Here we keep the numerical complications to a minimum and we concentrate on the underlying mathematical structure.

A review of matrices is in Chapter C and the notion of linear dependence is developed from a general point of view in Chapter D.

1. Linear independence

The following notation is used in Chapters 15, 16, and 17. By J we mean a bounded open interval containing the point t_0. Depending on context,

$$A, \ Y, \ F, \ U, \ V,$$

denote matrices or matrix-valued functions defined on J. However, C, K, and E are constant matrices, independent of t. All of these are square matrices of order n. If the meaning is clear from context, we sometimes say "matrix" instead of "matrix-valued function" and similarly for vectors. Unless the contrary is stated, it will not matter whether the elements of our vectors and matrices are real or complex.

The jth column of a square matrix is systematically denoted by the corresponding letter with a subscript. Thus,

$$A_j, \ Y_j, \ F_j, \ U_j, \ V_j, \ C_j, \ K_j, \ E_j, \ I_j$$

represent n-dimensional column vectors, the jth columns of the corresponding matrices A, Y, F, $\cdots$. We also use y, f, g, u, v to denote n-dimensional column vectors or vector-valued functions defined on J, while c and k are constant n-dimensional column vectors.

A matrix-valued function such as $A = A(t)$ is said to be continuous, bounded, or differentiable if each element a_{ij} in it is continuous, bounded, or differentiable, respectively. Differentiation and integration are performed elementwise; thus,

$$\frac{d}{dt}(a_{ij}) = \left(\frac{da_{ij}}{dt}\right), \qquad \int (a_{ij}) \, dt = \left(\int a_{ij} \, dt\right).$$

We shall be concerned with the equation $y' = Ay + f$, or more explicitly

$$\frac{dy}{dt} = A(t)y(t) + f(t), \qquad t \in J.$$

This is a condensed way of writing a linear system of differential equations in the n coordinates of y. For example, if y_j and f_j are real-valued functions defined on J, the system on the left below is equivalent to the matrix equation on the right:

$$
\begin{aligned}
y_1' &= a_{11}y_1 + a_{12}y_2 + f_1 \\
y_2' &= a_{21}y_1 + a_{22}y_2 + f_2
\end{aligned}
\qquad
\frac{d}{dt}\begin{pmatrix} y_1 \\ y_2 \end{pmatrix} = \begin{pmatrix} a_{11} & a_{12} \\ a_{21} & a_{22} \end{pmatrix}\begin{pmatrix} y_1 \\ y_2 \end{pmatrix} + \begin{pmatrix} f_1 \\ f_2 \end{pmatrix}.
$$

The equation $y' = Ay + f$ is the same as $Ty = f$ where

$$Ty = y' - Ay.$$

The domain of the operator T so defined consists of n-dimensional vector-valued functions differentiable on J. It is easily checked that T is linear. That is, if u and v are in the domain of T, and α is any scalar, then

$$T(u + v) = Tu + Tv, \qquad T(\alpha u) = \alpha Tu.$$

Proof of the second equation is left for you. The first follows from

$$T(u+v) = (u+v)' - A(u+v) = u' + v' - Au - Av = u' - Au + v' - Av = Tu + Tv.$$

By mathematical induction, this extends to three or more summands.

Since T is linear, the general principle of superposition, the principle of the complementary function, and other results established for linear operators in the foregoing pages apply to T. The following theorem indicates that the behavior of the equation $Ty = f$ is also similar to that in the scalar case:

THEOREM 1. *Let A and f be continuous and bounded for $t \in J$. Then for each k the initial-value problem*

$$y'(t) = A(t)y(t) + f(t), \qquad t \in J; \qquad y(t_0) = k$$

has one, and only one, solution. The solution exists on the whole interval J and $y(t)$ has a limit as t approaches either end of the interval.

Uniqueness is established in Chapter A, existence in DETA, page 341.

The assertion of uniqueness in Theorem 1 requires only that A be bounded on every closed subinterval contained in J, and it is known from advanced calculus that this holds if $A(t)$ is continuous. Existence of the solution on every closed subinterval of J, and hence on J, also follows from continuity alone. But the concluding statement about endpoint limits of $y(t)$ requires that A and f be bounded on J.

With these thoughts in mind, we drop statements about existence of endpoint limits and we make the following assumption once and for all:

BASIC ASSUMPTION. The functions A, F, and f are continuous on J.

The subject of matrix differential equations is one of those rare disciplines in which specific examples tend to be much more difficult than the general theorems they are supposed to illustrate. An early preoccupation with examples obscures the essential simplicity of the theory and does not do justice to its scope and power. In this chapter the basic results are presented without interruption. Later we give many examples, starting with the case $n = 2$ in Section 16.1.

(a) Linear dependence of vectors

The vectors Y_j are said to be *linearly dependent* if there are real or complex constants c_j, not all zero, such that

$$(1) \qquad c_1 Y_1 + c_2 Y_2 + \cdots + c_n Y_n = 0.$$

If no such constants exist the vectors are linearly independent. To interpret this condition, let the coordinates of the vector Y_j be y_{ij}. When $n = 3$ the condition for linear dependence takes the form

$$c_1 \begin{pmatrix} y_{11} \\ y_{21} \\ y_{31} \end{pmatrix} + c_2 \begin{pmatrix} y_{12} \\ y_{22} \\ y_{32} \end{pmatrix} + c_3 \begin{pmatrix} y_{13} \\ y_{23} \\ y_{33} \end{pmatrix} = \begin{pmatrix} 0 \\ 0 \\ 0 \end{pmatrix}.$$

These are equivalent to three linear homogeneous equations in the three unknowns c_1, c_2, c_3, namely

$$y_{11} c_1 + y_{12} c_2 + y_{13} c_3 = 0$$
$$y_{21} c_1 + y_{22} c_2 + y_{23} c_3 = 0.$$
$$y_{31} c_1 + y_{32} c_2 + y_{33} c_3 = 0$$

The coefficient matrix of this system is the matrix Y whose columns are the vectors Y_j. There is a nontrivial solution c_j if, and only if, the determinant $|Y|$ is 0. Hence the vectors are dependent when $|Y| = 0$, independent when $|Y| \neq 0$.

A similar result holds in general. For arbitrary n, Equation (1) leads to the following system of n linear equations in the n unknowns c_j:

(2)
$$\sum_{j=1}^{n} y_{ij} c_j = 0, \qquad i = 1, 2, \cdots, n.$$

These are equivalent to $Yc = 0$ where c is the column vector with elements c_j. Hence there is a nontrivial solution c if, and only if, $|Y| = 0$.

The column vector y is said to be a linear combination of the Y_j if there exist constants c_j such that

(3)
$$y = c_1 Y_1 + c_2 Y_2 + \cdots + c_n Y_n.$$

The above discussion shows that (3) is equivalent to $y = Yc$. Hence there is a solution c, for every choice of y, if and only if $|Y| \neq 0$. In that case the solution c is unique.

If every vector y can be represented uniquely in the form (3) the Y_j form a *basis*; more specifically, they form a basis for real n-dimensional space R^n when the vectors and constants are real, and for complex n-dimensional space C^n when they are complex. The above discussion can now be summarized as follows:

THEOREM 2. *The vectors Y_j are linearly dependent if and only if $|Y| = 0$. They form a basis for R^n or C^n, as the case may be, if and only if $|Y| \neq 0$.*

EXAMPLE 1. Test for linear dependence the four vectors on the left below:

$$\begin{pmatrix} 1 \\ 3 \\ 4 \\ 3 \end{pmatrix} \begin{pmatrix} 2 \\ 6 \\ 6 \\ 4 \end{pmatrix} \begin{pmatrix} 3 \\ 6 \\ 8 \\ 5 \end{pmatrix} \begin{pmatrix} 4 \\ 9 \\ 10 \\ 6 \end{pmatrix}, \qquad \begin{vmatrix} 1 & 2 & 3 & 4 \\ 3 & 6 & 6 & 9 \\ 4 & 6 & 8 & 10 \\ 3 & 4 & 5 & 6 \end{vmatrix}.$$

We see in Chapter E that the determinant on the right is 0. Hence the vectors are dependent.

(b) Linear dependence of vector-valued functions

We now extend the notion of linear dependence from vectors Y_j to vector-valued functions $Y_j(t)$. Such functions are linearly dependent on J if constants c_j, not all 0, can be found satisfying

$$c_1 Y_1(t) + c_2 Y_2(t) + \cdots + c_n Y_n(t) = 0, \qquad t \in J.$$

When no such constants exist the functions are linearly independent on J. For dependence, not only must the vectors $Y_j(t)$ be linearly dependent at each $t \in J$, but the *same* constants c_j must do the job at every t. This is an extremely restrictive condition. Nevertheless, we shall establish the following:

THEOREM 3.   *Let Y_j be n solutions of $y' = A(t)y$ on J. Suppose further that $|Y(t_0)| = 0$ at some point $t_0 \in J$. Then the functions $Y_j(t)$ are linearly dependent on J and $|Y(t)| = 0$ for $t \in J$.*

By Theorem 2 the vectors $Y_j(t_0)$ are linearly dependent. Hence we can find constants c_j, not all 0, such that the function

$$(4) \qquad\qquad y(t) = c_1 Y_1(t) + c_2 Y_2(t) + \cdots + c_n Y_n(t)$$

satisfies $y(t_0) = 0$. Linearity gives $y' = Ay$ and uniqueness then gives $y(t) = 0$ for $t \in J$. A second use of Theorem 2 completes the proof.

The functions Y_j in Theorem 3 form a *basis* for solutions of $Ty = 0$ if every solution of this equation can be represented uniquely in the form (4). We shall establish:

THEOREM 4.   *Let Y_j be n vector-valued functions satisfying $TY_j = 0$ on J. Then the Y_j form a basis for solutions of $Ty = 0$ if, and only if, $|Y(t_0)| \neq 0$ at some point $t_0 \in J$. In that case $|Y(t)| \neq 0$ at every point $t \in J$.*

For proof, note that (4) is the same as $y(t) = Y(t)c$. If $|Y(t_0)| \neq 0$ there is one, and only one, vector c such that

$$y(t_0) = Y(t_0)c.$$

By the uniqueness part of Theorem 1 we have $y(t) = Y(t)c$ on J and hence the Y_j form a basis. On the other hand if $|Y(t_0)| = 0$ the above discussion of (4) shows that the representation of the zero solution, $y = 0$, would not be unique. Hence the Y_j do not form a basis. The concluding statement follows from Theorem 3.

If the columns of Y are linearly independent solutions of $y' = Ay$ then Y is called a *fundamental matrix* for $y' = Ay$. By (4) and Theorem 4, a fundamental matrix yields both the general solution and the complete solution of $y' = Ay$. Further discussion is given in the following section.

PROBLEMS

1P. If α is constant check that the vectors on the left below are linearly dependent, but that the vector-valued functions on the right are linearly independent on every interval J, no matter how short:

$$\begin{pmatrix} \alpha \\ \alpha \end{pmatrix}, \; \begin{pmatrix} 2 \\ 2 \end{pmatrix}; \qquad \begin{pmatrix} t \\ t \end{pmatrix}, \; \begin{pmatrix} 2 \\ 2 \end{pmatrix}$$

2. The vectors $(t,\ 2t)^T$ and $(t,\ t^2)^T$ cannot be solutions of a 2-by-2 system $y' = Ay$, with $A(t)$ bounded, on any interval containing the point $t = 0$ or $t = 2$. Why not? Confirm this by finding the unique matrix A for which $y' = Ay$ has those solutions.

3. (abcde) For what values of the constant α, if any, are the columns of the following matrices linearly dependent?

$$\begin{pmatrix} 1 & 0 & 0 \\ 0 & 1 & \alpha \\ 0 & 0 & 1 \end{pmatrix} \begin{pmatrix} 1 & 1 & 0 \\ 3 & 1 & 1 \\ 4 & 2 & \alpha \end{pmatrix} \begin{pmatrix} 2 & 1 & 2 \\ 4 & 2 & 5 \\ 2 & 1 & \alpha \end{pmatrix} \begin{pmatrix} 1 & 1 & 0 \\ 1 & \alpha & 1 \\ -6 & 0 & \alpha \end{pmatrix} \begin{pmatrix} \alpha & -1 & 1 \\ 3 & \alpha & 1 \\ 3 & 1 & \alpha \end{pmatrix}$$

4. In Example 1 express the fourth vector as a linear combination of the other three.

---------------------------ANSWERS----------------------------

2. Because the vectors are dependent at $t = 0$ and $t = 2$, but are independent, as functions of t, on every interval. See Theorem 3.

$$A(t) = \frac{1}{t(t-2)} \begin{pmatrix} t-2 & 0 \\ -2t & -2+2t \end{pmatrix}$$

3. none, 1, all, $\{3, -2\}$, $\{-1, 0, 1\}$ 4. $V_4 = V_3 + V_2 - V_1$

---

2. Matrix differential equations

The linear operator $Ty = y' - Ay$ introduced in Section 1 has a natural extension from vectors to matrices. For example, the equation on the right below follows, *by definition*, from those on the left:

$$T\begin{pmatrix} y_1 \\ y_2 \end{pmatrix} = \begin{pmatrix} f_1 \\ f_2 \end{pmatrix}, \qquad T\begin{pmatrix} z_1 \\ z_2 \end{pmatrix} = \begin{pmatrix} g_1 \\ g_2 \end{pmatrix}, \qquad T\begin{pmatrix} y_1 & z_1 \\ y_2 & z_2 \end{pmatrix} = \begin{pmatrix} f_1 & g_1 \\ f_2 & g_2 \end{pmatrix}.$$

In a like manner, by definition,

$$TY = T(Y_1, Y_2, \cdots, Y_n) = (TY_1, TY_2, \cdots, TY_n).$$

The domain of T now consists of matrix-valued functions Y differentiable on J. This dual interpretation of T causes no difficulty. When we want to emphasize the distinction, we speak of $Ty = y' - Ay$ as the vector case and of $TY = Y' - AY$ as the matrix case.

In the matrix case linearity takes the form

(1) $T(U+V) = TU + TV,$ $T(UC) = (TU)C,$ $T(Uc) = (TU)c$

where U, V are n-by-n matrix-valued functions in the domain of T. Proof of the first equation is left for you. The second follows from

$$T(UC) = (UC)' - A(UC) = U'C - (AU)C = (U' - AU)C = (TU)C.$$

Instead of requiring C to be square as in our notational conventions, we could let C be any constant matrix with n rows. The choice $C = c$, a column vector of dimension n, then gives the third equation (1) as a special case of the second.

If we have n vector-valued functions F_j, with corresponding solutions Y_j, the equations

(2) $TY_1 = F_1,$ $TY_2 = F_2,$ $\cdots,$ $TY_n = F_n$

yield $T(Y_1, Y_2, \cdots, Y_n) = (TY_1, TY_2, \cdots, TY_n) = (F_1, F_2, \cdots, F_n)$. Hence $TY = F$. Conversely, $TY = F$ is equivalent to n vector equations, one for each column of Y and F. Applying Section 1, Theorem 1 to each equation (2), we get the following:

THEOREM 1.   *For each K, the following initial-value problem has one, and only one, solution:*

$$Y'(t) = A(t)Y(t) + F(t), \qquad t \in J; \qquad Y(t_0) = K.$$

(a) Fundamental solutions; variation of parameters

A *fundamental solution* of the equation $TY = 0$ is a solution $Y = U$ satisfying $|U(t_0)| \neq 0$ at some point t_0. A fundamental solution of $TY = 0$ is also a fundamental matrix for $Ty = 0$. By Section 1, Theorem 3 the condition $|U(t_0)| \neq 0$ implies $|U(t)| \neq 0$ on all of J, and hence the inverse $U(t)^{-1}$ of a fundamental matrix U exists at every point $t \in J$. This fact allows us to satisfy arbitrary initial conditions as in the following example.

EXAMPLE 1.   If U is a fundamental matrix for $Ty = 0$ (that is, a fundamental solution of $TY = 0$) solve the initial-value problems

$$y' = Ay, \quad y(t_0) = k; \qquad Y' = AY, \quad Y(t_0) = K.$$

In the first case let $y = Uc$. By linearity $y' = Ay$ and the initial condition holds if

$$U(t_0)c = k, \quad \text{hence if } c = U(t_0)^{-1}k.$$

The second case is similar, with c replaced by C and k by K. Thus the desired solutions are respectively

$$y(t) = U(t)U(t_0)^{-1}k, \qquad Y(t) = U(t)U(t_0)^{-1}K.$$

Another use of the fundamental solution is to solve the nonhomogeneous equation $Ty = f$. Instead of $y = Uc$ as above, we set $y = Uv$ where v is differentiable. Substituting $y = Uv$ into the equation $y' = Ay + f$ yields

$$Uv' + U'v = AUv + f, \quad Uv' + AUv = AUv + f, \quad Uv' = f.$$

Here the second equation follows from $U' = AU$ and the third is evident. The third equation gives $v' = U^{-1}f$, integration gives v, and the solution is $y = Uv$. In symbols,

$$(3) \qquad\qquad y(t) = U(t) \int U(t)^{-1} f(t)\, dt.$$

The constant of integration c implied in (3) produces an added term $U(t)c$. Since this satisfies the homogeneous equation, all choices of c are permissible. As in analogous cases presented earlier, this procedure is called the method of *variation of parameters*.

EXAMPLE 2. The *rest solution* of $Ty = f$ associated with the point t_0 is the solution satisfying $y(t_0) = 0$. To express this solution by a definite integral, all we have to do is to set $v(t_0) = 0$ in the above discussion. The result is

$$(4) \qquad\qquad y(t) = U(t) \int_{t_0}^{t} U(s)^{-1} f(s)\, ds \qquad \text{(rest solution)}.$$

(b) Liouville's equation

The *trace* of a matrix A is the sum of its diagonal elements,

$$\text{trace } A = a_{11} + a_{22} + \cdots + a_{nn}.$$

The following theorem of Liouville is a far-reaching generalization of Abel's identity for the Wronskian, to which it reduces when a single linear equation of order n is written as a first-order system:

THEOREM 2.   If $Y'(t) = A(t)Y(t)$ for $t \in J$ then for $t \in J$ and $t_0 \in J$

(5ab) $|Y(t)|' = |Y(t)| \operatorname{trace} A(t)$, $|Y(t)| = |Y(t_0)| e^{\int_{t_0}^{t} \operatorname{trace} A(s)\, ds}$.

When $n = 2$ the result can be established as suggested in Problem 9. The general case is settled in DETA, page 356.

PROBLEMS

1P. Show that the equation $T(CU) = CTU$ fails unless A and C commute.

2P. The terminology and algebra of complex numbers extends in a natural way to matrices and the extension will be taken for granted in this book. Check that the complex equation

$$(X + iY)' = (A + iB)(X + iY) + (F + iG)$$

is equivalent to the two real equations on the left below, and that these can be written (by block multiplication) in the matrix form on the right:

$$\begin{cases} X' = AX - BY + F \\ Y' = BX + AY + G \end{cases} \qquad \begin{pmatrix} X' \\ Y' \end{pmatrix} = \begin{pmatrix} A & -B \\ B & A \end{pmatrix} \begin{pmatrix} X \\ Y \end{pmatrix} + \begin{pmatrix} F \\ G \end{pmatrix}$$

3P. Give a sensible formulation of the principle of equating real parts for solutions of matrix differential equations and show that it follows from the special case $B = 0$ in Problem 2. No answer is provided.

4. A square matrix C is *nonsingular* if $|C| \neq 0$. If U is a fundamental solution and C is nonsingular show that UC is also a fundamental solution.

5. Theorem 1 (uniqueness) shows that (4) must give the same result no matter what fundamental solution U is used. If U is replaced by UC with C nonsingular as in Problem 4, show that C drops out of (4).

6. If V is a fundamental solution show that the following satisfies $U(t_0) = I$ and is also a fundamental solution: $U(t) = V(t)V(t_0)^{-1}$.

7. With U as in Problem 6, deduce from (4) that the following equation yields the unique solution of $y' = Ay + f$, $y(t_0) = k$:

$$y(t) = U(t) \left(\int_{t_0}^{t} U(s)^{-1} f(s)\, ds + k \right)$$

8. Show that the method of variation of parameters applies almost without change to $TY = F$; you merely set $Y = UV$ instead of $y = Uv$.

9. If $Y' = AY$ and $n = 2$ check that $|Y|' = D_1 + D_2$ where

$$D_1 = \begin{vmatrix} a_{11}y_{11} + a_{12}y_{21} & a_{11}y_{12} + a_{12}y_{22} \\ y_{21} & y_{22} \end{vmatrix} = a_{11}|Y|$$

Proceed similarly for D_2 and deduce Liouville's theorem when $n = 2$.

10. Referring to (5), let J be the interval (a, ∞) and let

$$R(t) = \text{Re} \left(\int_{t_0}^{t} (a_{11}(s) + a_{22}(s) + \cdots + a_{nn}(s))\, ds \right)$$

If all solutions of $Y'(t) = A(t)Y(t)$ are (a) bounded or (b) tend to 0 as $t \to \infty$, show (a) that $R(t) \leq$ const. or (b) that $R(t) \to -\infty$ as $t \to \infty$.

From a drop of water, a logician could infer the possibility of an Atlantic or a Niagara, without having seen or heard of one or the other.

—Sherlock Holmes

It has long been a complaint against mathematicians that they are hard to convince. But it is a far greater disqualification both for philosophy and for the affairs of life to be too easily convinced, to have too low a standard of proof.

—J. S. Hill

Nature's beauty dies. The day dawns when the nautilus is no more. The rainbow passes, the flower fades, the mountains crumble, the stars grow cold. But beauty in mathematics endures forever.

—H. E. Huntley

Chapter 16

EIGENVALUES AND EIGENVECTORS

WHEN the matrix A is constant, the equation $y' = Ay$ has nontrivial solutions of the form $ce^{\lambda t}$, as one might expect from the scalar case. These solutions lead to characteristic values λ_j, characteristic vectors E_j, and to canonical variables. The latter are related to y by a linear change of variables, $y = Ex$, and they satisfy an uncoupled system that can be solved by inspection. Canonical variables play a major role in engineering, economics, mechanics, and indeed in all fields that make extensive use of linear systems with constant coefficients.

Some acquaintance with Chapter 15 is desirable but not essential.

1. Introduction

In this chapter $A = (a_{ij})$ is a constant n-by-n matrix. As we know from Chapter 1, the scalar equation $y' = Ay$ for $n = 1$ has solutions of the form $ce^{\lambda t}$ where $\lambda = A$ and c is an arbitrary constant. It turns out that the vector equation $y' = Ay$ for $n \geq 2$ also has solutions of the form $ce^{\lambda t}$, but the vector c is no longer arbitrary and (unless $c = 0$) the exponent λ must be the root of a certain polynomial equation. Although our objective is to solve a differential equation, the computation of c and λ involves algebra rather than calculus.

Let us try, then, to choose λ and a vector $c \neq 0$ so that the function $y = ce^{\lambda t}$ satisfies $y' = Ay$. Substitution leads successively to

$$\lambda c e^{\lambda t} = A c e^{\lambda t}, \qquad Ac = \lambda c, \qquad (A - \lambda I)c = 0.$$

There is a nonzero solution c if, and only if, $|A - \lambda I| = 0$. This is the same as $P(\lambda) = 0$ where, by definition,

$$P(s) = |sI - A|.$$

It is easily seen that $P(s)$ is a polynomial in s. Since it behaves like $|Is| = s^n$ for large s, the leading term is s^n.

The above expression $P(s)$ is the *characteristic polynomial* of A, the equation $P(s) = 0$ is the *characteristic equation*, a solution λ of this equation is a *characteristic value*, and a nonzero vector c such that $Ac = \lambda c$ is a *characteristic vector*. The conditions are summarized by:

$$|\lambda I - A| = 0, \qquad Ac = \lambda c, \qquad c \neq 0.$$

Characteristic vectors and values are often called *eigenvectors* and *eigenvalues*, respectively. These words are hybrids of English and German, and following German usage, "ei" rhymes with π.

The following notation will be used throughout this chapter and the next. Since $P(s)$ has degree n there are n eigenvalues λ_j, not necessarily distinct. We denote corresponding eigenvectors by $E_j = (e_{ij})$. Thus, E_j is not 0 and satisfies each of the equivalent equations

$$(A - \lambda_j I)E_j = 0, \quad AE_j = \lambda_j E_j, \quad (\lambda_j I - A)E_j = 0.$$

By definition, the n-by-n matrix with columns E_j is E and the diagonal matrix with diagonal elements λ_j is Λ. When computing a particular eigenvector E_j, we use e_i as an abbreviation for e_{ij}.

The principal features of the theory are already present in the case $n = 2$, and this case is discussed next. When $n = 2$ the matrix $A = (a_{ij})$ is of order 2 and

$$\lambda I - A = \begin{pmatrix} \lambda & 0 \\ 0 & \lambda \end{pmatrix} - \begin{pmatrix} a_{11} & a_{12} \\ a_{21} & a_{22} \end{pmatrix} = \begin{pmatrix} \lambda - a_{11} & -a_{12} \\ -a_{21} & \lambda - a_{22} \end{pmatrix}.$$

By a short calculation the characteristic equation $|\lambda I - A| = 0$ is

$$\lambda^2 - (\text{trace } A)\lambda + (\det A) = 0.$$

Since the equation is quadratic there are two roots λ_j, not necessarily distinct. In this case

$$E_1 = \begin{pmatrix} e_{11} \\ e_{21} \end{pmatrix}, \quad E_2 = \begin{pmatrix} e_{12} \\ e_{22} \end{pmatrix}, \quad E = \begin{pmatrix} e_{11} & e_{12} \\ e_{21} & e_{22} \end{pmatrix}, \quad \Lambda = \begin{pmatrix} \lambda_1 & 0 \\ 0 & \lambda_2 \end{pmatrix}.$$

The matrix Λ is unique aside from the order in which the λ_i are written, but E is not unique, because any of its columns can be multiplied by a nonzero constant. For example, $5E_1$ is an eigenvector if E_1 is.

From the definitions it follows that the functions $E_1 e^{\lambda_1 t}$ and $E_2 e^{\lambda_2 t}$ each satisfy $y' = Ay$. If $\lambda_1 \neq \lambda_2$, these solutions are linearly independent, hence

$$U(t) = (E_1 e^{\lambda_1 t} \quad E_2 e^{\lambda_2 t})$$

is a fundamental matrix. The corresponding general solution is

$$y = \alpha_1 E_1 e^{\lambda_1 t} + \alpha_2 E_2 e^{\lambda_2 t}$$

where, as also below, the α_j denote real or complex constants. This illustrates one use of eigenvectors.

Another use depends on the identity $AE = E\Lambda$, which is verified as follows:

$$\begin{pmatrix} a_{11} & a_{12} \\ a_{21} & a_{22} \end{pmatrix} \begin{pmatrix} e_{11} & e_{12} \\ e_{21} & e_{22} \end{pmatrix} = \begin{pmatrix} \lambda_1 e_{11} & \lambda_2 e_{12} \\ \lambda_1 e_{21} & \lambda_2 e_{22} \end{pmatrix} = \begin{pmatrix} e_{11} & e_{12} \\ e_{21} & e_{22} \end{pmatrix} \begin{pmatrix} \lambda_1 & 0 \\ 0 & \lambda_2 \end{pmatrix}.$$

Here the first equality results from $AE_j = \lambda_j E_j$ and the second is evident. If the vectors E_1 and E_2 are linearly independent, as we now assume, then $|E| \neq 0$, and we can make the change of variables

$$y = Ex, \quad x = E^{-1}y.$$

With $y = Ex$ and $AE = E\Lambda$ the equation $y' = Ay$ gives, in succession,

$$y' = Ay, \quad Ex' = AEx, \quad Ex' = E\Lambda x, \quad x' = \Lambda x.$$

The last equation is equivalent to $x_1' = \lambda_1 x_1$, $x_2' = \lambda_2 x_2$ and hence, by inspection,

$$x_1 = \alpha_1 e^{\lambda_1 t}, \quad x_2 = \alpha_2 e^{\lambda_2 t}.$$

It is said that the original system $y' = Ay$ is uncoupled by the change of variable $y = Ex$ and the new variables x_i are called *canonical variables*.

     The success of the foregoing methods depends on the condition $|E| \neq 0$ which in turn depends on linear independence of the vectors E_1, E_2. To complete our discussion of the case $n = 2$ we show that, if the eigenvalues are distinct, the eigenvectors are independent.

     Suppose, then, that $\lambda_1 \neq \lambda_2$ and that

$$(1) \qquad \alpha_1 E_1 + \alpha_2 E_2 = 0.$$

We want to show that $\alpha_1 = \alpha_2 = 0$. To this end multiply (1) on the left by A. Since $AE_j = \lambda_j E_j$ the result is

$$(2) \qquad \alpha_1 \lambda_1 E_1 + \alpha_2 \lambda_2 E_2 = 0.$$

If (1) is multiplied by λ_2 and (2) is then subtracted we get

$$\alpha_1(\lambda_2 - \lambda_1)E_1 = 0.$$

This gives $\alpha_1 = 0$ and then $\alpha_2 = 0$ by (1).

EXAMPLE 1.   The matrix on the left below has the characteristic equation on the right:

$$A = \begin{pmatrix} 12 & 5 \\ -6 & 1 \end{pmatrix}, \qquad \begin{vmatrix} \lambda - 12 & -5 \\ 6 & \lambda - 1 \end{vmatrix} = \lambda^2 - 13\lambda + 42 = 0.$$

Hence $\lambda = 6$ and $\lambda = 7$. If $\lambda = 6$ the equation $(A - \lambda I)E_1 = 0$ reduces to

$$\begin{pmatrix} 6 & 5 \\ -6 & -5 \end{pmatrix} \begin{pmatrix} e_1 \\ e_2 \end{pmatrix} = \begin{pmatrix} 0 \\ 0 \end{pmatrix}.$$

This is equivalent to $6e_1 + 5e_2 = 0$. (We get only one equation because the determinant is 0. That was the reason for taking $\lambda = 6$.) We can use any nonzero value for e_1 and the equations then give e_2. To avoid fractions we choose $e_1 = 5$. The resulting characteristic vector E_1 and solution $u = E_1 e^{\lambda t}$ are on the left below:

$$E_1 = \begin{pmatrix} 5 \\ -6 \end{pmatrix}, \quad u = \begin{pmatrix} 5 \\ -6 \end{pmatrix} e^{6t}; \quad E_2 = \begin{pmatrix} 1 \\ -1 \end{pmatrix}, \quad v = \begin{pmatrix} 1 \\ -1 \end{pmatrix} e^{7t}.$$

It is left for you to verify that the choice $\lambda = 7$ yields $e_1 + e_2 = 0$ for the corresponding eigenvector E_2, and to get E_2 and v as above.

The matrix E and its inverse are, respectively,

$$E = \begin{pmatrix} 5 & 1 \\ -6 & -1 \end{pmatrix}, \quad E^{-1} = \begin{pmatrix} -1 & -1 \\ 6 & 5 \end{pmatrix}.$$

As a check, we write A, E, Λ side by side, thus:

$$\begin{pmatrix} 12 & 5 \\ -6 & 1 \end{pmatrix} \begin{pmatrix} 5 & 1 \\ -6 & -1 \end{pmatrix} \begin{pmatrix} 6 & 0 \\ 0 & 7 \end{pmatrix}.$$

The equation $AE = E\Lambda$ is easily verified and confirms the calculations.

EXAMPLE 2. Canonical variables x in Example 1 are given by

$$\begin{pmatrix} y_1 \\ y_2 \end{pmatrix} = \begin{pmatrix} 5 & 1 \\ -6 & -1 \end{pmatrix} \begin{pmatrix} x_1 \\ x_2 \end{pmatrix}, \quad \begin{pmatrix} x_1 \\ x_2 \end{pmatrix} = \begin{pmatrix} -1 & -1 \\ 6 & 5 \end{pmatrix} \begin{pmatrix} y_1 \\ y_2 \end{pmatrix}.$$

The equation $x' = \Lambda x$ is in this case $x_1' = 6x_1$, $x_2' = 7x_2$. As a check, we compute x_1' by use of the original system,

$$y_1' = 12y_1 + 5y_2, \quad y_2' = -6y_1 + y_2.$$

The equation $x_1 = -(y_1 + y_2)$ yields

$$x_1' = -y_1' - y_2' = -(12y_1 + 5y_2) - (-6y_1 + y_2) = -6y_1 - 6y_2 = 6x_1.$$

Verification that $x_2' = 7x_2$ is similar and is left for you.

EXAMPLE 3. For the problem

$$A = \begin{pmatrix} 6 & 1 \\ -1 & 8 \end{pmatrix}, \quad \begin{vmatrix} \lambda - 6 & -1 \\ 1 & \lambda - 8 \end{vmatrix} = \lambda^2 - 14\lambda + 49 = 0$$

there is only one characteristic value, $\lambda = 7$, and the equation for a corresponding characteristic vector is

$$\begin{pmatrix} -1 & 1 \\ -1 & 1 \end{pmatrix} \begin{pmatrix} e_1 \\ e_2 \end{pmatrix} = \begin{pmatrix} 0 \\ 0 \end{pmatrix}.$$

The sole solution is $e_1 = \alpha$, $e_2 = \alpha$ where α is constant, and there is no way to construct a fundamental matrix from this.

EXAMPLE 4. The following leads to complex eigenvalues $1 \pm 2i$:

$$\begin{pmatrix} 1 & 1 \\ -4 & 1 \end{pmatrix}, \qquad \begin{vmatrix} \lambda - 1 & -1 \\ 4 & \lambda - 1 \end{vmatrix} = \lambda^2 - 2\lambda + 5 = 0.$$

It is left for you to verify that, when $\lambda = 1 + 2i$, the equation $(A - \lambda I)E_1 = 0$ reduces to $e_2 = 2ie_1$. The form of the final solution depends on the choice of e_1, whether real or complex. For example, $e_1 = i$ gives the complex solution on the left below, which yields the real fundamental solution on the right:

$$\begin{pmatrix} i \\ -2 \end{pmatrix} e^t (\cos 2t + i \sin 2t), \qquad e^t \begin{pmatrix} -\sin 2t & \cos 2t \\ -2 \cos 2t & -2 \sin 2t \end{pmatrix}.$$

PROBLEMS

1. Using the method of eigenvectors, find a fundamental matrix in real form for the equation $y' = Ay$ associated with the following matrices A. Comparison with the answers is facilitated if you number real λ_j so $\lambda_1 < \lambda_2$. In (hjl) we have taken e_2 real, though other choices may be better:

(a) $\begin{pmatrix} -5 & 3 \\ -4 & 2 \end{pmatrix}$ (d) $\begin{pmatrix} 3 & 4 \\ 2 & 1 \end{pmatrix}$ (g) $\begin{pmatrix} -6 & 1 \\ 5 & -2 \end{pmatrix}$ (j) $\begin{pmatrix} -3 & 1 \\ -5 & -7 \end{pmatrix}$

(b) $\begin{pmatrix} 2 & -1 \\ -4 & -1 \end{pmatrix}$ (e) $\begin{pmatrix} 6 & 2 \\ 3 & 7 \end{pmatrix}$ (h) $\begin{pmatrix} 2 & 1 \\ -8 & 6 \end{pmatrix}$ (k) $\begin{pmatrix} -5 & 7 \\ 3 & -1 \end{pmatrix}$

(c) $\begin{pmatrix} 1 & 4 \\ 8 & 5 \end{pmatrix}$ (f) $\begin{pmatrix} 1 & 3 \\ -2 & -4 \end{pmatrix}$ (i) $\begin{pmatrix} 7 & -1 \\ 3 & 3 \end{pmatrix}$ (l) $\begin{pmatrix} -1 & 4 \\ -10 & -5 \end{pmatrix}$

2. (abcdefgik) Obtain canonical vectors x in Problem 1 and verify as in Example 2. (We omit (hjl) because complex canonical variables for real differential equations are seldom used.)

3P. If $P(s) = s^3 + p_2 s^2 + p_1 s + p_0$ is the characteristic polynomial for a 3-by-3 matrix $A = (a_{ij})$, verify that $p_2 = -$ trace A, $p_0 = -$ det A, and

$$p_1 = \begin{vmatrix} a_{11} & a_{12} \\ a_{21} & a_{22} \end{vmatrix} + \begin{vmatrix} a_{11} & a_{13} \\ a_{31} & a_{33} \end{vmatrix} + \begin{vmatrix} a_{22} & a_{23} \\ a_{32} & a_{33} \end{vmatrix}$$

Suggestion: The coefficient of s is $P'(0)$. Compute this derivative without expanding the determinant $|sI - A|$.

4P. By comparing the result of Problem 3 with the factored form

$$P(s) = (s - \lambda_1)(s - \lambda_2)(s - \lambda_3)$$

deduce that $\lambda_1 + \lambda_2 + \lambda_3 = \text{trace } A$, $\lambda_1\lambda_2\lambda_3 = \det A$.

5P. Following the pattern given in the text for $n = 2$, generalize the equation $AE = E\Lambda$ to the case $n = 3$. If $y = Ex$ this gives $x' = \Lambda x$ as before.

6P. The real eigenvectors E_i and E_j are said to be orthogonal if $E_j^T E_i = 0$. Assuming E_i and E_j orthogonal for all $i \neq j$, verify (a) for $n = 3$ and (b) in general that $E^T E$ is a diagonal matrix with diagonal entries $E_i^T E_i$. This is important in the study of real symmetric matrices.

7. Let $AE = E\Lambda$ where E is nonsingular. Check that the substitution $y = Ex$ in $y' = Ay + f$ gives the following uncoupled system:

$$x' = \Lambda x + g \quad \text{where} \quad g = E^{-1}f$$

If g is continuous deduce for $j = 1, 2, \cdots, n$ that

$$x_j(t) = e^{\lambda_j t} \int g_j(s) e^{-\lambda_j s}\, ds + c_j e^{\lambda_j t}$$

(handwritten: $\begin{matrix} \lambda+5 & -3 \\ 4 & \lambda-2 \end{matrix}$ $(\lambda+5)(\lambda-2)+12$ $\lambda^2+3\lambda-10+12$ $\lambda^2+3\lambda+2$ $(\lambda+1)(\lambda+2)$ $\lambda=-1 \;\; \lambda=-2$ $\begin{pmatrix} 4 & -3 \\ 4 & -3 \end{pmatrix}$ $\begin{pmatrix} 4 & -3 \\ 0 & 0 \end{pmatrix}$ $4c_1-3c_2=$ $\begin{matrix} 3 & -3 \\ & -4 \end{matrix}$ *)*

──────────────────────ANSWERS──────────────────────

1. For economy of layout, the consecutive letters run horizontally rather than vertically as in the problems.

   (a) $\begin{pmatrix} e^{-2t} & 3e^{-t} \\ e^{-2t} & 4e^{-t} \end{pmatrix}$ (b) $\begin{pmatrix} e^{-2t} & e^{3t} \\ 4e^{-2t} & -e^{3t} \end{pmatrix}$ (c) $\begin{pmatrix} -e^{-3t} & e^{9t} \\ e^{-3t} & 2e^{9t} \end{pmatrix}$

   (d) $\begin{pmatrix} -e^{-t} & 2e^{5t} \\ e^{-t} & e^{5t} \end{pmatrix}$ (e) $\begin{pmatrix} -e^{4t} & 2e^{9t} \\ e^{4t} & 3e^{9t} \end{pmatrix}$ (f) $\begin{pmatrix} -e^{-2t} & -3e^{-t} \\ e^{-2t} & 2e^{-t} \end{pmatrix}$

   (g) $\begin{pmatrix} -e^{-7t} & e^{-t} \\ e^{-7t} & 5e^{-t} \end{pmatrix}$ (h) $e^{4t}\begin{pmatrix} \cos 2t + \sin 2t & \cos 2t - \sin 2t \\ 4\cos 2t & -4\sin 2t \end{pmatrix}$

   (i) $\begin{pmatrix} e^{4t} & e^{6t} \\ 3e^{4t} & e^{6t} \end{pmatrix}$ (j) $e^{-5t}\begin{pmatrix} -2\cos t + \sin t & \cos t + 2\sin t \\ 5\cos t & -5\sin t \end{pmatrix}$

   (k) $\begin{pmatrix} -7e^{-8t} & e^{2t} \\ 3e^{-8t} & e^{2t} \end{pmatrix}$ (l) $e^{-3t}\begin{pmatrix} -\cos 6t + 3\sin 6t & 3\cos 6t + \sin 6t \\ 5\cos 6t & -5\sin 6t \end{pmatrix}$

2. Partial answer: A matrix E is obtained by setting $t = 0$ in the answers to Problem 1. Possible canonical variables are $x = |E|E^{-1}y$ where $|E|E^{-1}$ can be written by inspection. The factor $|E|$ is not necessary but simplifies the result.

2. Three-dimensional systems

If A is an n-by-n matrix, the conditions defining an eigenvalue λ_j and corresponding eigenvector E_j were stated in Section 1. Namely,

$$|\lambda_j I - A| = 0, \qquad AE_j = \lambda_j E_j, \qquad E_j \neq 0.$$

In the case $n = 3$ there are three eigenvalues λ_j, not necessarily distinct, and

$$\Lambda = \begin{pmatrix} \lambda_1 & 0 & 0 \\ 0 & \lambda_2 & 0 \\ 0 & 0 & \lambda_3 \end{pmatrix}.$$

The general theory for $n = 2$ was developed in Section 1 and was extended to $n = 3$ in the problems. Here the case $n = 3$ is illustrated in numerical examples. Since the calculations are decidedly error-prone, we give some suggestions for checking intermediate steps.

EXAMPLE 1. Find the eigenvalues of the matrix A on the left below:

$$\begin{pmatrix} 5 & 6 & 7 \\ 0 & -5 & 3 \\ 0 & -4 & 2 \end{pmatrix}, \qquad \begin{vmatrix} \lambda - 5 & -6 & -7 \\ 0 & \lambda + 5 & -3 \\ 0 & 4 & \lambda - 2 \end{vmatrix}.$$

$(\lambda - 5)(\lambda + 5)(\lambda - 2) + 2$

$(\lambda - 5)(\lambda^2 + 3\lambda + 2)$

The determinant on the right is $|\lambda I - A|$. Laplace expansion on the first column yields

$$|\lambda I - A| = (\lambda - 5)(\lambda^2 + 3\lambda + 2) = (\lambda - 5)(\lambda + 1)(\lambda + 2)$$

which is 0 for $\lambda = -2, -1$, or 5. Hence these are the eigenvalues. As a check, note that their sum is 2, which is the trace of A. See Section 1, Problem 4.

EXAMPLE 2. We shall now find the eigenvectors in Example 1. When $\lambda = -2, -1$, or 5 the matrices $A - \lambda I$ in Example 1 are, respectively,

(1)
$$\begin{pmatrix} 7 & 6 & 7 \\ 0 & -3 & 3 \\ 0 & -4 & 4 \end{pmatrix}, \qquad \begin{pmatrix} 6 & 6 & 7 \\ 0 & -4 & 3 \\ 0 & -4 & 3 \end{pmatrix}, \qquad \begin{pmatrix} 0 & 6 & 7 \\ 0 & -10 & 3 \\ 0 & -4 & -3 \end{pmatrix}.$$

In each case the determinant is 0; that is the reason for choosing λ as we did. By definition, the eigenvectors E_j are nontrivial solutions of the linear systems associated with the singular matrices (1).

We denote the eigenvector E_j under consideration by (e_i), so that e_i is an abbreviation for e_{ij}. An efficient procedure in the 3-by-3 case is to set $e_3 = 1$ and thus to get a system of two equations for e_1 and e_2. If this system has no solution, the correct value is $e_3 = 0$, and we are led to a 2-by-2 system for e_1 and e_2.

Let us illustrate in the case of the first matrix (1), which corresponds to $\lambda = -2$. When $e_3 = 1$ the equations are

$$7e_1 + 6e_2 + 7 = 0, \quad -3e_2 + 3 = 0.$$

The second of these gives $e_2 = 1$ and the first then gives $e_1 = -13/7$. We multiply by 7 to avoid fractions and get E_1 as shown below:

$$E_1 = \begin{pmatrix} -13 \\ 7 \\ 7 \end{pmatrix}, \quad E_2 = \begin{pmatrix} -23 \\ 9 \\ 12 \end{pmatrix}, \quad E_3 = \begin{pmatrix} 1 \\ 0 \\ 0 \end{pmatrix}.$$

In Problem 1 you are asked to derive the above vectors E_2 and E_3, using the second and third matrices (1) respectively.

The matrix E so obtained is written with A at its left and Λ at its right, thus:

$$\begin{pmatrix} 5 & 6 & 7 \\ 0 & -5 & 3 \\ 0 & -4 & 2 \end{pmatrix} \begin{pmatrix} -13 & -23 & 1 \\ 7 & 9 & 0 \\ 7 & 12 & 0 \end{pmatrix} \begin{pmatrix} -2 & 0 & 0 \\ 0 & -1 & 0 \\ 0 & 0 & 5 \end{pmatrix}.$$

The product of the first two is AE and of the last two is $E\Lambda$. In Section 1, Problem 5 it was seen that these must be equal, thus giving a check.

The point of the next example is to emphasize that the algebraic results lead to a complete solution of the analytic problems associated with the differential equation.

EXAMPLE 3. In Example 1 the general solution is

$$\alpha_1 E_1 e^{-2t} + \alpha_2 E_2 e^{-t} + \alpha_3 E_3 e^{5t}$$

where the E_j are given by Example 2. A fundamental matrix is

$$U(t) = (E_1 e^{-2t} \quad E_2 e^{-t} \quad E_3 e^{5t})$$

and the canonical variables $x = E^{-1}y$ satisfy

$$x_1' = -2x_1, \quad x_2' = -x_2, \quad x_3' = 5x_3.$$

¿From this the x_j can be found by inspection and the equation $y = Ex$ gives y. In Problem 1 you are asked to compute E^{-1} by reducing (E, I) or by the cofactor transpose, as you prefer, and to show that

$$7x_1 = 4y_2 - 3y_3, \quad 3x_2 = -y_2 + y_3, \quad 21x_3 = 21y_1 - 5y_2 + 44y_3.$$

The next example shows that there can be a set of linearly independent eigenvectors even if the eigenvalues are not distinct. The characteristic polynomial is less tractable than those considered hitherto, and we preface the example by some remarks on the problem of factorization.

Let p_j be the coefficient of λ^j in $P(\lambda)$ and note that $P(\lambda) = 0$ gives

$$p_0 = -\lambda(p_1 + \lambda p_2 + \cdots + \lambda^{n-1}).$$

If the coefficients and λ are integers this shows that the constant term p_0 is a multiple of λ. In other words, *any integer root must be a divisor of the constant term*. The same argument shows that *any rational root must be an integer*. Once a root λ_1 is found, we know that $\lambda - \lambda_1$ is a factor of $P(\lambda)$. The other factor is obtained by division.

EXAMPLE 4. In Problem 1 you are asked to show that the matrix on the left below has the characteristic equation on the right:

$$\begin{pmatrix} 11 & 12 & 6 \\ -4 & -5 & -2 \\ -4 & -4 & -3 \end{pmatrix}, \qquad \lambda^3 - 3\lambda^2 - 9\lambda - 5 = 0.$$

(It is advisable to expand directly as in Example 1 and to use Problem 3 of Section 1 as a check.) Integral roots, if any, must be divisors of the constant term, which is -5. Hence they can only be ± 1, ± 5. The root -1 is found by trial, and factorization yields

$$P(\lambda) = (\lambda + 1)(\lambda^2 - 4\lambda - 5) = (\lambda + 1)^2(\lambda - 5).$$

The solutions of $(A - \lambda I)E_i = 0$ are not affected when the rows of the coefficient matrix $A - \lambda I$ are divided by nonzero constants. After obvious simplification the matrices $A - \lambda I$ for $\lambda = -1$ and $\lambda = 5$ become, respectively,

$$\begin{pmatrix} 2 & 2 & 1 \\ 2 & 2 & 1 \\ 2 & 2 & 1 \end{pmatrix}, \qquad \begin{pmatrix} 1 & 2 & 1 \\ 2 & 5 & 1 \\ 1 & 1 & 2 \end{pmatrix}.$$

The first matrix leads to the single equation $2e_1 + 2e_2 + e_3 = 0$. An easy way to get linearly independent solutions is to take first $e_3 = 1$, $e_2 = 0$, and then $e_3 = 0$, $e_2 = 1$. The equations yield $e_1 = -1/2$ and $e_1 = -1$ in these two cases respectively. In Problem 1 you are asked to get a vector E_3 corresponding to $\lambda = 5$. When the E_j are multiplied by suitable constants, the matrices A, E, Λ are as follows, and the equation $AE = E\Lambda$ is easily verified:

$$\begin{pmatrix} 11 & 12 & 6 \\ -4 & -5 & -2 \\ -4 & -4 & -3 \end{pmatrix} \begin{pmatrix} -1 & -1 & -3 \\ 0 & 1 & 1 \\ 2 & 0 & 1 \end{pmatrix} \begin{pmatrix} -1 & 0 & 0 \\ 0 & -1 & 0 \\ 0 & 0 & 5 \end{pmatrix}.$$

In the case of a real symmetric matrix, it turns out that the eigenvalues are always real, and eigenvectors corresponding to different eigenvalues are always orthogonal. These results are discussed in Section 3. Here we illustrate in an example.

EXAMPLE 5. Find the eigenvalues for the symmetric matrix

$$A = \begin{pmatrix} 2 & 4 & -6 \\ 4 & 2 & -6 \\ -6 & -6 & -15 \end{pmatrix}.$$

In Problem 1 you are asked to determine the characteristic polynomial $P(\lambda)$ and, in particular, to verify that its constant term is

$$P(0) = -|A| = -324 = -(2^2)(3^4).$$

The rational eigenvalues, if any, must be integers and must be divisors of $P(0)$; but there are so many divisors that it is by no means obvious what to do next.

By sketching a graph of $P(s)$ we are led to surmise that it has a zero near -2, and if 2 is added to each diagonal element of A the resulting matrix has two equal rows. Hence its determinant is 0, and this shows that $\lambda = -2$ is indeed an eigenvalue. In Problem 1 you are asked to obtain the factored form

$$P(\lambda) = (\lambda + 2)(\lambda^2 + 9\lambda - 162)$$

and to get the remaining eigenvalues -18, 9.

EXAMPLE 6. With your cooperation, we shall verify that the eigenvectors in Example 5 are orthogonal. When $\lambda = -18$ the matrix $A - \lambda I$ is obtained by adding 18 to the diagonal elements of A and is displayed on the left below:

$$\begin{pmatrix} 20 & 4 & -6 \\ 4 & 20 & -6 \\ -6 & -6 & 3 \end{pmatrix} \begin{matrix} R_1/2 \\ R_2/2 \\ R_3/3 \end{matrix} \begin{pmatrix} 10 & 2 & -3 \\ 2 & 10 & -3 \\ -2 & -2 & 1 \end{pmatrix} \begin{matrix} R_1 + 3R_3 \\ R_2 + 3R_3 \end{matrix} \begin{pmatrix} 4 & -4 & 0 \\ -4 & 4 & 0 \\ -2 & -2 & 1 \end{pmatrix}.$$

The homogeneous system corresponding to the reduced matrix on the right has an obvious solution $E_1 = (1, 1, 4)^T$.

In Problem 1 you are asked to show that the choice $e_3 = 1$ is not possible when $\lambda = -2$, and that the choice $e_3 = 0$ (which is mandated by the failure of $e_3 = 1$) leads to a solution. You are also asked to solve the equations with $\lambda = 9$ by diagonalization, as was illustrated above for $\lambda = -18$. The final result is the matrix E in the center below:

$$(2) \qquad E^T E = \begin{pmatrix} 1 & 1 & 4 \\ -1 & 1 & 0 \\ -2 & -2 & 1 \end{pmatrix} \begin{pmatrix} 1 & -1 & -2 \\ 1 & 1 & -2 \\ 4 & 0 & 1 \end{pmatrix} = \begin{pmatrix} 18 & 0 & 0 \\ 0 & 2 & 0 \\ 0 & 0 & 9 \end{pmatrix}.$$

The product $E^T E$ is diagonal because the column vectors of E are orthogonal; see Section 1, Problem 6. The equation $Ex = y$ now yields $E^T Ex = E^T y$ and from this, x can be found by inspection. The variables x_j so defined satisfy $x_j' = \lambda_j x_j$ and y is found from $y = Ex$.

PROBLEMS

1P. In Examples 2, 3, 4, 5, and 6 we left a number of gaps, referring in each case to Problem 1. Study these examples and fill in the gaps.

2. The characteristic values are distinct integers. Find them:

(a) $\begin{pmatrix} 1 & -1 & -1 \\ 6 & 9 & 8 \\ -4 & -5 & -4 \end{pmatrix}$

(f) $\begin{pmatrix} 3 & 0 & 4 \\ 0 & 2 & 0 \\ -2 & 0 & -3 \end{pmatrix}$

(k) $\begin{pmatrix} 4 & -4 & 5 \\ 1 & 3 & 1 \\ -2 & 4 & -3 \end{pmatrix}$

(b) $\begin{pmatrix} -2 & -6 & -2 \\ -9 & -3 & -9 \\ -1 & -3 & -1 \end{pmatrix}$

(g) $\begin{pmatrix} 14 & 13 & -14 \\ -4 & -1 & 6 \\ 2 & 3 & 2 \end{pmatrix}$

(l) $\begin{pmatrix} 2 & 4 & 3 \\ 5 & 3 & 5 \\ -4 & -4 & -5 \end{pmatrix}$

(c) $\begin{pmatrix} -1 & 3 & 0 \\ 3 & -1 & 0 \\ -1 & -1 & -2 \end{pmatrix}$

(h) $\begin{pmatrix} 3 & 4 & 1 \\ -3 & -4 & 3 \\ -2 & -2 & 6 \end{pmatrix}$

(m) $\begin{pmatrix} 3 & -6 & -2 \\ 2 & 3 & 2 \\ -2 & 6 & 3 \end{pmatrix}$

(d) $\begin{pmatrix} 5 & 0 & -1 \\ -1 & 3 & -1 \\ -2 & -1 & 4 \end{pmatrix}$

(i) $\begin{pmatrix} -8 & -11 & 16 \\ 2 & 3 & -4 \\ -4 & -5 & 8 \end{pmatrix}$

(n) $\begin{pmatrix} -7 & -3 & -8 \\ 2 & 2 & 2 \\ 7 & 3 & 8 \end{pmatrix}$

(e) $\begin{pmatrix} 0 & 2 & 0 \\ 3 & 0 & 3 \\ 0 & 1 & 0 \end{pmatrix}$

(j) $\begin{pmatrix} 5 & 0 & 6 \\ -1 & 1 & -1 \\ -3 & 0 & -4 \end{pmatrix}$

(o) $\begin{pmatrix} -7 & -12 & -6 \\ 2 & 2 & 2 \\ 7 & 12 & 6 \end{pmatrix}$

3. (abcdefghijklmno) Find eigenvectors for the eigenvalues in Problem 2 and check that $AE = E\Lambda$.

4. (abcdefghijklmno) In Problem 3 find canonical variables $x = E^{-1}y$ and verify at least one equation $dx_j/dt = \lambda_j x_j$.

5. (abcdef) The following matrices have integral eigenvalues not all distinct. Find three linearly independent eigenvectors and check that $AE = E\Lambda$:

(a) $\begin{pmatrix} -3 & -1 & -2 \\ 0 & -2 & 0 \\ 1 & 1 & 0 \end{pmatrix}$

(c) $\begin{pmatrix} 3 & 0 & -1 \\ 0 & 4 & 0 \\ -1 & 0 & 3 \end{pmatrix}$

(e) $\begin{pmatrix} 1 & 3 & 6 \\ 0 & 0 & -2 \\ 0 & 1 & 3 \end{pmatrix}$

(b) $\begin{pmatrix} -1 & -4 & 2 \\ 0 & 1 & 0 \\ -1 & -2 & 2 \end{pmatrix}$

(d) $\begin{pmatrix} 0 & -2 & -2 \\ -3 & -1 & -3 \\ -1 & -1 & 1 \end{pmatrix}$

(f) $\begin{pmatrix} -5 & -4 & 8 \\ 0 & -1 & 0 \\ -4 & -4 & 7 \end{pmatrix}$

6P. Check that the matrix E^T in (2) gives, with very little calculation,

$$18x_1 = y_1 + y_2 + 4y_3, \qquad 2x_2 = -y_1 + y_2, \qquad 9x_3 = -2y_1 - 2y_2 + y_3$$

7. The following symmetric matrices have distinct integral eigenvalues. Find the eigenvalues and the matrix E of eigenvectors. Check that $AE = E\Lambda$, and also that $E^T E$ is diagonal. Thus find canonical variables $x = E^T y$ as in Problem 6:

(a) $\begin{pmatrix} 0 & 0 & -2 \\ 0 & 0 & 0 \\ -2 & 0 & 0 \end{pmatrix}$ (b) $\begin{pmatrix} 3 & 0 & -1 \\ 0 & 3 & 0 \\ -1 & 0 & 3 \end{pmatrix}$ (c) $\begin{pmatrix} -7 & 0 & 2 \\ 0 & -5 & 2 \\ 2 & 2 & -6 \end{pmatrix}$

────────────────ANSWERS────────────────

2. (abcde) (1, 2, 3) (−12, 0, 6) (−4, −2, 2) (2, 4, 6) (−3, 0, 3)
   (fghij) (−1, 1, 2) (4, 5, 6) (−1, 2, 4) (0, 1, 2) (−1, 1, 2)
   (klmno) (−1, 2, 3) (−2, −1, 3) (1, 3, 5) (0, 1, 2) (−1, 0, 2)

3. Here are possible matrices E for eigenvalues as above.

(a) $\begin{pmatrix} 0 & 1 & 1 \\ -1 & -2 & -5 \\ 1 & 1 & 3 \end{pmatrix}$ (f) $\begin{pmatrix} -1 & -2 & 0 \\ 0 & 0 & 1 \\ 1 & 1 & 0 \end{pmatrix}$ (k) $\begin{pmatrix} -1 & -3 & -1 \\ 0 & 1 & 1 \\ 1 & 2 & 1 \end{pmatrix}$

(b) $\begin{pmatrix} 2 & -1 & 2 \\ 3 & 0 & -3 \\ 1 & 1 & 1 \end{pmatrix}$ (g) $\begin{pmatrix} 4 & 3 & 5 \\ -2 & -1 & -2 \\ 1 & 1 & 1 \end{pmatrix}$ (l) $\begin{pmatrix} -1 & -1 & -1 \\ 1 & 0 & -1 \\ 0 & 1 & 1 \end{pmatrix}$

(c) $\begin{pmatrix} -1 & 0 & -2 \\ 1 & 0 & -2 \\ 0 & 1 & 1 \end{pmatrix}$ (h) $\begin{pmatrix} -1 & 3 & 1 \\ 1 & -1 & 0 \\ 0 & 1 & 1 \end{pmatrix}$ (m) $\begin{pmatrix} -1 & -3 & -1 \\ -1 & -1 & 0 \\ 2 & 3 & 1 \end{pmatrix}$

(d) $\begin{pmatrix} 1 & 1 & -1 \\ 4 & -2 & 0 \\ 3 & 1 & 1 \end{pmatrix}$ (i) $\begin{pmatrix} 2 & 3 & 7 \\ 0 & -1 & -2 \\ 1 & 1 & 3 \end{pmatrix}$ (n) $\begin{pmatrix} -5 & -1 & -3 \\ 1 & 0 & 1 \\ 4 & 1 & 3 \end{pmatrix}$

(e) $\begin{pmatrix} 2 & -1 & 2 \\ -3 & 0 & 3 \\ 1 & 1 & 1 \end{pmatrix}$ (j) $\begin{pmatrix} -1 & 0 & -2 \\ 0 & 1 & 1 \\ 1 & 0 & 1 \end{pmatrix}$ (o) $\begin{pmatrix} -1 & -6 & -4 \\ 0 & 1 & 1 \\ 1 & 5 & 4 \end{pmatrix}$

4. Before finding E^{-1} check that your matrix E is consistent with the book's answers to Problem 3; it need not be identical. Since verification is part of the problem, no further answer is provided.

5. (abcdef) $(-2, -2, -1)$ $(0, 1, 1)$ $(2, 4, 4)$ $(-4, 2, 2)$ $(1, 1, 2)$ $(-1, -1, 3)$

Possible matrices of eigenvectors E with eigenvalues as above:

$$(a) \begin{pmatrix} -2 & -1 & -1 \\ 0 & 1 & 0 \\ 1 & 0 & 1 \end{pmatrix} \quad (c) \begin{pmatrix} 1 & -1 & 0 \\ 0 & 0 & 1 \\ 1 & 1 & 0 \end{pmatrix} \quad (e) \begin{pmatrix} 0 & 1 & 3 \\ -2 & 0 & -1 \\ 1 & 0 & 1 \end{pmatrix}$$

$$(b) \begin{pmatrix} 2 & 1 & -2 \\ 0 & 0 & 1 \\ 1 & 1 & 0 \end{pmatrix} \quad (d) \begin{pmatrix} 2 & -1 & -1 \\ 3 & 0 & 1 \\ 1 & 1 & 0 \end{pmatrix} \quad (f) \begin{pmatrix} 2 & -1 & 1 \\ 0 & 1 & 0 \\ 1 & 0 & 1 \end{pmatrix}$$

7. Eigenvalues: (a) $(-2, 0, 2)$ (b) $(2, 3, 4)$ (c) $(-9, -6, -3)$

Canonical variables x are given in terms of y by $x = E^T y$ where possible values of E^T are:

$$(a) \begin{pmatrix} 1 & 0 & 1 \\ 0 & 1 & 0 \\ -1 & 0 & 1 \end{pmatrix} \quad (b) \begin{pmatrix} 1 & 0 & 1 \\ 0 & 1 & 0 \\ -1 & 0 & 1 \end{pmatrix} \quad (c) \begin{pmatrix} -2 & -1 & 2 \\ 2 & -2 & 1 \\ 1 & 2 & 2 \end{pmatrix}$$

3. Supplementary topics

Results illustrated for $n = 2$ and $n = 3$ in the foregoing pages are now summarized for arbitrary n. We begin with a brief review. If A is a square matrix of order n, there are n eigenvalues λ_j, not necessarily distinct, and corresponding eigenvectors E_j. By definition, $E_j \neq 0$ and

$$(1) \qquad AE_1 = \lambda_1 E_1, \qquad AE_2 = \lambda_2 E_2, \qquad \cdots, \qquad AE_n = \lambda_n E_n.$$

We set $E = (E_1 \ E_2 \ \cdots \ E_n)$ and $\Lambda = \text{diag}\,(\lambda_1 \ \lambda_2 \cdots \lambda_n)$. Equation (1) shows that each column of the matrix

$$(2) \qquad U(t) = (E_1 e^{\lambda_1 t} \quad E_2 e^{\lambda_2 t} \ \cdots \ E_n e^{\lambda_n t})$$

satisfies $y' = Ay$; that is the way the notion of characteristic values and characteristic vectors arose in the first place. In many cases the vectors E_j can be chosen to be linearly independent, even if the λ_j are not all distinct. Then $U(t)$ gives a fundamental matrix for $y' = Ay$. By Section 15.1, Theorem 2, the condition of linear independence is equivalent to $|E| \neq 0$.

The method of canonical variables depends on the equation $AE = E\Lambda$, which was established for $n = 2$ and $n = 3$ in Section 1. Here is a concise proof for arbitrary n:

$$AE = (AE_1 \ AE_2 \ \cdots \ AE_n) = (\lambda_1 E_1 \ \lambda_2 E_2 \ \cdots \ \lambda_n E_n) = E\Lambda.$$

If $|E| \neq 0$ the substitution $y = Ex$ in $y' = Ay$ yields $Ex' = AEx = E\Lambda x$, hence $x' = \Lambda x$. We summarize as follows:

THEOREM 1. *If $|E| \neq 0$ then $U(t)$ in (2) is a fundamental matrix for $y' = Ay$ and the change of variables $y = Ex$ leads to the uncoupled system $x' = \Lambda x$.*

The equation $AE = E\Lambda$ was found above under the assumption that the columns E_j of E are eigenvectors. Conversely, if $AE = E\Lambda$, where Λ is diagonal, then the columns of E satisfy $AE_j = \lambda_j E_j$. Hence the nonzero columns are eigenvectors and the corresponding entries of Λ are eigenvalues. In most applications, the condition $|E| \neq 0$ of Theorem 1 is ensured by the following. For proof, see DETA page 383:

THEOREM 2. *If the n eigenvalues λ_i are distinct, the eigenvectors E_i are linearly independent.*

Let us recall that the matrix A is symmetric if $A = A^T$, Hermitian if $A = A^*$. Here A^* denotes the conjugate transpose, or adjoint, of A. When A is real $A^* = A^T$ and the terms "Hermitian" and "symmetric" mean the same thing. In fields such as biomathematics and economics the matrices are usually real. But in electrical engineering and quantum mechanics they are usually complex. To accommodate the latter subjects, our theorems are stated for Hermitian matrices.

If u and v are column vectors of dimension n, the expression v^*u is a matrix of one element, namely,

$$v^*u = (\alpha) \quad \text{where} \quad \alpha = u_1\bar{v}_1 + u_2\bar{v}_2 + \cdots + u_n\bar{v}_n.$$

By definition, the vectors u and v are *orthogonal* if $v^*u = (0)$. The special case $v = u$ gives

$$u^*u = (\beta) \quad \text{where} \quad \beta = |u_1|^2 + |u_2|^2 + \cdots + |u_n|^2.$$

The quantity $\sqrt{\beta}$ so defined is the length of u and is denoted by $\|u\|$. Note that $\|u\|$ is real and is 0 only when $u = 0$. The following theorem is established in DETA pages 384–385:

THEOREM 3. *Let $A = A^*$. Then all eigenvalues λ of A are real and eigenvectors u, v belonging to different eigenvalues λ, μ are orthogonal.*

If there are n eigenvectors E_j, and if they are mutually orthogonal as in Theorem 3, the method of Section 1, Problem 6 shows that

(3) $$E^*E = \operatorname{diag}\,(\|E_i\|^2).$$

Since the eigenvector E_j is not zero we can divide E_j by $\|E_j\|$ and assume without loss of generality that $\|E_j\| = 1$. In that case equation (3) yields

$$E^*E = I, \quad \text{hence} \quad E^{-1} = E^*.$$

As stated in Chapter C, matrices with this property are called *unitary*.

If E is unitary the equation $AE = E\Lambda$ can be multiplied on the right or left by E^* to yield the equivalent equations

$$A = E\Lambda E^*, \quad E^*AE = \Lambda.$$

Thus we are led to the following theorem when A has n linearly independent eigenvectors. In DETA, page 406 it is seen that the result holds without this restriction:

THEOREM 4.   *If $A^* = A$ there is a unitary matrix E such that E^*AE is diagonal.*

Theorem 4 has an interesting geometric interpretation. The column vectors $x = (x_j)$ and $y = (y_j)$ can be written

$$x = x_1 I_1 + x_2 I_2 + \cdots + x_n I_n, \quad y = y_1 I_1 + y_2 I_2 + \cdots + y_n I_n$$

where I_j are columns of the identity matrix I. It is said that the ordered set $\{I_j\}$ is the *standard basis* and that x_j and y_j are coordinates of x and y relative to the standard basis.

Instead of I_j we could use the columns B_j of any nonsingular matrix B. Since the columns of such a matrix form a basis, x and y can be written uniquely in the form

$$x = \xi_1 B_1 + \xi_2 B_2 + \cdots + \xi_n B_n, \quad y = \eta_1 B_1 + \eta_2 B_2 + \cdots + \eta_n B_n.$$

Here ξ_j and η_j are the coordinates of x and y relative to the new basis. The above equations are equivalent to

$$x = B\xi, \quad y = B\eta.$$

In these coordinates the transformation $y = Ax$ becomes $B\eta = AB\xi$, hence

$$\eta = B^{-1}AB\xi.$$

It is said that A and $B^{-1}AB$ are *similar*, because they represent the same transformation to two different bases. Theorem 4 indicates that every Hermitian matrix is similar to a diagonal matrix, and that if the columns of E are used as a basis, the transformation $y = Ax$ takes the simple form $\eta = \Lambda\xi$ in the new coordinates.

PROBLEMS

1. As in Section 1, Problem 4, derive the identities

$$\sum_{i=1}^{n} \lambda_i = \sum_{i=1}^{n} a_{ii}, \qquad \prod_{i=1}^{n} \lambda_i = \det A$$

2. This problem and the next show how higher-order equations can be studied by means of eigenvectors. Let A be an n-by-n matrix with distinct negative eigenvalues $\lambda_j = -\omega_j^2$. Check that the function $y = c\cos(\omega t - \theta)$ satisfies $y'' = Ay$ if, and only if, $\omega^2 c = Ac$. Thus get the solution

$$y = E_1 \cos(\omega_1 t - \theta_1) + E_2 \cos(\omega_2 t - \theta_2) + \cdots + E_n \cos(\omega_n t - \theta_n)$$

where the E_j are characteristic vectors for λ_j and θ_j are arbitrary constants. The values ω_j, or sometimes $\omega_j/2\pi$, are referred to as *characteristic frequencies* and the corresponding solutions $E_j \cos(\omega_j t - \theta_j)$ are the *normal modes* of oscillation.

3. The vibrations of an oscillating mass-spring system are governed by the equations

$$m_1 y_1'' = -k_1 y_1 + k(y_2 - y_1), \qquad m_2 y_2'' = -k_2 y_2 - k(y_2 - y_1)$$

(a) Write as a 2-by-2 system $y'' = Ay$ and, if $m_1 = k = k_1 = 2$, $m_2 = k_2 = 1$ solve in the form

$$\begin{pmatrix} y_1 \\ y_2 \end{pmatrix} = \alpha_1 \begin{pmatrix} 1 \\ 1 \end{pmatrix} \cos(t - \theta_1) + \alpha_2 \begin{pmatrix} 1 \\ -2 \end{pmatrix} \cos(2t - \theta_2)$$

In the first term on the right the two masses oscillate with the same amplitude and phase; in the second the ratio of amplitudes is 2 to 1 and the oscillations are 180° out of phase. These solutions are the normal modes of oscillation referred to in Problem 2.

Chapter 17

THE MATRIX EXPONENTIAL

THE SET of eigenvalues of a square matrix A, with their multiplicities, is called the spectrum of A. If a complex-valued function ψ is defined on the spectrum of A, it is possible to compute $\psi(A)$, and a single formula will do for the whole class of matrices with given spectrum. Applying this with $\psi(s) = e^{st}$ we find that $Y = e^{At}$ satisfies $Y' = AY$, $Y(0) = I$. A major advantage of this approach is that multiple eigenvalues require only minor modification and may actually lead to simplification. Once the spectrum is known, the evaluation of e^{At} requires little more than matrix multiplication.

Prerequisite: Chapters 15 and 16.

1. Distinct eigenvalues

If a is a number, we know from Chapter 1 that the initial-value problem

$$\frac{dy}{dt} = ay, \qquad y(0) = 1 \quad \text{has the solution } y = e^{at}.$$

Our objective here is to obtain a similar result when a is replaced by a constant n-by-n matrix A. We will define e^{At} in such a way that it is easy to compute, and furthermore,

$$\frac{dY}{dt} = AY, \qquad Y(0) = I \quad \text{has the solution } Y = e^{At}.$$

Throughout Section 1, including the problems, we make the following

SIMPLIFYING ASSUMPTION. The n eigenvalues λ_j of A are distinct.

The numbered theorems and definitions are worded, however, so that they are correct in the case of multiple eigenvalues and they will be applied to that case in the next section.

Some basic relations were derived in Chapter C, Problems 8, 9, and 10, and you may find it helpful to read these problems if you have not already done so. They show that results that are derived here only briefly, or only for $n = 2$, actually hold in full generality.

Before discussing e^{At} we consider polynomials in A. Suppose

$$\phi(s) = q_0 + q_1 s + q_2 s^2 + \cdots + q_m s^m$$

where the coefficients and s are real or complex. Then, by definition,

$$\phi(A) = q_0 I + q_1 A + q_2 A^2 + \cdots + q_m A^m.$$

Note that the constant term q_0 is replaced by $q_0 I$.

Evaluation of a polynomial $\phi(\Lambda)$ is easy when the matrix Λ is diagonal. As an illustration, let

$$\Lambda = \begin{pmatrix} \lambda_1 & 0 \\ 0 & \lambda_2 \end{pmatrix}$$

and let q_j be any real or complex number. Then

$$q_j \begin{pmatrix} \lambda_1 & 0 \\ 0 & \lambda_2 \end{pmatrix}^j = q_j \begin{pmatrix} \lambda_1^j & 0 \\ 0 & \lambda_2^j \end{pmatrix} = \begin{pmatrix} q_j \lambda_1^j & 0 \\ 0 & q_j \lambda_2^j \end{pmatrix}$$

for $j = 0, 1, 2, \cdots$. Summation on j from 0 to m yields

$$\phi(\Lambda) = \begin{pmatrix} \phi(\lambda_1) & 0 \\ 0 & \phi(\lambda_2) \end{pmatrix}$$

with ϕ as above. In condensed notation

(1) $$\phi(\Lambda) = \text{diag } \phi(\lambda_j).$$

Under our simplifying assumption there is an invertible matrix E such that $AE = E\Lambda$, or $A = E\Lambda E^{-1}$. This gives

$$q_j A^j = q_j E\Lambda^j E^{-1} = E q_j \Lambda^j E^{-1}$$

for $j = 0, 1, 2, \cdots$. Summation on j from 0 to m yields

(2) $$\phi(A) = E\phi(\Lambda)E^{-1}.$$

By the results of Chapter C cited above, (1) and (2) hold for arbitrary n. Both equations are basic in the rest of this chapter.

Equations (1) and (2) show that if two polynomials ϕ_1 and ϕ_2 have the same value at each λ_j, then they give the same value for $\phi(A)$. The set of values λ_j is so important that it has a special name:

DEFINITION 1. *The set of eigenvalues of A together with their multiplicities is called the spectrum of A.*

In this section the multiplicities are all 1 and the spectrum is the set of distinct numbers

$$\lambda_1, \lambda_2, \cdots, \lambda_n.$$

The result stated informally above can now be restated as follows:

THEOREM 1. *Let ϕ_1 and ϕ_2 be two polynomials such that $\phi_1 = \phi_2$ on the spectrum of A. Then $\phi_1(A) = \phi_2(A)$.*

(a) Functions of matrices

Theorem 1 allows us to compute $\psi(A)$ for much more general functions than polynomials. Throughout the following discussion, ψ denotes an arbitrary complex-valued function that is defined on the spectrum of A. It need not even be continuous.

DEFINITION 2. *Let ϕ be any polynomial such that $\phi(\lambda) = \psi(\lambda)$ on the spectrum of A. Then $\psi(A) = \phi(A)$.*

By Theorem 1, all such polynomials ϕ yield the same value for $\psi(A)$.

To evaluate $\psi(A)$ we must construct a polynomial that agrees with ψ on the spectrum of A. An easy way of doing this was found by Lagrange. Starting with the characteristic polynomial

$$P(s) = (s - \lambda_1)(s - \lambda_2) \cdots (s - \lambda_n)$$

Lagrange forms additional polynomials

$$L_j(s) = \frac{N_j(s)}{N_j(\lambda_j)}$$

where $N_j(s)$ is obtained by omitting the factor $(s - \lambda_j)$ from $P(s)$. It vanishes at each λ_i for $i \neq j$, but not at λ_j. Hence

$$(3) \qquad L_j(\lambda_i) = 0 \quad \text{for} \quad i \neq j, \qquad L_j(\lambda_j) = 1.$$

A moment's thought shows that the *Lagrange polynomial*

$$\phi(s) = L_1(s)\psi(\lambda_1) + L_2(s)\psi(\lambda_2) + \cdots + L_n(s)\psi(\lambda_n)$$

has degree $\leq n - 1$ and satisfies $\phi(\lambda_j) = \psi(\lambda_j)$ for all j. The following examples show how Lagrange polynomials are used to compute e^{At}.

EXAMPLE 1. Let A be a 3-by-3 matrix with eigenvalues 0, 1, 4. To find $\psi(A)$, we start with $P(s) = s(s - 1)(s - 4)$. The three numerators $N_i(s)$ are

$$(s - 1)(s - 4), \quad s(s - 4), \quad s(s - 1)$$

and their values at 0, 1, 4 are 4, -3, 12 respectively. Hence the Lagrange polynomial ϕ for ψ is

$$\phi(s) = \frac{(s - 1)(s - 4)}{4}\psi(0) + \frac{s(s - 4)}{-3}\psi(1) + \frac{s(s - 1)}{12}\psi(4).$$

Definition 2 with this choice of ϕ yields

$$(4) \qquad \psi(A) = \frac{(A - I)(A - 4I)}{4}\psi(0) + \frac{A(A - 4I)}{-3}\psi(1) + \frac{A(A - I)}{12}\psi(4).$$

EXAMPLE 2.   To find e^{At} in Example 1, we set $\psi(s) = e^{st}$ where t is a constant parameter. Thus

$$\psi(0) = 1, \quad \psi(1) = e^t, \quad \psi(4) = e^{4t}.$$

Substitution into (4) yields

$$e^{At} = \frac{(A-I)(A-4I)}{4} + \frac{A(A-4I)}{-3}e^t + \frac{A(A-I)}{12}e^{4t}.$$

In Part (b) below, we see that this yields a fundamental solution of $Y' = AY$ for the whole class of 3-by-3 matrices A with eigenvalues 0, 1, 4. Also, by Part (c), the nonzero columns of the products

$$(A-I)(A-4I), \quad A(A-4I), \quad A(A-I)$$

are characteristic vectors associated with $\lambda = 0, 1, 4$ respectively.

(b) Analytic properties

It is a remarkable fact that functions such as $\sin At$ or e^{At} have analytic properties similar to those of $\sin \lambda t$ or $e^{\lambda t}$. We now explain why this is so. If ϕ is a polynomial, the relations

$$\phi(\Lambda) = \operatorname{diag} \phi(\lambda_j), \quad \phi(A) = E\phi(\Lambda)E^{-1}$$

were verified by direct calculation. Since we can always find a polynomial ϕ that agrees with ψ on the spectrum of A, Definition 2 has the effect of replacing ϕ by corresponding equations with ψ. In other words,

(5) $$\psi(\Lambda) = \operatorname{diag} \psi(\lambda_j), \quad \psi(A) = E\psi(\Lambda)E^{-1}.$$

For example, when $n = 2$

(6) $$\psi(\Lambda) = \begin{pmatrix} \psi(\lambda_1) & 0 \\ 0 & \psi(\lambda_2) \end{pmatrix}.$$

By definition, the *matrix exponential* e^{At} is the result obtained when $\psi(s) = e^{st}$ where t is a constant parameter. Substitution into (6) yields

$$e^{\Lambda t} = \begin{pmatrix} e^{\lambda_1 t} & \cdot \ 0 \\ 0 & e^{\lambda_2 t} \end{pmatrix}.$$

The right-hand side is differentiable, hence the left-hand side is also. The result of differentiation is

$$\frac{d}{dt}e^{\Lambda t} = \begin{pmatrix} e^{\lambda_1 t}\lambda_1 & 0 \\ 0 & e^{\lambda_2 t}\lambda_2 \end{pmatrix} = \begin{pmatrix} e^{\lambda_1 t} & 0 \\ 0 & e^{\lambda_2 t} \end{pmatrix}\begin{pmatrix} \lambda_1 & 0 \\ 0 & \lambda_2 \end{pmatrix} = e^{\Lambda t}\Lambda.$$

Thus $(d/dt)e^{\Lambda t} = e^{\Lambda t}\Lambda$, as one would expect from the behavior of $e^{\lambda t}$. In Problem 12 you are asked to generalize this equation to diagonal matrices of order n. The calculation is similar to that for $n = 2$.

Setting $\psi(s) = e^{st}$ in (5) we get

$$e^{At} = Ee^{\Lambda t}E^{-1}$$

and, differentiating with the aid of the result found above,

$$\frac{d}{dt}e^{At} = Ee^{\Lambda t}\Lambda E^{-1}.$$

Now comes an important point. If a factor $E^{-1}E$ is inserted between $e^{\Lambda t}$ and Λ, the above product can be written

$$\frac{d}{dt}e^{At} = (Ee^{\Lambda t}E^{-1})(E\Lambda E^{-1}) = e^{At}A.$$

Thus, $(d/dt)e^{At} = e^{At}A$. When $t = 0$ both $e^{\Lambda t}$ and e^{At} reduce to I, and we summarize as follows:

THEOREM 2. *The matrix exponential satisfies* $(d/dt)e^{At} = e^{At}A$, $e^0 = I$.

In Problem 2 it is seen that the inverse of e^{At} is obtained by simply writing $-t$ for t. For example, the formula for solving $y' = Ay + f$ by variation of parameters is

$$y = e^{At}\int e^{-At}f(t)\,dt.$$

This property of the exponential is very convenient in applications.

(c) A connection with eigenvectors

As in Chapter 16, we use E_j to denote an eigenvector for λ_j. Under our simplifying assumption the functions $e^{\lambda_j t}$ are linearly independent. Let L_j be the coefficient polynomials in the Lagrange polynomial for $\psi(s) = e^{st}$. The equation for e^{At} has the form

$$(7) \qquad e^{At} = M_1 e^{\lambda_1 t} + M_2 e^{\lambda_2 t} + \cdots + M_n e^{\lambda_n t}$$

where the M_j are square matrices of order n given by $M_j = L_j(A)$. (Here the subscript does not denote the jth column, as elsewhere in this chapter, but denotes the jth matrix.) The sum satisfies $dU/dt = AU$ by Theorem 2 and hence each term in it satisfies the same equation. That is,

$$(8) \qquad \frac{d}{dt}M_j e^{\lambda_j t} = AM_j e^{\lambda_j t}, \qquad j = 1, 2, \cdots, n.$$

This follows from linear independence of the functions $e^{\lambda_j t}$. Equation (8) is equivalent to $\lambda_j M_j = A M_j$, which in turn is equivalent to the following:

(i) The columns of M_j are scalar multiples of E_j.

Thus we can get a characteristic vector E_j by looking at any nonzero column of M_j. Since $e^0 = I$ we have also

(ii) $M_1 + M_2 + \cdots + M_n = I$.

These two conditions show that the sum satisfies $Y' = AY$, $Y(0) = I$ and, by uniqueness, it must agree with e^{At}. Under our simplifying assumption, (i) and (ii) give a definitive check on numerical calculations. When there are multiple eigenvalues, however, it will be seen that Equation (7) requires modification, and this leads to a corresponding modification of (i) and (ii).

(d) The Cayley-Hamilton theorem

If we choose $\phi_1(s) = P(s)$, where P is the characteristic polynomial of A, then $\phi_1(\lambda) = 0$ on the spectrum of A, and Theorem 1 with $\phi_2 = 0$ gives $\phi_1(A) = \phi_2(A) = 0$. This suggests the following interesting result, which is known as the Cayley-Hamilton theorem. The above proof assumes distinct eigenvalues, but the general case is settled in DETA, pages 402–403:

THEOREM 3. *Every square matrix satisfies its own characteristic equation.*

The following example shows how Theorem 3 can be used to verify that e^{At} actually satisfies the expected differential equation.

EXAMPLE 3. In Example 2 we found e^{At} as a sum of three terms. These were constant multiples of

$$U = (A - I)(A - 4I), \quad V = A(A - 4I)e^t, \quad W = A(A - I)e^{4t}$$

respectively. We want to check the work by showing that each of these satisfies $dY/dt = AY$. After obvious simplification in the case of V and W, the equations $dY/dt = AY$ for U, V, W reduce to

$$0 = A(A - I)(A - 4I), \quad A(A - 4I) = A^2(A - 4I), \quad 4A(A - I) = A^2(A - I).$$

To check this, recall that the characteristic polynomial in Example 1 was

$$P(s) = s(s - 1)(s - 4) = s^3 - 5s^2 + 4s.$$

Theorem 3 gives $A^3 - 5A^2 + 4A = 0$, which yields each equation above.

PROBLEMS

1P. (a) If α and β are real or complex constants check that

$$(A + \alpha I)(A + \beta I) = A^2 + (\alpha + \beta)A + \alpha\beta I = (A + \beta I)(A + \alpha I)$$

(b) Let ϕ_1 and ϕ_2 be polynomials with real or complex coefficients α_j and β_k, respectively. Sum the identity

$$(\alpha_j A^j)(\beta_k A^k) = (\beta_k A^k)(\alpha_j A^j)$$

on j and k and deduce $\phi_1(A)\phi_2(A) = \phi_2(A)\phi_1(A)$. Definition 2 now yields

$$\psi_1(A)\psi_2(A) = \psi_2(A)\psi_1(A)$$

when ψ_i are defined on the spectrum of A. For example, $e^{At}A = Ae^{At}$.

(c) Show that $(A + B)(A - B) \neq (A - B)(A + B)$ if A and B are square matrices that do not commute. In Part (b) it is important that only a single matrix is involved. The reason why the problem worked out as it did is that A^j and A^k commute.

2P. Let A be a square matrix of order 2 with distinct eigenvalues λ_1 and λ_2.

(a) Find the Lagrange polynomial $\phi(s)$ that agrees with $\psi(s)$ at λ_1 and λ_2 and thus derive the formula

$$\psi(A) = \frac{A - \lambda_2 I}{\lambda_1 - \lambda_2}\psi(\lambda_1) + \frac{A - \lambda_1 I}{\lambda_2 - \lambda_1}\psi(\lambda_2)$$

(b) If A is real and λ is complex, let $\psi(s) = e^{st}$, $\lambda_1 = \lambda$, $\lambda_2 = \bar{\lambda}$ and deduce from (a) that

$$e^{At} = \frac{\text{Im } (A - \bar{\lambda}I)e^{\lambda t}}{\text{Im } \lambda}$$

3. (abcdefghijkl) Using Problem 2 with $\psi(s) = e^{st}$ write down e^{At} for some of the 2-by-2 systems in Section 16.1, Problem 1. You can check by verifying that $Y(t) = e^{At}Y(0)$ where $Y(t)$ is the answer from Chapter 16.

4P. A 3-by-3 matrix A has characteristic polynomial $P(s) = s(s^2 - 1)$. Find the Lagrange polynomial ϕ for ψ, then $\psi(A)$, and deduce that

$$2e^{At} = 2(I - A^2) + (A^2 - A)e^{-t} + (A^2 + A)e^t$$

5. Theorem 3 gives $A^3 = A$ for the matrix of Problem 4. Using this verify that the formula there given satisfies $(d/dt)e^{At} = Ae^{At}$.

6P. Check that the following matrix A has $P(s) = s(s^2 - 1)$:

$$A = \begin{pmatrix} -1 & -1 & 1 \\ -2 & -3 & 2 \\ -4 & -5 & 4 \end{pmatrix}$$

Compute A^2 and, by the result of Problem 4, get

$$e^{At} = \begin{pmatrix} 2 & 1 & -1 \\ 0 & 0 & 0 \\ 2 & 1 & -1 \end{pmatrix} + \begin{pmatrix} 0 & 0 & 0 \\ 1 & 2 & -1 \\ 1 & 2 & -1 \end{pmatrix} e^{-t} + \begin{pmatrix} -1 & -1 & 1 \\ -1 & -1 & 1 \\ -3 & -3 & 3 \end{pmatrix} e^{t}$$

7. (abcdefghijklmno) This is best done by means of a hand calculator programmed to compute matrix products. Find e^{At} for some of the 3-by-3 matrices in Section 16.2, Problem 2. Check that $e^0 = I$ and that the columns of M_j are multiples of the corresponding eigenvector E_j given in the answers.

8. A real 3-by-3 matrix A has eigenvalues $1, -i, i$. Check that the part of e^{At} associated with $-i$ and i is

$$(A - I)\text{Re}\,\frac{A - iI}{-1 + i}e^{-it}$$

and thus get $2e^{At} = (A^2 + I)e^t - (A^2 - I)\cos t - (A - I)^2 \sin t$.

9. A square matrix of order 4 has $P(s) = s(s^2 - 1)(s - 2)$. Find $\phi(s)$, then $\psi(A)$, and deduce that e^{At} is

$$\frac{(A^2 - I)(A - 2I)}{2} - \frac{(A^2 - A)(A - 2I)}{6}e^{-t} - \frac{(A^2 + A)(A - 2I)}{2}e^t + \frac{A^3 - A}{6}e^{2t}$$

10. Choose $\psi(s) = \cos\sqrt{s}\,t$ in Example 1 to get

$$\cos\sqrt{A}\,t = \frac{(A - I)(A - 4I)}{4} + \frac{A(A - 4I)}{-3}\cos t + \frac{A(A - I)}{12}\cos 2t$$

It can be shown that this satisfies $Y'' + AY = 0$, $Y(0) = I$, $Y'(0) = 0$.

11. Prove that $(d/dt)e^{\Lambda t} = e^{\Lambda t}\Lambda$ for diagonal matrices Λ of arbitrary order.

12P. Prove the first equation below, then the second, then the third:

$$e^{\Lambda s}e^{\Lambda t} = e^{\Lambda(s+t)}, \qquad e^{As}e^{At} = e^{A(s+t)}, \qquad e^{-At}e^{At} = I$$

2. Multiple eigenvalues

If the characteristic polynomial $P(s)$ has $(s - \lambda)^m$ as a factor, and no higher power of $s - \lambda$, then λ is said to be an eigenvalue of multiplicity m. There is a general principle in the theory of multiple eigenvalues, which can be stated informally as follows: In addition to $\psi(\lambda)$, use all derivatives up to order $m - 1$.

Let us explain how the above principle applies to Section 1. A function ψ is *defined on the spectrum of* A if the values

$$\psi(\lambda), \qquad \psi'(\lambda), \qquad \cdots, \qquad \psi^{(m-1)}(\lambda)$$

exist at each eigenvalue $\lambda = \lambda_j$ of multiplicity $m = m_j$. Two functions ψ_1 and ψ_2 are such that $\psi_1 = \psi_2$ *on the spectrum of* A if, for each λ as above,

$$\psi_1(\lambda) = \psi_2(\lambda), \qquad \psi_1'(\lambda) = \psi_2'(\lambda), \qquad \cdots, \qquad \psi_1^{(m-1)}(\lambda) = \psi_2^{(m-1)}(\lambda).$$

A central theorem in the subject states that if two polynomials ϕ_1 and ϕ_2 are such that $\phi_1 = \phi_2$ on the spectrum of A, then $\phi_1(A) = \phi_2(A)$. The proof is in DETA, page 403. This extends Section 1, Theorem 1 to the case of multiple eigenvalues.

Definitions 1 and 2 of Section 1 are already worded to allow multiple eigenvalues. With the above conventions, Theorems 1, 2, and 3 of Section 1 retain their validity, though the proofs do not. Instead of discussing proofs we show how problems with multiple eigenvalues are actually solved. Surprising as it may seem, the details of calculation are often easier than when all eigenvalues are distinct.

Our assumption that ψ is defined on the spectrum of A requires existence of at least one derivative at each multiple eigenvalue λ. The notion of derivative is readily extended to allow a complex variable; the defining limit is the same, and for a polynomial or a power series the formula is the same, as in the real case.

If ϕ is any polynomial such that $\phi = \psi$ on the spectrum of A, then $\psi(A) = \phi(A)$ by definition. The only new problem is construction of the polynomial ϕ. The method of Lagrange does not work, because the denominator $N_j(\lambda_j)$ is 0 at a multiple eigenvalue λ_j.

To deal with this problem, recall the equation

$$\psi(A) = L_1(A)\psi(\lambda_1) + L_2(A)\psi(\lambda_2) + \cdots + L_n(A)\psi(\lambda_n)$$

which holds when the λ_j are distinct. The coefficients $L_j(A)$ are polynomials in A, hence each of them is a matrix M_j of order n. Thus the above equation becomes

$$\psi(A) = M_1\psi(\lambda_1) + M_2\psi(\lambda_2) + \cdots + M_n\psi(\lambda_n).$$

It is important that the M_j are independent of ψ.

In the case of multiple eigenvalues the only change is as follows: Instead of the m terms

$$M_{j\,0}\psi(\lambda) + M_{j\,1}\psi(\lambda) + \cdots + M_{j\,m-1}\psi(\lambda)$$

corresponding to an eigenvalue $\lambda = \lambda_j$ of multiplicity $m = m_j$, we must use

$$M_{j\,0}\psi(\lambda) + M_{j\,1}\psi'(\lambda) + \cdots + M_{j\,m-1}\psi^{(m-1)}(\lambda).$$

Note that this agrees with the principle stated informally above. If $\psi(s) = e^{st}$ the nonzero columns of $M_{j\,m-1}$ must be eigenvectors corresponding to λ_m, though no such result holds for the other matrices $M_{j\,k}$.

EXAMPLE 1. A 3-by-3 matrix A has $P(s) = (s-4)^2(s-7)$, so that the spectrum is 4, 4, 7. In this case

$$(1) \qquad\qquad \psi(A) = M_1\psi(4) + M_2\psi'(4) + M_3\psi(7)$$

where the M_j are 3-by-3 matrices independent of ψ. To find the M_j we make three independent choices of ψ. The most obvious choices are

$$\psi(s) = 1, \ s, \ s^2$$

for which $\psi(A) = I$, A, A^2 respectively. But simpler equations are obtained if we choose ψ from among the functions

$$\psi(s) = 1, \quad s-4, \quad (s-4)^2, \quad s-7.$$

The choices $\psi(s) = 1$, $s-4$, $(s-4)^2$ in (1) yield, respectively,

$$I = M_1 + M_3, \quad A - 4I = M_2 + 3M_3, \quad (A-4I)^2 = 9M_3.$$

The third equation gives M_3, the first then gives M_1, the second gives M_2, and the choice $s - 7$ provides a check. It is left for you to verify that

$$\psi(A) = \frac{(A-I)(A-7I)}{-9}\psi(4) - \frac{(A-4I)(A-7I)}{3}\psi'(4) + \frac{(A-4I)^2}{9}\psi(7).$$

EXAMPLE 2. To find e^{At} in Example 1, take

$$\psi(s) = e^{st}, \quad \psi'(s) = te^{st}.$$

Substitution into the result of Example 1 yields

$$e^{At} = \frac{(A-I)(A-7I)}{-9}e^{4t} - \frac{(A-4I)(A-7I)}{3}te^{4t} + \frac{(A-4I)^2}{9}e^{7t}.$$

EXAMPLE 3.   A matrix A has spectrum λ, λ, λ, λ. To get e^{At} we start from

$$M_1\psi(\lambda) + M_2\psi'(\lambda) + M_3\psi''(\lambda) + M_4\psi'''(\lambda) = \psi(A).$$

The choices $\psi(s) = (s - \lambda)^j$ for $j = 0$, 1, 2, 3 yield

$$M_1 = I, \quad M_2 = A - \lambda I, \quad 2!M_3 = (A - \lambda I)^2, \quad 3!M_4 = (A - \lambda I)^3.$$

¿From $\psi(s) = e^{st}$ follows $\psi^{(j)}(s) = t^j e^{st}$, so that finally

$$e^{At} = \left(I + (A - \lambda I)t + (A - \lambda I)^2 \frac{t^2}{2!} + (A - \lambda I)^3 \frac{t^3}{3!} \right) e^{\lambda t}.$$

The result obviously generalizes.

It is a remarkable fact that when A is real and symmetric, or more generally when A is Hermitian, we need to consider only the distinct values of λ_j. There are no derivatives in the equation for $\psi(A)$ and no terms of the form $Mte^{\lambda t}$ in e^{At}. The following theorem is established in DETA, page 406:

THEOREM 1.   If $A = A^*$ there is a unitary matrix E such that $E^*AE = \Lambda$ where Λ is diagonal. The entries of Λ are the eigenvalues of A, repeated according to their multiplicity.

The unitary matrix of Theorem 1 satisfies $E^*E = I$ by definition. When any one of the equivalent equations

$$A = E\Lambda E^{-1}, \quad E^{-1}AE = \Lambda, \quad E^*AE = \Lambda$$

holds, the construction of $\psi(A)$ can be based upon the distinct eigenvalues λ_j, as if they had multiplicity 1. This fact leads to astonishing simplification, as illustrated in the following example.

EXAMPLE 4.   A real symmetric matrix of order n has two distinct eigenvalues, λ and μ, with combined multiplicity $m_1 + m_2 = n$. Hence

$$M_1\psi(\lambda) + M_2\psi(\mu) = \psi(A).$$

Choosing $\psi(s) = s - \lambda$ and $s - \mu$ we get

$$M_2(\mu - \lambda) = A - \lambda I, \quad M_1(\lambda - \mu) = A - \mu I.$$

¿From this M_1 and M_2 are found by inspection. The choice $\psi(s) = e^{st}$ yields

$$e^{At} = \frac{A - \mu I}{\lambda - \mu} e^{\lambda t} - \frac{A - \lambda I}{\lambda - \mu} e^{\mu t}.$$

This satisfies $Y' = AY$, $Y(0) = I$. The nonzero columns of $A - \mu I$ are eigenvectors for λ and those of $A - \lambda I$ are eigenvectors for μ. If $n = 1000$, say, the superiority of this method over that of Chapter 16 is obvious.

PROBLEMS

1P. Let A be a 2-by-2 matrix with two equal eigenvalues λ. Show that

$$\psi(A) = I \, \psi(\lambda) + (A - \lambda I) \, \psi'(\lambda)$$

2. The following matrices do not have two linearly independent eigenvectors. Taking $\psi(s) = e^{st}$ in Problem 1, find e^{At}:

(a) $\begin{pmatrix} 1 & 1 \\ -1 & 3 \end{pmatrix}$ (b) $\begin{pmatrix} -3 & 2 \\ -2 & 1 \end{pmatrix}$ (c) $\begin{pmatrix} 5 & 1 \\ -1 & 3 \end{pmatrix}$ (d) $\begin{pmatrix} -5 & -2 \\ 2 & -1 \end{pmatrix}$

3. The numbers in parentheses give the spectra for real symmetric matrices A. As in Example 4, find e^{At}:

(a) (0, 0, 1, 1)　　(b) (1, 1, 1, 2)　　(c) (3, 3, 4, 4, 4, 4)

4. Verify that the symmetric matrix on the left has spectrum 4, 4, 7 and show that the result of Example 4 gives the formula on the right:

$$A = \begin{pmatrix} 5 & 1 & 1 \\ 1 & 5 & 1 \\ 1 & 1 & 5 \end{pmatrix} \qquad 3e^{At} = \begin{pmatrix} 2 & -1 & -1 \\ -1 & 2 & -1 \\ -1 & -1 & 2 \end{pmatrix} e^{4t} + \begin{pmatrix} 1 & 1 & 1 \\ 1 & 1 & 1 \\ 1 & 1 & 1 \end{pmatrix} e^{7t}$$

5. Check that the matrix of Problem 4 satisfies $(A - 4I)(A - 7I) = 0$ and hence $A^2 = 11A - 28I$. Thus show that the formula of Example 1 simplifies to yield the result of Problem 4.

6. Here are six symmetric matrices each of which has a multiple eigenvalue. Using the result of Example 4, find e^{At}:

(a) $\begin{pmatrix} -2 & 0 & 0 \\ 0 & 0 & 0 \\ 0 & 0 & -2 \end{pmatrix}$ (c) $\begin{pmatrix} 1 & 1 & -2 \\ 1 & 1 & -2 \\ -2 & -2 & 4 \end{pmatrix}$ (e) $\begin{pmatrix} -5 & -2 & -2 \\ -2 & -2 & 4 \\ -2 & 4 & -2 \end{pmatrix}$

(b) $\begin{pmatrix} -1 & 2 & -4 \\ 2 & 2 & 2 \\ -4 & 2 & -1 \end{pmatrix}$ (d) $\begin{pmatrix} 1 & 0 & 0 \\ 0 & 2 & 0 \\ 0 & 0 & 1 \end{pmatrix}$ (f) $\begin{pmatrix} 1 & -1 & -1 \\ -1 & 1 & -1 \\ -1 & -1 & 1 \end{pmatrix}$

7. (abcdef) Review. This takes a good deal of time. In Problem 6 obtain a nonsingular matrix E of eigenvectors by the methods of Chapter 16.

8. (abcdef) Find e^{At} if the spectrum of A is as follows:

$$(0, 1, 1) \quad (3, 3, 3) \quad (2, 2, 5) \quad (0, 0, 0, 1) \quad (1, 1, 0, 0) \quad (0, 1, 1, 2)$$

9. (abcdefghijkl) The following matrices have integral eigenvalues, one of which is of multiplicity 2. Proceeding as in Examples 1 and 2, express e^{At} in terms of A:

(a) $\begin{pmatrix} 0 & 0 & -1 \\ 1 & 0 & -1 \\ 0 & 0 & -1 \end{pmatrix}$
(e) $\begin{pmatrix} 3 & 6 & 3 \\ -1 & 2 & -1 \\ 1 & -3 & 2 \end{pmatrix}$
(i) $\begin{pmatrix} 7 & -3 & -3 \\ 2 & 1 & -2 \\ 3 & -3 & 1 \end{pmatrix}$

(b) $\begin{pmatrix} -1 & 2 & -3 \\ 7 & 4 & 7 \\ -1 & -1 & 2 \end{pmatrix}$
(f) $\begin{pmatrix} 0 & 0 & -3 \\ -1 & 2 & -4 \\ -2 & 0 & -1 \end{pmatrix}$
(j) $\begin{pmatrix} 5 & 2 & 2 \\ 1 & 5 & -1 \\ -2 & 2 & 9 \end{pmatrix}$

(c) $\begin{pmatrix} -3 & -4 & -4 \\ 2 & 3 & 2 \\ 5 & 2 & 7 \end{pmatrix}$
(g) $\begin{pmatrix} 5 & 6 & 5 \\ 1 & 2 & 1 \\ -3 & -3 & -2 \end{pmatrix}$
(k) $\begin{pmatrix} 5 & 9 & -15 \\ 0 & 5 & -2 \\ 1 & 3 & -3 \end{pmatrix}$

(d) $\begin{pmatrix} 4 & 4 & 4 \\ 1 & 2 & 1 \\ -2 & -2 & -1 \end{pmatrix}$
(h) $\begin{pmatrix} -2 & 3 & 3 \\ 2 & -2 & 2 \\ -3 & -3 & -8 \end{pmatrix}$
(l) $\begin{pmatrix} 3 & 1 & 1 \\ -4 & 7 & 2 \\ 1 & -1 & 3 \end{pmatrix}$

10. (abcdefghijkl) Find two independent eigenvectors in Problem 9.

11. (abcdefghijkl) Preferably using a calculator capable of computing matrix products, work out some of the results of Problem 9 explicitly.

12. Let A be a 3-by-3 matrix with eigenvalues λ_1, λ_2, λ_3. Show that the nonzero columns of $(A - \lambda_2 I)(A - \lambda_3 I)$ are eigenvectors for λ_1.

13. Let A be a 3-by-3 matrix with spectrum (λ, λ, μ) where $\lambda \neq \mu$. Show that

$$e^{At} = Bte^{\lambda t} + Ce^{\mu t} + (I - C)e^{\lambda t}$$

where $(\lambda - \mu)^2 C = (A - \lambda I)^2$, $(\lambda - \mu)B = (A - \lambda I)(A - \mu I)$. By Problem 12 the nonzero columns of B and C are eigenvectors for λ and μ respectively.

14. Let Λ be a diagonal matrix of order n with only the two distinct eigenvalues λ, μ, and let $A = E^{-1}\Lambda E$. Verify the first of the following by direct computation and deduce the second by use of Section 1, Equation (2):

$$(\Lambda - \lambda I)(\Lambda - \mu I) = 0, \qquad (A - \lambda I)(A - \mu I) = 0$$

---ANSWERS---

2. $I e^{\lambda t} + B t e^{\lambda t}$ where $\lambda = 2, -1, 4, -3$. Check that the nonzero columns of B are eigenvectors.

3. (a) $Ae^t - A + I$, (b) $(A - I)e^{2t} - (A - 2I)e^t$, (c) $(A - 3I)e^{4t} - (A - 4I)e^{3t}$

6. (abcdef) Partial answer: Here are the eigenvalues. Example 4 then gives the answer almost by inspection:

$$(-2, -2, 0), \quad (-6, 3, 3), \quad (0, 0, 6), \quad (1, 1, 2), \quad (-6, -6, 3), \quad (-1, 2, 2)$$

7. With eigenvalues as above, possible matrices E are:

(a) $\begin{pmatrix} 0 & 1 & 0 \\ 0 & 0 & 1 \\ 1 & 0 & 0 \end{pmatrix}$ (c) $\begin{pmatrix} 2 & -1 & -1 \\ 0 & 1 & -1 \\ 1 & 0 & 2 \end{pmatrix}$ (e) $\begin{pmatrix} 2 & 2 & -1 \\ 0 & 1 & 2 \\ 1 & 0 & 2 \end{pmatrix}$

(b) $\begin{pmatrix} 2 & -1 & 1 \\ -1 & 0 & 2 \\ 2 & 1 & 0 \end{pmatrix}$ (d) $\begin{pmatrix} 0 & 1 & 0 \\ 0 & 0 & 1 \\ 1 & 0 & 0 \end{pmatrix}$ (f) $\begin{pmatrix} 1 & -1 & -1 \\ 1 & 0 & 1 \\ 1 & 1 & 0 \end{pmatrix}$

8. (a) $e^{At} = (A - I)^2 + (2A - A^2)e^t + (A^2 - A)te^t$

(b) $e^{At} = Ie^{3t} + (A - 3I)te^{3t} + (A - 3I)^2 t^2 e^{3t}/2$

(c) $9e^{At} = -(A^2 - 4A - 5I)e^{2t} - 3(A^2 - 7A + 10I)te^{2t} + (A - 2I)^2 e^{5t}$

(d) $e^{At} = I - A^3 + (A - A^3)t + (A^2 - A^3)t^2/2 + A^3 e^t$

(e) $e^{At} = (A - I)^2(I + 2A) + (A - I)^2 At + (3A^2 - 2A^3)e^t + (A^3 - A^2)te^t$

(f) $2e^{At} = (A - I)^2(2 - A + Ae^{2t}) + 2(2A - A^2)(e^t - te^t + Ate^t)$

9. The spectrum is λ, λ, μ with (λ, μ) as follows. The answers are then given by Problem 13:

(a) $(0, -1)$ (c) $(3, 1)$ (e) $(2, 3)$ (g) $(2, 1)$ (i) $(4, 1)$ (k) $(2, 3)$

(b) $(4, -3)$ (d) $(2, 1)$ (f) $(2, -3)$ (h) $(-5, -2)$ (j) $(7, 5)$ (l) $(4, 5)$

10. Possible eigenvectors with (λ, μ) as above are the columns of the following matrices. These are for (abcdef), those below for (ghijkl):

$\begin{pmatrix} 0 & 1 \\ 1 & 0 \\ 0 & 1 \end{pmatrix}$ $\begin{pmatrix} -1 & -1 \\ -1 & 1 \\ 1 & 0 \end{pmatrix}$ $\begin{pmatrix} -2 & -4 \\ 1 & 1 \\ 2 & 3 \end{pmatrix}$ $\begin{pmatrix} -2 & 0 \\ -1 & -1 \\ 2 & 1 \end{pmatrix}$ $\begin{pmatrix} -3 & -1 \\ -1 & -1 \\ 3 & 2 \end{pmatrix}$ $\begin{pmatrix} 0 & 1 \\ 1 & 1 \\ 0 & 1 \end{pmatrix}$

$\begin{pmatrix} -3 & -1 \\ -1 & -1 \\ 3 & 2 \end{pmatrix}$ $\begin{pmatrix} -1 & -1 \\ 0 & -1 \\ 1 & 1 \end{pmatrix}$ $\begin{pmatrix} 1 & 1 \\ 0 & 1 \\ 1 & 1 \end{pmatrix}$ $\begin{pmatrix} 1 & 1 \\ 0 & -1 \\ 1 & 1 \end{pmatrix}$ $\begin{pmatrix} 9 & 3 \\ 2 & 1 \\ 3 & 1 \end{pmatrix}$ $\begin{pmatrix} -1 & -1 \\ -2 & -3 \\ 1 & 1 \end{pmatrix}$

11. Check that $e^0 = I$ and, by the above answers, that nonzero columns of the coefficients of $te^{\lambda t}$ and of $e^{\mu t}$ are eigenvectors for λ and μ.

Chapter 18

BOUNDARY-VALUE PROBLEMS

W ITH HISTORICAL roots going back to early work of Fourier, Sturm, and Liouville, the results of this chapter are an important aid in solving the boundary-value problems of mathematical physics. After a concise introduction to Fourier series we discuss a number of classical problems involving partial differential equations. The principal object is to get the correct form of the solutions and questions of convergence (which are difficult) are not considered. The exposition is further simplified by systematic use of dimensionless variables and by an efficient method for computing Fourier coefficients. These aspects of the presentation make it possible to outline a variety of interesting applications in limited space.

Knowledge of Section 9.2 is desirable but not essential.

1. Fourier series

It is assumed throughout that all functions are integrable over the relevant intervals. We begin with a brief introduction to Fourier series, which were first used systematically in Joseph Fourier's book of 1822 and are named after him.

If m and n are real constants, the functions $\sin mx$, $\cos mx$ satisfy the first of the following equations and $\sin nx$, $\cos nx$ satisfy the second:

$$u'' = -m^2 u, \qquad v'' = -n^2 v.$$

These equations yield

$$(uv' - vu')' = uv'' - vu'' = (m^2 - n^2)uv,$$

hence

$$uv' - vu' \Big|_{-\pi}^{\pi} = (m^2 - n^2) \int_{-\pi}^{\pi} uv \, dx.$$

Here, and in analogous cases below, $u, u', \cdots$ stand for $u(x), u'(x), \cdots$.

So far, m and n have been unrestricted. We now assume that m and n are integers which, without loss of generality, are nonnegative. In that case $uv' - vu'$ has period 2π, the left side of the above equation is 0, and therefore

$$\int_{-\pi}^{\pi} uv dx = 0, \qquad m^2 \neq n^2.$$

When $m = n$ the procedure gives no information, but obviously

$$\int_{-\pi}^{\pi} \sin mx \cos mx \, dx = \frac{1}{2} \int_{-\pi}^{\pi} \sin 2mx \, dx = 0.$$

In general, two functions u and v are *orthogonal* over a given interval if the integral of their product over that interval is 0. What has been shown is that any two different members of the sequence

$$(1) \qquad 1, \quad \sin x, \quad \cos x, \quad \sin 2x, \quad \cos 2x, \quad \cdots, \quad \sin nx, \quad \cos nx, \quad \cdots$$

are orthogonal over the interval $-\pi \le x \le \pi$.

It turns out that broad classes of functions f can be represented by a convergent series

$$(2) \qquad f(x) = \frac{a_0}{2} + \sum_{n=1}^{\infty}(a_n \cos nx + b_n \sin nx), \qquad -\pi \le x \le \pi.$$

Assuming that this holds, and that term-by-term integration is permissible, we shall derive a formula for the coefficients.

To get a_0, integrate from $-\pi$ to π. The only surviving term on the right is πa_0 and division by π gives a_0. To find a_n with $n \ge 1$, multiply by $\cos nx$. The result is

$$f(x) \cos nx = \cdots + a_n \cos^2 nx + \cdots$$

where the terms not written involve a product of $\cos nx$ with some other member of the sequence (1). If we integrate from $-\pi$ to π these products integrate to 0 by orthogonality and the surviving term is

$$\int_{-\pi}^{\pi} a_n \cos^2 nx \, dx = \int_{-\pi}^{\pi} a_n \frac{1 + \cos 2nx}{2} \, dx = \pi a_n.$$

A similar calculation gives πb_n, with the result

$$(3) \qquad a_n = \frac{1}{\pi} \int_{-\pi}^{\pi} f(x) \cos nx \, dx, \quad b_n = \frac{1}{\pi} \int_{-\pi}^{\pi} f(x) \sin nx \, dx.$$

The constant term in (2) is written $a_0/2$ so that the formula for a_n remains valid when $n = 0$.

The constants a_n, b_n in (3) are called *Fourier coefficients* and the equations (3) are the Euler-Fourier formulas; they were stated by Euler as early as 1750. The series (2) with coefficients given by (3) is denoted by $S(f)$ and we write $f \sim S(f)$ to express the fact that $S(f)$ is the Fourier series of f. An equation such as $S(f) = S(g)$ means that $S(f)$ and $S(g)$ have the same coefficients; it says nothing about convergence.

If the Fourier cosine and sine coefficients are denoted by $a_n(f)$ and $b_n(f)$, respectively, it is obvious from the definition that

$$a_n(f + g) = a_n(f) + a_n(g), \quad a_n(cf) = ca_n(f).$$

Using this and the corresponding equations for b_n, we get

$$S(f + g) = S(f) + S(g), \quad S(cf) = cS(f).$$

In other words, the Fourier series for $f + g$ can be obtained by termwise addition, and the series for cf is obtained when the series for f is multiplied termwise by c.

The calculation of a Fourier series is especially simple when f is even or odd. Recall that "even" and "odd" are defined as follows:

$$f \text{ is even if } f(-x) = f(x); \quad f \text{ is odd if } f(-x) = -f(x).$$

In both cases, the equation is to hold for all x for which $f(x)$ is defined. It is left for you to check that the sum of two even functions is even, the sum of two odd functions is odd, the product of two even or two odd functions is even, and the product of an even and an odd function is odd. Symbolically,

$$e + e = e, \quad o + o = o, \quad ee = e, \quad oo = e, \quad eo = o.$$

Furthermore if F is integrable on $(-\pi, \pi)$ then

$$(4) \qquad \int_{-\pi}^{\pi} F(x)\, dx = \gamma \int_{0}^{\pi} F(x)\, dx$$

where $\gamma = 2$ if F is even and $\gamma = 0$ if F is odd. These remarks lead to the following:

THEOREM 1. *If $f(x)$ is even its Fourier series has cosine terms only and is*

$$(5a) \qquad f(x) \sim \frac{a_0}{2} + \sum_{k=1}^{\infty} a_k \cos kx, \quad a_k = \frac{2}{\pi} \int_{0}^{\pi} f(x) \cos kx\, dx.$$

If $f(x)$ is odd its Fourier series has sine terms only and is

$$(5b) \qquad f(x) \sim \sum_{k=1}^{\infty} b_k \sin kx, \quad b_k = \frac{2}{\pi} \int_{0}^{\pi} f(x) \sin kx\, dx.$$

For proof, suppose f is even. Then $f(x) \cos nx$ is even and $f(x) \sin nx$ is odd. By (4) the first integral (3) agrees with that in (5a) and the second integral (3) is 0. This gives (5a). Proof of (5b) is similar.

The coefficients (5) are determined when f is given only on the interval $(0, \pi)$. For this reason the series in (5) are called the *half-range cosine series* and the *half-range sine series*, respectively. The full-range series may involve both sines and cosines and it requires that f be defined on the whole interval $(-\pi, \pi)$. The following theorem applies to half-range as well as full-range series. For proof see DETA, pages 448–449:

THEOREM 2.   *Let* $(-\pi, \pi)$ *be divided by points* x_k *into a finite number of open intervals on each of which* $f(x)$ *has a bounded derivative. Suppose further that* f *has period* 2π. *Then the full-range Fourier series for* f *converges at each value of* x *to*

(6)
$$\frac{f(x-) + f(x+)}{2}.$$

As explained more fully in DETA, page 421, it is customary to assume (and we do assume) that f is periodic and is defined by (6) at all points where (6) is meaningful. For most functions of interest in applications the relation $f \sim S$ can then be replaced by $f = S$.

EXAMPLE 1.   Let $f(x) = \pi/2$ for $0 \le x \le \pi/2$ and $f(x) = 0$ elsewhere on $(0, \pi)$. The half-range cosine coefficients satisfy

$$a_0 = \frac{2}{\pi} \int_0^{\pi/2} \frac{\pi}{2}\, dx = \frac{\pi}{2}, \quad a_n = \int_0^{\pi/2} \cos nx\, dx = \frac{\sin (n\pi/2)}{n}.$$

The function $\sin (n\pi/2)$ has the values

$$\sin \frac{n\pi}{2} = 1,\ 0,\ -1,\ 0 \quad | \quad \text{repeat}$$

for $n = 1, 2, 3, 4, \cdots$, where the word "repeat" means that the next four values are the same, the next four are the same, and so on. These give the numerators for a_n and the denominators n are $1, 2, 3, \cdots$. Taking account of the term $a_0/2$ we get the half-range cosine series

$$\frac{\pi}{4} + \frac{\cos x}{1} - \frac{\cos 3x}{3} + \frac{\cos 5x}{5} - \frac{\cos 7x}{7} + \cdots.$$

The sum of the series is even because the terms $\cos nx$ are even, and it has period 2π because the terms do. By Theorem 2, the series converges to the function of period 2π shown graphically in Figure 1(a).

EXAMPLE 2.   It is easily checked that the sine coefficients for the function in Example 1 satisfy $nb_n = 1 - \cos(n\pi/2)$. Here

$$1 - \cos \frac{n\pi}{2} = 1,\ 2,\ 1,\ 0 \quad | \quad \text{repeat}$$

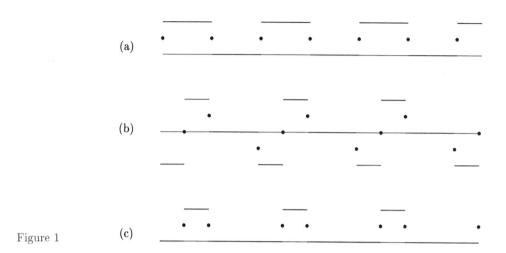

Figure 1

and the half-range sine series is

$$\frac{\sin x}{1} + 2\frac{\sin 2x}{2} + \frac{\sin 3x}{3} + \frac{\sin 5x}{5} + 2\frac{\sin 6x}{6} + \frac{\sin 7x}{7} + \frac{\sin 9x}{9} + \cdots$$

The function to which this series converges is shown in Figure 1(b). It is odd because the terms $\sin nx$ are odd and it clearly has period 2π.

EXAMPLE 3. Let $f(x) = \pi/2$ for $0 \le x \le \pi/2$ and $f(x) = 0$ elsewhere on $(-\pi, \pi)$. As discussed more fully later, the full-range Fourier series for this function is half the sum of the two series obtained in Examples 1 and 2. See Figure 1(c).

The following useful result is established in DETA, page 427:

THEOREM 3. *Let* $f(x)$ *coincide with a polynomial on* $[0, \pi]$ *and let*

$$\alpha_k = f^{(k)}(0), \quad \beta_k = f^{(k)}(\pi), \qquad k = 0, 1, 2, \cdots.$$

Then for $n \ge 1$ *the coefficient* a_n *in the half-range cosine series satisfies*

$$\frac{\pi}{2}a_n = (-1)^n \left(\frac{\beta_1}{n^2} - \frac{\beta_3}{n^4} + \frac{\beta_5}{n^6} - \cdots \right) - \left(\frac{\alpha_1}{n^2} - \frac{\alpha_3}{n^4} + \frac{\alpha_5}{n^6} - \cdots \right)$$

and the coefficient b_n *in the half-range sine series satisfies*

$$\frac{\pi}{2}b_n = \left(\frac{\alpha_0}{n} - \frac{\alpha_2}{n^3} + \frac{\alpha_4}{n^5} - \cdots \right) - (-1)^n \left(\frac{\beta_0}{n} - \frac{\beta_2}{n^3} + \frac{\beta_4}{n^5} - \cdots \right).$$

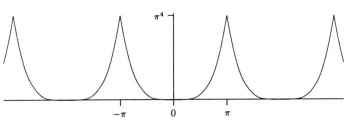

Figure 2

EXAMPLE 4.   Let us find the half-range cosine series for x^4. The relevant derivatives and their values at 0 and π are

$$4x^3, \qquad 24x; \qquad \alpha_1 = 0, \qquad \alpha_3 = 0; \qquad \beta_1 = 4\pi^3, \qquad \beta_3 = 24\pi.$$

By direct integration

$$a_0 = \frac{2}{\pi} \int_0^\pi x^4 \, dx = \frac{2\pi^4}{5}.$$

It is left for you to check that the formula for a_n in Theorem 3 gives

$$x^4 \sim \frac{\pi^4}{5} + \sum_{n=1}^{\infty} (-1)^n \left(\frac{8\pi^2}{n^2} - \frac{48}{n^4} \right) \cos nx.$$

By Theorem 2 the series converges to x^4 for $-\pi \le x \le \pi$ and to the periodic extension of x^4 outside this interval. See Figure 2. An idea of the rate at which the Fourier series converges is given by Figure 3, which shows the function together with two partial sums of the Fourier cosine series. The sums have 8 and 24 terms, respectively.

It turns out that the half-range Fourier series can be used to get practically all the formulas normally associated with the full-range series. Here are two examples.

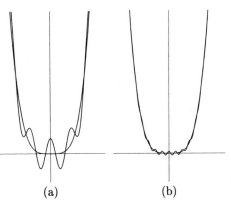

(a)   (b)

Figure 3

EXAMPLE 5.   If C and S are the half-range cosine and sine series such that

$$C(x) \sim \frac{f(x) + f(-x)}{2}, \qquad S(x) \sim \frac{f(x) - f(-x)}{2},$$

then the full-range Fourier series for f is $C(x) + S(x)$. This follows from the fact that the functions generating $C(x)$ and $S(x)$ are even and odd, respectively, and from the fact that

$$f(x) = \frac{f(x) + f(-x)}{2} + \frac{f(x) - f(-x)}{2}.$$

As an illustration, the equation

$$1 + x + x^2 + x^3 + x^4 + x^5 = (1 + x^2 + x^4) + (x + x^3 + x^5)$$

allows use of Theorem 3 to get the full-range Fourier series of the function on the left, even though this function is neither even nor odd. All we have to do is expand the two functions on the right into cosine and sine series, respectively.

EXAMPLE 6. Let $C(x)$ be the half-range cosine series and $S(x)$ the half-range sine series for $f(x)$ on the range $(0, \pi)$. Then

$$\frac{C(x) + S(x)}{2}$$

is the full-range Fourier series for the function that agrees with f on $(0, \pi)$ and is 0 on $(-\pi, 0)$. This is an immediate consequence of the Euler-Fourier formulas.

PROBLEMS

1P. We shall get the half-range cosine and sine series for $\pi - x$.

(a) Show by integration that $a_0 = \pi$ and by differentiation that, in the notation of Theorem 3, $\alpha_0 = \pi$, $\alpha_1 = -1$, $\beta_0 = 0$, $\beta_1 = -1$.

(b) Get a_n and b_n from Theorem 3 and conclude that

$$\pi - x = \frac{\pi}{2} + \frac{4}{\pi} \sum_{n=1}^{\infty} \frac{\cos(2n-1)x}{(2n-1)^2}, \qquad \pi - x = 2 \sum_{n=1}^{\infty} \frac{\sin nx}{n}$$

(c) Confirm the results of Part (b) by (5).

(d) By Example 6, half the sum of the series of Part (b) gives the full-range Fourier series for the function defined by $f(x) = 0$ for $-\pi < x < 0$, and by $f(x) = \pi - x$ for $0 < x < \pi$. Sketch the graphs of the functions to which the three Fourier series in Parts (bd) converge. No answer is provided.

2. (a) Using Theorem 3 obtain the half-range cosine and sine expansions for π and x and check that their difference agrees with Problem 1.

(b) Let $f(x) = 0$ for $-\pi < x < 0$ and $f(x) = x$ for $0 < x < \pi$. Find the full-range Fourier series by the Euler-Fourier formulas and verify agreement with the result of Part (a). Here and below, when a problem is self-checking no answer is provided.

3. Using integral tables if you wish, show that

$$|\sin x| \sim \frac{2}{\pi} - \frac{4}{\pi} \sum_{n=1}^{\infty} \frac{\cos 2nx}{(2n)^2 - 1}$$

This represents a fully rectified wave and $(|\sin x| + \sin x)/2$ represents a half-rectified wave. For $x \geq 0$ the graphs of these functions are given in Section 22.1, Figure 2.

4. (ab) Obtain the following expansions for x^2 and x^6, respectively:

$$\frac{\pi^2}{3} + \sum_{n=1}^{\infty} (-1)^n \frac{4\cos nx}{n^2} \qquad \frac{\pi^6}{7} + \sum_{n=1}^{\infty} (-1)^n \left(\frac{12\pi^4}{n^2} - \frac{240\pi^2}{n^4} + \frac{1440}{n^6} \right) \cos nx$$

5. (ab) Expand x^3 and x^5 in half-range sine series and verify that the results agree with those obtained by differentiating the half-range cosine series for $x^4/4$ and $x^6/6$ as given by Example 4 and Problem 4.

6. Consider the functions $x(\pi - x)$, $(\pi - x)^2$, $\pi(\pi - x)$.

(a) Find the half-range cosine series for each function and check that the third series is the sum of the other two.

(b) Do Part (a) with the word "cosine" replaced by "sine".

7. (ab) Proceed as in Problem 6: $x^3(\pi^2 - x^2)$, $x^2(x^3 - \pi^3)$, $(\pi x)^2(x - \pi)$.

8. Graph the function to which the Fourier series converges (a) in Problems 4 and 5, (b) in Problem 6, and (c) in Problem 7. (Four, six, and six graphs, respectively.)

9. This pertains to $y'' + \lambda y = 0$, $0 \leq x \leq \pi$ with the boundary conditions

$$\text{(a) } y(0) = y(\pi) = 0 \qquad \text{(b) } y'(0) = y'(\pi) = 0$$

If there are nontrivial solutions show that $\lambda = m^2$ where m is a positive integer in (a) and a nonnegative integer in (b). This gives the solutions $\sin mx$ and $\cos mx$. Show that each set $\{\sin mx\}$ and $\{\cos mx\}$ is orthogonal on the interval $(0, \pi)$.

2. Separation of variables

One of the main uses of Fourier series is to solve partial differential equations. Three typical examples are

$$u_t = c^2 u_{xx}, \qquad u_{xx} + u_{yy} = 0, \qquad u_{tt} = c^2 u_{xx}$$

where the subscripts denote partial differentiation. The first of these equations describes heat flow in a uniform insulated bar. The constant c depends on the conductivity, density, and heat capacity of the material and the solution $u(x,t)$ denotes the temperature at point x of the bar at time t. The second equation describes the steady-state temperature $u(x,y)$ at point (x,y) of a uniform sheet, and the third describes wave motion of small amplitude on a uniform stretched string. Here the constant c is determined by the tension and density and is the same as the speed of propagation of the waves.

By a change of scale, constants such as c in the above equations can be reduced to 1, and a length ℓ can be reduced to 1 or π. This use of *dimensionless variables* does not diminish the mathematical content of our investigation, and as explained later, it does not entail any loss of generality.

From now on we assume that all functions are sufficiently well-behaved to justify the calculations. The word "solution" is used in the sense of "formal solution." This means that the solution has the correct form but the question of convergence is left open. A discussion of the theoretical foundations is in DETA, Chapter 18.

The connection between Fourier series and partial differential equations rests on the following fact. Let F be a function of x alone and G a function of t alone, each with a nonempty domain of definition. If

$$F(x) = G(t)$$

for all values of the variables for which the two sides are defined, then both F and G are constant. To see why, choose any value t_0 in the domain of G. The resulting equation

$$F(x) = G(t_0)$$

shows that F is constant. Similarly, G is constant. It is said that the variables in $F(x) = G(t)$ are *separated*, the common constant value of F and G is the *separation constant*, and the procedure described in this section is called the method of *separation of variables*.

The following preliminary remark helps in determining the sign of the separation constant:

(1) $\qquad y'' + \lambda y = 0, \quad y(a) = y(b) = 0, \quad y \neq 0 \ \Rightarrow \ \lambda > 0.$

More precisely, it is assumed that y is continuous for $a \leq x \leq b$, that the differential equation holds on the interior of this interval, and that y is not identically zero. The conclusion is that $\lambda > 0$.

For proof, if $y(x) > 0$ at any point a theorem of advanced calculus shows that y has a positive maximum at some point c, and at this point

$$\lambda y(c) = -y''(c) \geq 0.$$

This gives $\lambda \geq 0$. Similarly, $\lambda \geq 0$ if $y(x) < 0$ at any point, and hence $\lambda \geq 0$ unless y is identically 0. Since the case $\lambda = 0$ is trivial, this completes the proof.

Naturally, (1) can also be obtained by solving the differential equation. But the above procedure applies to the general equation

$$y'' + p(x)y' + \lambda q(x) = 0, \qquad y(a) = y(b) = 0$$

which is often encountered in advanced work. If $q > 0$, and if there is a nontrivial solution y, the conclusion $\lambda \geq 0$ follows as in the simpler case above. No hypothesis on p is needed.

(a) Transient heat flow in a rod

A uniform insulated bar extending from $x = 0$ to $x = \pi$ has its ends kept at temperature 0. The initial temperature is $f(x)$ at point x of the bar. We want to find the temperature $u(x, t)$ at x at any subsequent time t. See Figure 4, which shows the bar with the relevant part of the (x, t) plane above it.

This problem can be formulated as follows:

$$u_t = u_{xx}, \qquad u(0, t) = 0$$

$$u(\pi, t) = 0, \qquad u(x, 0) = f(x).$$

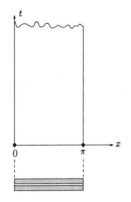

The differential equation holds for $0 < x < \pi$ and $t > 0$. It expresses the law of heat flow. The other equations describe the boundary conditions and the hypothesis that the initial temperature is $f(x)$.

Figure 4

Let us try $u(x, t) = R(x)S(t)$. Substitution into the differential equation gives successively

(2) $$RS' = R''S, \qquad \frac{R''}{R} = \frac{S'}{S}, \qquad \frac{R''}{R} = \frac{S'}{S} = -\lambda.$$

The separation constant is denoted by $-\lambda$ because λ will then turn out to be positive. In fact,

$$R'' + \lambda R = 0, \qquad R(0) = R(\pi) = 0$$

where the boundary conditions are inherited from those in the original problem. The condition $\lambda > 0$ follows from (1).

We set $\lambda = \omega^2$ and find in succession

$$R = a \cos \omega x + b \sin \omega x, \quad R = b \sin \omega x, \quad \omega = n$$

where a, b are constant and n is an integer. The solutions are

$$R(x) = b_n \sin nx, \quad \lambda = n^2, \quad n = 1, 2, 3, \cdots.$$

The notation $b = b_n$ indicates that the constants can depend on n. We get $S(t) = e^{-\lambda t}$ from the last equation (2), so that

$$S(t)R(x) = b_n e^{-n^2 t} \sin nx.$$

Each of these functions is called a *partial solution* because it satisfies part of the problem. Namely, it satisfies both the differential equation and the boundary conditions.

To satisfy the initial condition we try a superposition of partial solutions,

$$(3) \qquad\qquad u(x,t) = \sum_{n=1}^{\infty} b_n e^{-n^2 t} \sin nx.$$

The series takes the value 0 when $x = 0$ or $x = \pi$, so that the boundary conditions hold for all choices of $\{b_n\}$. Setting $t = 0$ yields

$$f(x) = \sum_{n=1}^{\infty} b_n \sin nx, \quad \text{hence} \quad b_n = \frac{2}{\pi} \int_0^{\pi} f(\xi) \sin n\xi \, d\xi.$$

We have written the integration variable as ξ to avoid confusion with x. If the above value b_n is substituted into the series for $u(x,t)$ we get

$$u(x,t) = \int_0^{\pi} K(t,x,\xi) f(\xi) \, d\xi, \quad K(t,x,\xi) = \frac{2}{\pi} \sum_{n=1}^{\infty} e^{-n^2 t} \sin nx \sin n\xi.$$

The function K is called the *kernel* and the formula is the *kernel representation* of the solution.

(b) The Dirichlet problem

The Laplace equation in two dimensions is $u_{xx} + u_{yy} = 0$ and solutions of Laplace's equation are called *harmonic functions*. The *Dirichlet problem* is to find a function harmonic in a given region with prescribed values on the boundary. It arises in connection with such topics as heat conduction, ideal fluid flow, diffusion, electrostatics, and gravitation. Aside from physical applications, the problem plays a prominent role in a branch of mathematics known as potential theory.

We now solve the Dirichlet problem for a strip. The conditions for steady-state heat flow in the semi-infinite strip of Figure 5 are

$$u_{xx} + u_{yy} = 0, \quad u(0, y) = 0, \quad u(\pi, y) = 0, \quad u(x, 0) = f(x).$$

The first equation holds for $0 < x < \pi$ and $0 < y < \infty$. It describes the law of steady-state heat flow. The next three equations give boundary conditions on the three edges of the strip.

Figure 5

If $f(x) = \sin x$ the problem has two distinct solutions,

$$u(x, y) = e^y \sin x, \quad u(x, y) = e^{-y} \sin x.$$

Uniqueness fails because the other end of the strip must be taken into account even though it is infinitely far away. We therefore impose the additional condition

$$(4) \qquad \lim_{y \to \infty} u(x, y) = 0 \text{ uniformly in } x.$$

A brief argument, which we omit, now shows that there is at most one solution.

The trial solution $X(x)Y(y)$ yields

$$X''Y + Y''X = 0, \quad -\frac{X''}{X} = \frac{Y''}{Y}$$

and, setting the two terms in the second equation equal to a separation constant λ, we get

$$X'' + \lambda X = 0, \quad Y'' = \lambda Y.$$

The boundary conditions $X(0) = X(\pi) = 0$ are inherited from those in the original problem.

The choice $\lambda = \omega^2 > 0$ is justified by (1) and leads to

$$X(x) = a \cos \omega x + b \sin \omega x.$$

The condition $X(0) = 0$ rules out the cosine and $X(\pi) = 0$ shows that $\omega = n$, an integer. Since $Y'' = \lambda Y$ this gives the partial solutions

$$b_n e^{ny} \sin nx, \quad b_n e^{-ny} \sin nx, \qquad n = 1, 2, \cdots.$$

Each of these satisfies the differential equation and the lateral boundary conditions.

In view of (4) the partial solutions involving e^{ny} are rejected. We form a superposition of the others as in Part (a) and we determine the coefficients by setting $y = 0$. The result is

$$(5) \qquad u(x,y) = \sum_{n=1}^{\infty} b_n e^{-ny} \sin nx, \quad b_n = \frac{2}{\pi} \int_0^{\pi} f(\xi) \sin n\xi \, d\xi.$$

(c) Change of scale

On an interval $(-\ell, \ell)$, rather than $(-\pi, \pi)$, an argument similar to that at the beginning of Section 1 gives the full-range Fourier series

$$g(t) \sim \frac{a_0}{2} + \sum_1^{\infty} \left(a_n \cos \frac{n\pi t}{\ell} + b_n \sin \frac{n\pi t}{\ell} \right), \qquad -\ell \le t \le \ell$$

where

$$a_n = \frac{1}{\ell} \int_{-\ell}^{\ell} g(t) \cos \frac{n\pi t}{\ell} \, dt, \quad b_n = \frac{1}{\ell} \int_{-\ell}^{\ell} g(t) \sin \frac{n\pi t}{\ell} \, dt.$$

If f is even or odd we get corresponding half-range series. The following theorem reduces all three series to the case $\ell = \pi$:

THEOREM 1.   *The Fourier series for $g(t)$ on the range $(-\ell, \ell)$ or $(0, \ell)$ is the same as the series for the function $f(x) = g(\ell x/\pi)$ on the range $(-\pi, \pi)$ or $(0, \pi)$, respectively.*

For proof, the substitution $t = \ell x / \pi$ gives

$$\frac{1}{\ell} \int_{-\ell}^{\ell} g(t) \cos \frac{n\pi t}{\ell} \, dt = \frac{1}{\pi} \int_{-\pi}^{\pi} g\left(\frac{\ell x}{\pi} \right) \cos nx \, dx.$$

The right side is a_n for the function $f(x) = g(\ell x/\pi)$. Discussion of b_n and of the half-range cases is similar.

When Section 1, Theorem 3 is applied to $f(x) = g(\ell x/\pi)$ the chain rule gives

$$f^{(j)}(x) = \left(\frac{\ell}{\pi} \right)^j g^{(j)}(t), \qquad t = \frac{\ell x}{\pi},$$

and hence the quantities α_j and β_j in Section 1, Theorem 3 are

$$(6) \qquad \alpha_j = \left(\frac{\ell}{\pi} \right)^j g^{(j)}(0), \quad \beta_j = \left(\frac{\ell}{\pi} \right)^j g^{(j)}(\ell).$$

EXAMPLE 1.   If $g(t) = t(1-t)$ on the interval $(0,1)$ the above formulas give $\alpha_0 = \beta_0 = 0$, $\alpha_2 = \beta_2 = -2/\pi^2$. Hence

$$\frac{\pi}{2} b_n = \frac{2}{\pi n^3}(1 - (-1)^n).$$

Upon writing the series for $f(x)$ with these coefficients and then setting $x = \pi t$, we get the half-range sine series,

$$t(1-t) = \frac{8}{\pi^3}\left(\frac{\sin \pi t}{1^3} + \frac{\sin 3\pi t}{3^3} + \frac{\sin 5\pi t}{5^3} + \cdots\right), \qquad 0 < t < 1.$$

A change of variable is also useful in connection with partial differential equations. For example, consider the problem

$$v_t = c^2 v_{xx}, \quad v(0,t) = v(\ell, t) = 0, \quad v(x,0) = g(x)$$

where c is constant. The substitution $v(x,t) = u(\alpha x, \beta t)$ gives

$$\beta u_t = c^2 \alpha^2 u_{xx}, \quad u(0, \beta t) = u(\alpha \ell, \beta t) = 0, \quad u(\alpha x, 0) = g(x).$$

We determine α by setting $\alpha \ell = \pi$ and then we determine β setting $\beta = c^2 \alpha^2$. The resulting problem is the same as that in Part (a), with $f(x) = g(\ell x/\pi)$. Hence (3) gives

$$(7) \qquad v(x,t) = \sum_{n=1}^{\infty} b_n e^{-(n\pi c/\ell)^2 t} \sin \frac{n\pi x}{\ell}, \quad b_n = \frac{2}{\pi} \int_0^\pi g\left(\frac{\ell \xi}{\pi}\right) \sin n\xi \, d\xi$$

in this seemingly more general case. The discussion illustrates an important method of reducing the number of parameters in physical problems. Further development leads to the study of dimensional analysis and dimensionless variables, which was initiated toward the end of the last century by the mathematical physicist Lord Rayleigh.

PROBLEMS

1. In (3) suppose the initial temperature is $f(x) = x(\pi - x)$; note that this satisfies the boundary conditions. Find $u(x,t)$ and check that the temperature at the midpoint of the bar is

$$u\left(\frac{\pi}{2}, t\right) = \frac{8}{\pi}\left(\frac{e^{-t}}{1^3} - \frac{e^{-3^2 t}}{3^3} + \frac{e^{-5^2 t}}{5^3} - \frac{e^{-7^2 t}}{7^3} + \cdots\right)$$

2. If both ends of the bar in Part (a) are insulated the boundary conditions are $u_x(0,t) = u_x(\pi,t) = 0$. Obtain the partial solutions $a_n e^{-n^2 t} \cos nx$ and thus show that

$$u(x,t) = \frac{a_0}{2} + \sum_{n=1}^{\infty} a_n e^{-n^2 t} \cos nx, \quad a_n = \frac{2}{\pi} \int_0^{\pi} f(\xi) \cos n\xi \, d\xi$$

3. If $f(x) = \pi/4$ in (5), the boundary conditions at the corners are mutually contradictory. Ignoring this difficulty, obtain the solution

$$u(x,y) = e^{-y} \sin x + e^{-3y} \frac{\sin 3x}{3} + e^{-5y} \frac{\sin 5x}{5} + \cdots$$

4P. The vertical displacement $u(x,t)$ of a vibrating string extending from 0 to π satisfies

$$u_{xx} = u_{tt}, \quad u(0,t) = 0, \quad u(\pi,t) = 0$$

The first equation holds for $0 < x < \pi$, $0 < t < \infty$ and describes the physics of small oscillations. The next two equations are boundary conditions stating that the endpoints of the string remain fixed throughout the motion. Obtain the partial solutions

$$b_n \sin nx \cos nt, \quad b_n \sin nx \sin nt$$

where n is an integer.

(a)

5. The plucked string. Let $u(x,0) = f(x)$ and $u_t(x,0) = 0$ in Problem 4. This means that the string is given an initial displacement and is gently released, as in a harp; see Figure 6(a). Obtain the solution

(b)

Figure 6

$$u(x,t) = \sum_{n=1}^{\infty} b_n \cos nt \sin nx, \quad b_n = \frac{2}{\pi} \int_0^{\pi} f(\xi) \sin n\xi \, d\xi$$

6. The struck string. In Problem 4 let $u(x,0) = 0$, $u_t(x,0) = g(x)$. This means that the string has initial displacement zero but is given an initial velocity, as in a piano; see Figure 6(b). Obtain the solution

$$u(x,t) = \sum_{n=1}^{\infty} b_n \sin nt \sin nx, \quad b_n = \frac{2}{n\pi} \int_0^{\pi} g(\xi) \sin n\xi \, d\xi$$

The sum of this function and that in Problem 5 solves the string problem with arbitrary initial displacement and velocity.

7. A uniform rectangle of width π and height b has its top edge at temperature $f(x)$ and the remaining three edges at temperature 0 as shown in Figure 7. The steady-state temperature u satisfies $u_{xx} + u_{yy} = 0$ for $0 < x < \pi$ and $0 < y < b$, with the boundary conditions

$$u(x, b) = f(x), \quad u(0, y) = 0, \quad u(x, 0) = 0, \quad u(\pi, y) = 0$$

Separate variables and express the solution by means of the kernel

$$K(x, y, \xi) = \frac{2}{\pi} \sum_{n=1}^{\infty} \frac{\sinh ny}{\sinh nb} \sin nx \sin n\xi$$

8. Explain how to solve the Dirichlet problem for a square by four uses of the result of Problem 7 with $b = \pi$.

Figure 7

9. Suppose the two lateral edges of the rectangle in Problem 7 are insulated, so that

$$u_x(0, y) = u_x(\pi, y) = 0$$

while the other two boundary conditions are still $u(x, 0) = 0$, $u(x, b) = f(x)$. Obtain the solution

$$u(x, y) = \frac{a_0}{2} y + \sum_{n=1}^{\infty} a_n \cos nx \sinh ny$$

Express the coefficients a_n in terms of f, evaluate when $f(x) = c$, a constant, and verify that the resulting solution satisfies both the differential equation and the boundary conditions.

10. Rewrite the formula for b_n in (7) by setting $\eta = \ell\xi/\pi$ and obtain this solution directly by separation of variables; do not use (3).

11. Show that the half-range cosine and sine series for $g(t) = (t - 1)^2$ on the interval $(0, 1)$ are respectively

$$\frac{1}{3} + \frac{4}{\pi^2} \sum_{n=1}^{\infty} \frac{\cos n\pi t}{n^2}, \quad \frac{2}{\pi} \sum_{n=1}^{\infty} \frac{\sin n\pi t}{n} - \frac{8}{\pi^3} \sum_{n=1}^{\infty} \frac{\sin(2n - 1)\pi t}{(2n - 1)^3}$$

Half the sum of these series gives the full-range Fourier series for the function that equals $(t - 1)^2$ on the interval $(0, 1)$ and equals 0 on $(-1, 0)$.

Chapter 19

THE PHASE PLANE

MANY PROBLEMS involve two unknown functions, $x(t)$ and $y(t)$, where dx/dt and dy/dt are given in terms of x and y. In such cases a solution can be described very conveniently by plotting the locus of the point $(x(t), y(t))$ in the (x, y) plane. For linear equations with constant coefficients, the resulting curves fall into three main classes, and graphical methods yield detailed information with little effort. A particularly simple case is that in which $x = u(t)$ and $y = u'(t)$ where u satisfies a second-order linear equation with constant coefficients. This case is investigated in detail and more general cases are then reduced to it.

1. A second-order equation

Throughout this chapter p, q, a, b, c, d, k are real constants and primes denote differentiation with respect to t. We begin by considering the linear equation

(1) $$u'' + pu' + qu = 0,$$

which has already been studied extensively in this book. If $u(t)$ gives the position at time t of a point moving on the u axis, the function

$$v(t) = \frac{du}{dt} = u'(t)$$

denotes the velocity.

The behavior of the system can be described not only by a graph of $u(t)$ versus t but also by giving the locus of the point $(u(t), v(t))$ in the (u, v) plane. When the (u, v) plane is associated with a differential equation in this way, it is called the *phase plane*. The curves given by $(u(t), v(t))$ as t increases from $-\infty$ to ∞ are called *trajectories* and the set of points occupied by a trajectory is the *orbit* or *trace*. For example, in Chapter 7 it was seen that the phase-plane orbits for $u'' + u = 0$ are circles centered at the origin. The main difference between an orbit and a trajectory is that the latter is associated with the specific parameter t. The parameterization gives the trajectory an orientation, as shown by arrows in the accompanying figures.

Throughout this chapter (u, v) gives the position of a point moving in the phase plane. The velocity of this point is not u' but is the two-dimensional vector

$$(u', v') = \left(\frac{du}{dt}, \frac{dv}{dt} \right) = \mathbf{i}u' + \mathbf{j}v'.$$

The calculation

$$\text{slope of vector} = \frac{dv/dt}{du/dt} = \frac{dv}{du} = \text{slope of curve}$$

shows that the velocity vector is tangent to the orbit; more specifically, the vector gives the direction in which the point (u, v) on the trajectory is moving at any given time. This fact is an aid in sketching the curves.

The analysis of orbits in the phase plane can be simplified by a change of variables known as an affine transformation. An *affine transformation* is defined by

$$U = Au + Bv \qquad \begin{vmatrix} A & B \\ C & D \end{vmatrix} \neq 0$$
$$V = Cu + Dv$$

where A, B, C, D are real constants. Since the determinant of the system is not 0, we can solve for (u, v) in terms of (U, V). As an illustration, each of the three transformations

$$U = 4u - v \qquad U = 5u + v \qquad U = u + v$$
$$V = u - v \qquad V = 2u - v \qquad V = -4u$$

is affine because the corresponding determinants are $-3, -7, 4$, none of which is 0. These transformations will be used in Examples 1, 2, and 3.

The main features of an affine transformation are best described by saying what happens to a given geometric figure in the (u, v) plane when the figure is mapped into the (U, V) plane. It can be shown that under an affine transformation:

(a) parallel lines go into parallel lines

(b) intersecting lines go into intersecting lines

(c) a square goes into a parallelogram

(d) a circle goes into an ellipse

(e) tangent curves go into tangent curves

(f) any conic goes into a conic of the same type.

The proof of these facts is not essential to your understanding of differential equations and is omitted.

When one geometric figure is obtained from another by an affine transformation, the two figures are said to be *affine equivalent*. The two families of curves in Figure 1 are affine equivalent, so are those in Figure 2, and so are those in Figure 3. The following examples show where these figures came from.

EXAMPLE 1. Let it be required to find the phase-plane orbits of

$$u'' - 5u' + 4u = 0.$$

The characteristic polynomial is

$$s^2 - 5s + 4 = (s-1)(s-4)$$

so that, by the theory of Chapter 6 together with $v = u'$,

$$(2) \quad u = c_1 e^t + c_2 e^{4t}, \quad v = c_1 e^t + 4c_2 e^{4t}.$$

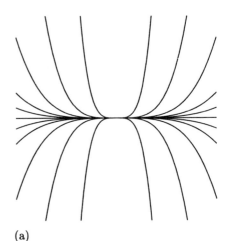

(a)

These are parametric equations of the trajectories. By inspection

$$4u - v = 3c_1 e^t, \quad u - v = -3c_2 e^{4t}.$$

If we set $U = 4u - v$, $V = u - v$ these equations become

$$(3) \qquad U = C_1 e^t, \quad V = C_2 e^{4t}$$

where $C_1 = 3c_1$ and $C_2 = -3c_2$. From this follows

$$|U|^4 = |C_1|^4 e^{4t}, \quad |V| = |C_2| e^{4t}.$$

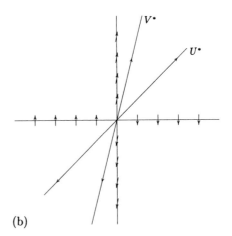

(b)

Assuming $C_1 \neq 0$ we see that $|V|/|U|^4$ is constant and hence

$$(4) \quad |V| = k|U|^4, \qquad k \geq 0 \quad \text{const.}$$

We are interested not just in a single curve (3) but in the whole family of curves obtained for various choices of C_1 and C_2. If (U, V) is a point on one of these curves then each of the points

$$(-U, V), \quad (U, -V), \quad (-U, -V)$$

is also on one of these curves. For example, $(U, -V)$ is on the curve (3) obtained when C_2 is changed to $-C_2$.

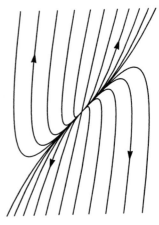

(c)

Figure 1

Thus the family of curves (3), considered as a whole, has the same symmetry properties as the family defined by (4). Either set of curves can be plotted by finding the graphs in the first quadrant and extending to other quadrants by symmetry. In the first quadrant (4) is equivalent to

$$V = kU^4$$

which is plotted with ease. The result after extension by symmetry is shown in Figure 1(a). These are the orbits in the (U, V) plane.

The orbits in the (u, v) plane are obtained by the affine transformation $U = 4u - v$, $V = u - v$ and each of them satisfies the equation

$$|u - v| = k|4u - v|^4, \qquad k = \text{const.}$$

This is not very helpful in sketching the curves, however, and another method is used here.

The U axis in the (U, V) plane is described by the equation $V = 0$, hence its image in the (u, v) plane is the line with equation

$$u - v = 0.$$

This line is labeled U^* in Figure 1(b). In a like manner, the image of the V axis is the line V^* with the equation

$$4u - v = 0.$$

If $v = u'$ the equation $u'' - 5u' + 4u = 0$ yields

$$u' = v, \quad v' = -4u + 5v$$

and this gives the tangent vector (u', v'). In particular, on the two axes in the (u, v) plane the velocity vector is

$$(0, -4u), \qquad v = 0; \qquad (v, 5v), \qquad u = 0$$

respectively. This information is shown by arrows in Figure 1(b). (The arrows ought to get shorter as they approach the origin, but we are interested only in their direction.) On the lines U^* and V^* the velocity vectors are

$$(u, u), \qquad v = u; \qquad (v, 4v), \qquad v = 4u$$

respectively. They have slopes 1 and 4, the same as the slopes of the lines $v = u$ and $v = 4u$ on which they lie. Hence these lines must be orbits of the differential equation. Indeed, we get

$$u = c_1 e^t, \quad v = c_1 e^t; \qquad u = c_2 e^{4t}, \quad v = 4c_2 e^{4t}$$

by setting $c_2 = 0$ or $c_1 = 0$, respectively, in (2). These parametric equations give $v = u$ and $v = 4u$, and they also confirm the fact that the point (u, v) moves away from the origin as t increases.

The orbits in the (u, v) plane can now be sketched with ease. Those in the (U, V) plane are tangent to the U axis at the origin and are almost parallel to the V axis at distant points. Hence those in the (u, v) plane are tangent to U^* at the origin and are almost parallel to V^* at distant points. Taking account of the arrows in Figure 1(b), we get the curves in Figure 1(c).

Before giving more examples we introduce some additional terminology. The origin $(0, 0)$ in phase space is called a *critical point* for the equation (1). Critical points are classified into three main types according to the behavior of the characteristic equation

$$s^2 + ps + q = 0.$$

If both roots are real, unequal, and of the same sign the critical point is a *node*. As an illustration, the critical point in Example 1 is a node because the roots are 2, 5. When the roots are real and of opposite signs the critical point is a *saddle*. When they are complex but not purely imaginary the critical point is a *focus*. A node or focus is *stable* if the point $(u(t), v(t))$ moves to the node or focus as a limit for $t \to \infty$ and it is *unstable* if the point moves arbitrarily far away as $t \to \infty$. It will be seen that a node is stable if, and only if, the roots are negative, and a focus is stable if, and only if, the roots have negative real parts. The critical point in Example 1 is an unstable node.

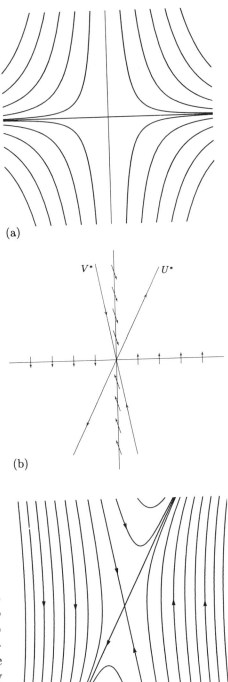

(a)

(b)

(c)   Figure 2

EXAMPLE 2. For the equation $u'' + 3u' - 10u = 0$ the roots of the characteristic equation are $2, -5$. Since these are real and of opposite signs, the critical point is a saddle. The solution is

$$u = c_1 e^{2t} + c_2 e^{-5t}, \quad v = 2c_1 e^{2t} - 5c_2 e^{-5t}.$$

By inspection

$$5u + v = 7c_1 e^{2t}, \quad 2u - v = 7c_2 e^{-5t}.$$

With $U = 5u + v$ and $V = 2u - v$ this gives

$$U = C_1 e^{2t}, \quad V = C_2 e^{-5t}$$

where $C_1 = 7c_1$, $C_2 = 7c_2$. Hence

$$|U|^5 = |C_1|^5 e^{10t}, \quad |V|^2 = |C_2|^2 e^{-10t}.$$

These equations yield $|U|^5 |V|^2 = k$ where k is constant. As in the previous case one can plot the locus in the first quadrant of the (U, V) plane and extend to other quadrants by symmetry. The result is shown in Figure 2(a).

The images of the U and V axes are the lines U^* and V^* with the respective equations

$$2u - v = 0, \quad 5u + v = 0$$

and the tangent vector is $(u', v') = (v, 10u - 3v)$. It is left for you to check the details of Figure 2(b) and to obtain further verification by setting $c_2 = 0$ or $c_1 = 0$ in the formulas with which we began. The final result is shown in Figure 2(c).

EXAMPLE 3. As our last example we consider

$$u'' + 2u' + 17u = 0.$$

By the quadratic formula the characteristic equation has roots $-1 \pm 4i$. Hence the solution can be written in the form

$$u(t) = Ae^{-t} \cos(4t - \phi)$$

where A and ϕ are constant; see Chapter 6, Section 3, Problem 7. The derivative is

$$v = -Ae^{-t} \cos(4t - \phi) - 4Ae^{-t} \sin(4t - \phi).$$

By inspection

$$u + v = -4Ae^{-t}\sin(4t - \phi),$$

$$-4u = -4Ae^{-t}\cos(4t - \phi).$$

If we set $U = u + v$, $V = -4u$
the equations become

$$U = -4Ae^{-t}\sin(4t - \phi),$$

$$V = -4Ae^{-t}\cos(4t - \phi).$$

Squaring and adding yield

$$U^2 + V^2 = k^2 e^{-2t}$$

where $k = -4A$ is constant.

If the factor e^{-2t} were absent the
locus would be a circle of radius $|k|$
centered at the origin. The effect of
the factor e^{-2t} is to diminish the distance to the origin as t increases, so
that the actual curve is a spiral. See
Figure 3(a). In this case it is not important to get the images U^* and V^*
of the U and V axes, but the tangent
vector

$$(u', v') = (v, -17u - 2v)$$

shows whether the spiral is traversed
in a clockwise or counterclockwise direction. On the axes the vector is

$$(0, -17u), \qquad v = 0$$

$$(v, -2v), \qquad u = 0$$

and the direction is clockwise as in Figure 3(c).

The factor e^{-2t} in the U, V equations or e^{-t} in the original equations
shows that the point $(u(t), v(t))$ tends
to the origin as $t \to \infty$. The origin
is a *stable focus*. It is stable because

(a)

(b)

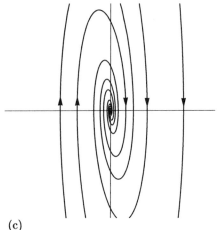

(c)

Figure 3

the point (u, v) tends to $(0, 0)$ and it is a focus because the characteristic equation has complex roots not purely imaginary.

PROBLEMS

1P. We shall find the phase-plane orbits of $u'' - 7u' + 10u = 0$.

(a) Check that the roots of the characteristic equation are 2 and 5 (and hence the critical point is an unstable node.) Set $v = u'$ to get

$$u = c_1 e^{2t} + c_2 e^{5t} \qquad v = 2c_1 e^{2t} + 5c_2 e^{5t}$$

(b) Compute $U = 5u - v$, $V = 2u - v$ and check that $|V|^2 = K|U|^5$.

(c) Sketch the curves $V = kU^{5/2}$ and, by symmetry, the curves of Part (b).

(d) The image U^* of the U axis has the equation $v = 2u$. Similarly get the equation of V^* and sketch both images U^* and V^* in the (u, v) plane. Suggestion: Since the v, V^* axes are nearly parallel, it is advisable to use a full $8\frac{1}{2}$-by-11 inch sheet of paper.

(e) Using the tangent vector $(v, -10u + 7v)$, make appropriate arrows on the u, v, U^*, V^* axes and sketch the curves.

(f) Obtain a partial check by the formulas of Part (a).

2. Proceeding as in Examples 1, 2, and 3 analyze the phase-plane orbits:

(a) $u'' + 3u' + 2u = 0$ (d) $u'' + u' - 2u = 0$ (g) $u'' + 4u' + 13u = 0$

(b) $u'' - u = 0$ (e) $u'' + 2u' + 5u = 0$ (h) $u'' + 9u' + 20u = 0$

(c) $u'' - 2u' + 2u = 0$ (f) $u'' - 9u' + 14u = 0$ (i) $u'' + u' - 20u = 0$

-----------------------------------ANSWERS-----------------------------------

Abbreviations: N,S,F = node, saddle, focus; u,s,c,cc = unstable, stable, clockwise, counterclockwise. A saddle is always regarded as unstable, so that the phrase "unstable saddle" is redundant.

2. Partial answers:

(a) sN, $|2u + v|^2 = k|u + v|$ (f) uN, $|2u - v|^2 = k|7u - v|^7$

(b) S, $|u^2 - v^2| = k$ (g) scF, $(2u + v)^2 + 9u^2 = k^2 e^{-4t}$

(c) ucF, $(u - v)^2 + u^2 = k^2 e^{2t}$ (h) sN, $|5u + v|^5 = k|4u + v|^4$

(d) S, $|u - v||2u + v|^2 = k$ (i) S, $|4u - v|^4|5u + v|^5 = k$

(e) scF, $(u + v)^2 + 4u^2 = k^2 e^{-2t}$

2. The general case

In the previous examples and problems it was repeatedly observed that the orbits in the case of a node all have the same general character. So do those for a saddle, and those for a focus. We now confirm this behavior in the general case

$$u'' + pu' + qu = 0, \qquad v = u'.$$

If the roots λ and μ of the characteristic equation are real and distinct the solution u and its derivative v are

$$u = c_1 e^{\lambda t} + c_2 e^{\mu t}, \quad v = \lambda c_1 e^{\lambda t} + \mu c_2 e^{\mu t}.$$

These equations give

$$\mu u - v = (\mu - \lambda) c_1 e^{\lambda t}, \quad \lambda u - v = (\lambda - \mu) c_2 e^{\mu t}.$$

With $V = \mu u - v$ and $U = \lambda u - v$ we get, as in Section 1,

(1) $$|V|^{\mu} = K|U|^{\lambda}, \qquad K \geq 0 \text{ const.}$$

(i) Node. Let λ and μ be unequal real numbers of the same sign, where without loss of generality $\lambda/\mu > 1$. Equation (1) with $k = K^{\lambda/\mu}$ yields

$$|V| = k|U|^{\lambda/\mu}, \qquad k \geq 0 \text{ const.}$$

The curves in the (U, V) plane resemble the parabolas $|V| = k|U|^2$ and the curves in the (u, v) plane are given by an affine transformation. This is the case of a node. The analysis shows that the orbits always have the general character of those shown in Section 1, Figure 1.

(ii) Saddle. If λ and μ have opposite signs let $\mu > 0$ and $\lambda = -\rho < 0$. The curves now take the form
$$|U|^{\rho} |V|^{\mu} = K$$

and resemble the hyperbolas $|U||V| = K$. This is the case of a saddle as shown in Section 1, Figure 2.

(iii) Focus. For a focus the roots have the form $\sigma \pm i\omega$ where neither σ nor ω is 0. The solution u and its derivative $v = u'$ are now

$$u = A e^{\sigma t} \cos(\omega t - \phi), \quad v = \sigma A e^{\sigma t} \cos(\omega t - \phi) - \omega A e^{\sigma t} \sin(\omega t - \phi)$$

where A and ϕ are constant. These equations give

$$\sigma u - v = \omega A e^{\sigma t} \sin(\omega t - \phi), \quad \omega u = \omega A e^{\sigma t} \cos(\omega t - \phi).$$

The variables $U = \sigma u - v$, $V = \omega u$ satisfy

$$U^2 + V^2 = k^2 e^{2\sigma t}$$

where $k = |\omega A|$. This would represent a circle if $\sigma = 0$; since $\sigma \neq 0$, it represents a spiral. When t increases by $2\pi/\omega$ the point (U, V) returns to the same radial line from which it started, but the exponential factor multiplies its distance to the origin by $e^{2\pi\sigma/\omega}$. The same factor intervenes when we go once around the spiral in the (u, v) plane. The critical point is a focus. The analysis shows that the orbits always have the general character of those shown in Section 1, Figure 3.

In the sequel, the following result is used to reduce more general cases to those considered here:

THEOREM 1. *If two affine transformations are carried out one after the other, the result is again an affine transformation.*

We give the proof in matrix notation, though familiarity with matrices is not assumed elsewhere in this chapter. The matrix form of an affine transformation is

$$\begin{pmatrix} X \\ Y \end{pmatrix} = \begin{pmatrix} A & B \\ C & D \end{pmatrix} \begin{pmatrix} x \\ y \end{pmatrix}, \qquad \begin{vmatrix} A & B \\ C & D \end{vmatrix} \neq 0.$$

If two such transformations are carried out one after the other, the matrix of the resulting transformation is the product of the individual matrices and its determinant is the product of the individual determinants. This shows that the determinant is nonzero and Theorem 1 follows.

PROBLEMS

1P. With $D = d/dt$ the system $x' = ax + by$, $y' = cx + dy$ can be written

$$(D - a)x = by, \qquad (D - d)y = cx$$

(a) Operate on the first equation with $D - d$ and on the second with $D - a$ to get

$$x'' - (a + d)x' + (ad - bc)x = 0, \qquad y'' - (a + d)y' + (ad - bc)y = 0$$

(b) Show that the original system has a nonzero solution satisfying $y = mx$, with m constant, if, and only if, $c + dm = m(a + bm)$.

(c) The characteristic polynomial for the second-order equations in (a) and the quadratic for m in (b) are respectively

$$s^2 - (a+d)s + (ad - bc) = 0, \qquad bm^2 + (a-d)m - c = 0$$

Check that these have the same discriminant.

(d) If m satisfies the second equation of Part (c) show that $a + bm$ satisfies the first.

2P. In the case of a focus let us suppose that you have made a rough sketch of a spiral. According to the text a full turn of the spiral changes the distance to the origin by a factor $e^{2\pi\sigma/\omega}$. If $\lambda = 1 + 8i$ check that this factor is about 2.2, so your rough sketch is probably about right. But if $\lambda = 8 + i$ the factor is

$$e^{16\pi} \doteq 6,761,000,000,000,000,000,000$$

and the trajectory goes off the paper so fast that your rough sketch is probably quite wrong. This matter will not be emphasized but you should be aware of it.

3. If the complex number z is written in polar form and ϕ is real the transformation $Z = e^{i\theta}z$ becomes

$$Z = e^{i\theta}re^{i\phi} = re^{i(\theta+\phi)}$$

This represents a *rotation* about the origin through the angle ϕ. Writing

$$X + iY = (\cos\phi + i\sin\phi)(x + iy)$$

and separating into real and imaginary parts, obtain the real form of the rotation in the (x, y) plane:

$$X = x\cos\phi - y\sin\phi, \qquad Y = x\sin\phi + y\cos\phi$$

4. Let S denote the square in the (x, y) plane with vertices at $(0,0)$, $(0,1)$, $(1,0)$, $(1,1)$. Sketch the image of S in the (X, Y) plane under the affine transformations

$$\text{(a)} \begin{cases} X = 2x \\ Y = 3y \end{cases} \qquad \text{(b)} \begin{cases} X = -x \\ Y = y \end{cases} \qquad \text{(c)} \begin{cases} \sqrt{2}X = x - y \\ \sqrt{2}Y = x + y \end{cases}$$

The first of these represents a change of the x and y scales, the second a reflection, and the third a rotation.

5. The triangle (P_1, P_2, P_3) formed by three noncollinear points in the (x, y) plane is *positively oriented* if the interior of the triangle lies on your left when you traverse the boundary from P_1 to P_2 to P_3 to P_1, in that order. If the interior lies on your right, the triangle (P_1, P_2, P_3) is *negatively oriented*. See Figure 4.

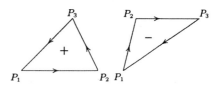

Figure 4

(a) One of the following triangles (P_1, P_2, P_3) is positively oriented and one is negatively oriented. Plot the points and decide which is which:

$$(2, -1), \ (3, 4), \ (-1, 2); \qquad (5, 2), \ (0, 1), \ (-1, 6)$$

(b) It can be shown that an affine transformation preserves orientation if its determinant is positive, and reverses orientation if its determinant is negative. Check this statement by finding the image of each triangle in Part (a) under each of the following affine transformations. No answer is given:

$$X = 2x + 3y, \ Y = x - y; \qquad X = -3x + y, \ Y = x - 2y$$

6. For the system $x' = x$, $y' = -y$ check that the equation $xy = 1$ represents a single hyperbola and two orbits, while $|x||y| = 1$ represents two hyperbolas and four orbits. The fact that a single Cartesian equation may represent the union of two or more orbits causes no practical difficulty, but you should be aware of it.

7. The purpose of this problem is to remind you that parametric equations $x = \xi(t)$, $y = \eta(t)$, and a seemingly equivalent Cartesian equation $\theta(x, y) = 0$, need not describe the same locus. When there is a discrepancy, the parametric equations take precedence.

(a) Given $x' = x$, $y' = y$; $x \, dy = y \, dx$ check that the respective solutions are

$$x = c_1 e^t, \quad y = c_2 e^t; \qquad y = cx \quad \text{or} \quad x = cy$$

The t-system has $x = y = 0$ as sole solution through the origin while the (x, y)-equation has infinitely many.

(b) Given $x' = -y$, $y' = x$; $x \, dx + y \, dy = 0$ check that the respective solutions are

$$x = c \cos(t - t_0), \quad y = c \sin(t - t_0); \qquad x^2 + y^2 = c^2$$

Again the t-system has $x = y = 0$ as sole solution through the origin but the (x, y)-equation has none. Further discussion is given in Chapter 2, Section 2.

3. A two-dimensional linear system

The following *equivalence theorem* reduces a general linear system to the simpler cases we have already considered:

THEOREM 1. *Unless $b = c = 0$ and $a = d$, the phase-plane orbits of*

$$(1) \qquad\qquad x' = ax + by, \quad y' = cx + dy$$

are affinely equivalent to those of the second-order equation

$$(2) \qquad\qquad u'' + pu' + qu = 0$$

where $p = -(a + d)$ and $q = ad - bc$. Futhermore both x and y satisfy (2).

When the theorem applies, a second affine transformation takes us from (u, u') in (2) to the functions (U, V) of Section 2. Hence the (x, y) trajectories are affinely equivalent to the (U, V) trajectories; see Section 2, Theorem 1.

Readers familiar with 2-by-2 matrices will recognize that (1) can be written in matrix form as

$$\begin{pmatrix} x' \\ y' \end{pmatrix} = \begin{pmatrix} a & b \\ c & d \end{pmatrix} \begin{pmatrix} x \\ y \end{pmatrix}.$$

The characteristic polynomial of the matrix is

$$\begin{vmatrix} a - s & b \\ c & d - s \end{vmatrix} = s^2 - (a + d)s + (ad - bc)$$

which is the same as the characteristic polynomial for (2). In the excluded case $b = c = 0$, $a = d$ the matrix is a scalar multiple of the identity matrix. It is seen in Section 4 that this case is trivial.

We now give the proof. That x and y satisfy (2) was shown in Section 2, Problem 1. To prove the affine equivalence suppose first that $b \neq 0$. Since the phase-plane orbits of (2) are given by

$$u' = v, \quad v' = -qu - pv$$

we try the affine transformation $u = x$, $v = ax + by$. The first equation (1) is then $u' = v$, as desired. Since $u = x$ satisfies (2), as noted above, the equation $v' = u'' = -qu - pv$ also holds.

If $c \neq 0$ the affine transformation $u = y, v = cx + dy$ reduces the second equation (1) to $u' = v$, and since $u = y$ satisfies (2), the result follows again. In the remaining case (1) is $x' = ax, y' = dy$ with $a \neq d$. The affine transformation

$$u = x + y, \quad v = ax + dy$$

yields $u' = v$. Since x and y both satisfy (2) the same is true of $u = x + y$ by linearity. This completes the proof.

By definition, the critical point in (1) is a node, saddle, or focus when the critical point in (2) is a node, saddle, or focus, respectively. This equivalence leads to a simple decision procedure as explained next.

(a) Classification of critical points

If λ and μ are the roots of the characteristic equation associated with (2) the identity

$$(s - \lambda)(s - \mu) = s^2 + ps + q$$

shows that

$$\lambda + \mu = -p, \quad \lambda\mu = q.$$

Furthermore, the discriminant of the quadratic is $p^2 - 4q$. Using these facts, we shall justify the classification of critical points shown in Figure 5.

The roots are complex if $4q > p^2$, and real and unequal if $4q < p^2$. This leads to the classification as "focus" above the parabola of the figure and "node or saddle" below. The positive q axis is excluded in the case of a focus because if $p = 0$ the roots are purely imaginary.

Since $\lambda\mu = q$ the two real roots λ and μ have the same sign if $q > 0$, and they have opposite signs if $q < 0$. This leads to the classification as "node" above the p axis and "saddle" below.

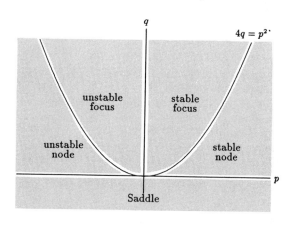

Figure 5

In the case of complex roots the quadratic formula shows that the real part of each root is $-p/2$. Hence the roots both have negative real parts if $p > 0$, positive real parts if $p < 0$. The formula $\lambda + \mu = -p$ gives a similar result for real roots, both of which have the same sign. Namely, the roots are negative if $p > 0$, positive if $p < 0$. These remarks yield the classification "stable" for $p > 0$, "unstable" for $p < 0$, as shown in Figure 5. Our justification of Figure 5 is now complete.

EXAMPLE 1. Classify the critical point for each system:

$$
\begin{aligned}
x' &= 2x + 4y \\
y' &= x + 3y
\end{aligned}
\qquad\qquad
\begin{aligned}
x' &= 2x + 3y \\
y' &= x - 4y
\end{aligned}
\qquad\qquad
\begin{aligned}
x' &= -2x - 3y \\
y' &= x - 4y
\end{aligned}
$$

It is convenient to abbreviate each differential equation by giving the matrix of coefficients, thus:

$$\begin{pmatrix} 2 & 4 \\ 1 & 3 \end{pmatrix} \qquad \begin{pmatrix} 2 & 3 \\ 1 & -4 \end{pmatrix} \qquad \begin{pmatrix} -2 & -3 \\ 1 & -4 \end{pmatrix}$$

In the first case $p = -5$ and the determinant is $q = 2$. Since

$$4q < p^2, \quad p < 0, \quad q > 0$$

Figure 5 shows that the origin is an unstable node. In the second case $p = 2$ and $q = -11$. Since $q < 0$ the critical point is a saddle. In the third case $4q > p^2$, $p > 0$ and the critical point is a stable focus.

EXAMPLE 2.   The classification can be supplemented by plotting the velocity vectors on the x and y axes, namely,

$$(ax, cx) \quad \text{at} \quad (x, 0), \qquad (by, dy) \quad \text{at} \quad (0, y).$$

For example in the third matrix of Example 1 these vectors are $(-2x, x)$ and $(-3y, -4y)$ and the direction of the spiral is counterclockwise. The same method shows that the spiral at a focus is always counterclockwise when c is positive, clockwise when c is negative.

Despite a word of caution in Section 2, Problem 2, the trajectories at a focus can usually be sketched with tolerable accuracy by use of Figure 5 and Example 2. The next example explains what to do at a node or saddle.

EXAMPLE 3.   For the two systems

$$\begin{aligned} x' &= 6x + y \\ y' &= 4x + 3y \end{aligned} \qquad\qquad\qquad \begin{aligned} x' &= -3x + 2y \\ y' &= -3x + 4y \end{aligned}$$

it is left for you to check that the origin is an unstable node and a saddle, respectively, and that the roots (λ, μ) are $(2, 7)$ in the first case, $(-2, 3)$ in the second. We write x corresponding to each root and we get y from $y = x' - 6x$ in the first system, from $2y = x' + 3x$ in the second. The respective results are

$$\begin{aligned} x &= ke^{2t}, & y &= -4x \\ x &= ke^{7t}, & y &= x \end{aligned} \qquad\qquad \begin{aligned} x &= ke^{-2t}, & 2y &= x \\ x &= ke^{3t}, & y &= 3x \end{aligned}$$

where k is constant. These are parametric equations for the straight-line orbits.

In the first case $|x|$ and $|y|$ increase with t, since the exponents are positive, and the lines are directed away from the origin. As $t \to -\infty$ the term e^{2t} is much larger than e^{7t}. Hence the orbits are tangent at the origin to

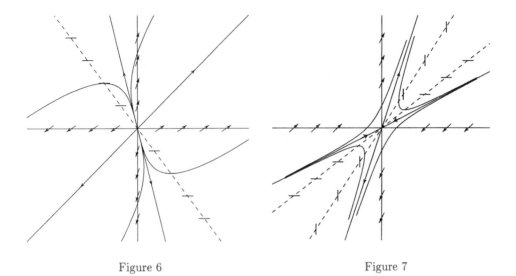

Figure 6   Figure 7

the line $y = -4x$ associated with $\lambda = 2$. As $t \to \infty$ the term e^{7t} dominates and the orbits behave like $y = x$ at distant points. See Figure 6, which contains additional information not considered here.

In the second case the line $2y = x$ is associated with the negative exponent -2 and is directed toward the origin, while the other line is directed away. The trajectories are easily sketched and are shown, with additional information, in Figure 7.

The general character of the trajectories was already given by Theorem 1 and Section 2, and that is the reason why so little calculation was required. For independent confirmation (which is not really needed) one can use a general solution (x, y). In the first system, for example, x and $y = x' - 6x$ yield

$$(3) \qquad \frac{y}{x} = \frac{-4c_1 e^{2t} + c_2 e^{7t}}{c_1 e^{2t} + c_2 e^{7t}} = \frac{-4c_1 + c_2 e^{5t}}{c_1 + c_2 e^{5t}}.$$

If $c_1 c_2 \neq 0$, so that we are not on a straight-line trajectory, it is clear that $x \to 0$, $y \to 0$, $y/x \to -4$ as $t \to -\infty$. This gives an independent proof that the trajectories come in to the origin tangent to the line $y = -4x$. Equation (3) for $t \to \infty$ also confirms the behavior at distant points.

(b) Isoclines and fixed points

Dividing the second equation (1) by the first gives

$$(4) \qquad \frac{dy}{dx} = \frac{cx + dy}{ax + by} = \frac{c + d(y/x)}{a + b(y/x)}.$$

The first equality (4) is valid if $ax + by \neq 0$ and the second if also $x \neq 0$. The expression on the right shows that dy/dx is constant when y/x is constant.

In other words, if m is constant the orbits have the same slope at all points of the line $y = mx$. The line $y = mx$ is an example of an isocline. In general, an *isocline* is a locus on which all solutions of a given differential equation have the same slope. The prefix "iso" means "same".

Among the most helpful isoclines are the x and y axes and the two lines

$$ax + by = 0, \quad cx + dy = 0$$

on which the tangent to the orbit is vertical or horizontal, respectively. Some of these are shown as dotted lines in Figures 6, 7, and 8.

As a rule, an isocline is *not a solution* of the differential equation. However, if the slope m of the line $y = mx$ agrees with dy/dx in (4) then the isocline is a solution. By inspection of (4) the condition for this to happen is

(5)
$$m = \frac{c + dm}{a + bm}.$$

With an obvious choice of F the equation has the form $m = F(m)$. The number m is not changed by F and is called a *fixed point*.

In Section 2, Problem 1 it was seen that the quadratic for m given by (5) and the characteristic equation for λ, μ have the same discriminant. Hence if either equation has complex roots so does the other. If either equation has real and distinct roots so does the other. This confirms the fact that there are two straight-line orbits $y = mx$ in the case of a node or saddle and none in the case of a focus.

The line $x = 0$ is an orbit when $b = 0$ and corresponds to $m = \infty$. This case is taken up in Problem 6 and is excluded from the following theorem:

THEOREM 2. *In the case of a node or saddle with $b \neq 0$:*

(a) *The slopes m of the two orbits $y = mx$ are given by (5).*

(b) *The two values of $a + bm$ satisfy the characteristic equation.*

(c) *The trajectory with orbit $y = mx$ is oriented toward the origin or away from the origin if $a + bm < 0$ or $a + bm > 0$, respectively.*

(d) *For a node, the line of tangency at the origin is the line $y = mx$ for which $|a + bm|$ is smaller.*

Part (a) was established in the above discussion, and both (a) and (b) in Section 2, Problem 1. The substitution $y = mx$ into (1) yields

$$x' = x(a + bm), \quad \text{hence} \quad x = ke^{(a+bm)t}$$

where k is constant. This gives (c) and reconfirms (b). Part (d) is obtained from an equation like (3) with general exponents (λ, μ); the details are easy and are omitted. For examples, see Problems 2 and 3.

PROBLEMS

1. Each matrix stands for a system of differential equations. Thus (a) pertains to the system $x' = -5x + 3y$, $y' = -4x + 2y$. Classify the critical point:

(a) $\begin{pmatrix} -5 & 3 \\ -4 & 2 \end{pmatrix}$　(d) $\begin{pmatrix} 3 & 4 \\ 2 & 1 \end{pmatrix}$　(g) $\begin{pmatrix} -6 & 1 \\ 5 & -2 \end{pmatrix}$　(j) $\begin{pmatrix} -3 & 1 \\ -5 & -7 \end{pmatrix}$

(b) $\begin{pmatrix} 2 & -1 \\ -4 & -1 \end{pmatrix}$　(e) $\begin{pmatrix} 6 & 2 \\ 3 & 7 \end{pmatrix}$　(h) $\begin{pmatrix} 2 & 1 \\ -8 & 6 \end{pmatrix}$　(k) $\begin{pmatrix} -5 & 7 \\ 3 & -1 \end{pmatrix}$

(c) $\begin{pmatrix} 1 & 4 \\ 8 & 5 \end{pmatrix}$　(f) $\begin{pmatrix} 1 & 3 \\ -2 & -4 \end{pmatrix}$　(i) $\begin{pmatrix} 7 & -1 \\ 3 & 3 \end{pmatrix}$　(l) $\begin{pmatrix} -1 & 4 \\ -10 & -5 \end{pmatrix}$

2P. For $x' = 6x + y$, $y' = 4x + 3y$ verify that the origin is an unstable node and draw velocity vectors on the axes as in Example 2. Then:

(a) Check that the fixed-point relation on the left below yields the equations on the right, and plot the corresponding orbits $y = mx$:

$$m = \frac{4 + 3m}{6 + m}, \qquad m^2 + 3m - 4 = 0, \qquad (m-1)(m+4) = 0$$

(b) Obviously $y = mx$ yields $x' = (6 + m)x$. Thus get

$$x = ce^{7t}, \qquad m = 1; \qquad x = ce^{2t}, \qquad m = -4$$

This shows that the trajectories are oriented away from the origin. Since $|2|$ is smaller than $|7|$, Theorem 2 says that the other orbits are tangent to the line $y = -4x$ at the origin and they behave like $y = x$ at distant points. Sketch the trajectories.

(c) Verify that the exponents 7 and 2 obtained in Part (b) satisfy

$$\begin{vmatrix} 6 - s & 1 \\ 4 & 3 - s \end{vmatrix} = 0$$

3P. For $x' = -3x + 2y$, $y' = -3x + 4y$, verify that the origin is a saddle and draw velocity vectors on the axes as in Example 2. Then:

(a) From the fixed-point equation, obtain the orbits $2y = x$ and $y = 3x$.

(b) Obviously $y = mx$ yields $x' = (-3 + 2m)x$. Using this, orient the trajectories corresponding to $y = mx$ in (a) and sketch some other trajectories.

(c) Check that the exponents $-2, 3$ obtained in Part (b) satisfy the characteristic equation.

4P. In this problem

$$x' = x - 4y, \quad y' = 4x + y.$$

Check that the origin is an unstable counterclockwise focus and verify the additional features shown in Figure 8.

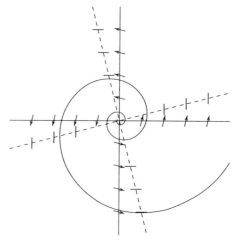

5. (abcdefghijkl) Sketch the trajectories in Problem 1. That is, sketch the orbits, and provide them with arrows showing orientation. For a node or saddle, you will find it instructive to do some as in Example 3, others by use of Theorem 2.

Figure 8

6. For the system $x' = ax$, $y' = cx + dy$ with $a \neq d$ check that

$$x = c_1 e^{at}, \qquad cx + (d - a)y = c_2 e^{dt}$$

Hence $U = x$, $V = cx + (d - a)y$ satisfy $|U|^d = |V|^a k$ and the trajectories can be plotted as in Section 1.

7. Eigenvectors. Problems 7, 8, and 9 depend on Chapter 16, Section 1. We set $X = (x, y)^T$ and we denote eigenvectors for λ, μ by E_1, E_2.

(a) If λ, μ are real and distinct, the lines $y = mx$ of Theorem 2 are obtained by the choices $c_1 = 0$ or $c_2 = 0$ in

$$X = c_1 E_1 e^{\lambda t} + c_2 E_2 e^{\mu t}$$

With this as a clue, deduce Theorem 2.

(b) By means of canonical variables show that the (x, y) orbits in the case of a node or saddle are affinely equivalent to the (U, V) orbits for those cases in Section 2.

(c) Obtain the trajectories in Example 3 by eigenvectors.

8. Using complex eigenvectors, show that the orbits for a focus are affinely equivalent to the (U, V) orbits obtained for that case in Section 2.

9. (abcdefghijkl) Use eigenvectors to sketch the trajectories in Problem 1.

---ANSWERS---

1. (afg) sN (bcdk) S (ei) uN (h) ucF (jl) scF
2. See Figure 6.   3. See Figure 7.
5. Partial answer: The equations are the same as those in Chapter 16, Section 1, Problem 1 and answers are given there. Here is a clue to their interpretation:
   (a) $x = c_1 e^{-2t} + 3c_2 e^{-t}$, $y = c_1 e^{-2t} + 4c_2 e^{-t}$

4. Degenerate cases

If (p, q) is in any one of the five regions of Figure 5 the point (p_1, q_1) is in the same region provided its distance to (p, q) is sufficiently small. This means that the classification of the critical point as node, saddle, or focus is unchanged by sufficiently small perturbations of the coefficients.

We now consider limiting or degenerate cases that do not exhibit this insensitivity. On the contrary, the classification depends on certain exact equalities, and can change if a single coefficient is altered by an arbitrarily small amount. Since the coefficients are seldom known with infinite precision, the classification as node, saddle, or focus is more fundamental than the cases considered here.

(i) Center. If $4q > p^2$, as for a focus, but $p = 0$, the roots are purely imaginary and the critical point is called a *center*. Instead of being a spiral the orbits are ellipses centered at the origin as shown in Figure 9. To see why, note that the equivalent second-order equation is in this case

$$u'' + qu = 0, \quad q > 0.$$

Setting $q = \omega^2$ and recalling that $v = u'$ we get

$$u = A\cos(\omega t - \phi), \quad v = -\omega A \sin(\omega t - \phi)$$

where A and ϕ are constant. If $A \neq 0$ then

$$\frac{u^2}{A^2} + \frac{v^2}{(\omega A)^2} = 1$$

Figure 9

and this is the equation of an ellipse. A different proof was given in Chapter 7 and still another is given in DETA, page 473. The orbits are plotted by the same method as for a focus.

(ii) Degenerate node. Here we assume $4q = p^2$, $p \neq 0$ and we also assume that we do not have the case $b = c = 0$, $a = d$ that was excluded in the equivalence theorem. The roots of the characteristic equation are both equal to $-p/2$. Since they have the same sign one would expect the orbits to resemble those for a node, and they do. The only significant difference is that the fixed-point equation $m = F(m)$ has two equal roots, and the two straight-line solutions are now coincident. Thus, the orbits are tangent to a line at the origin and almost parallel to the same line at distant points. The proof requires only minor modification of the proof of Section 3, Theorem 2(d). The critical point in this case is a *degenerate node*. It is stable if $p > 0$, unstable if $p < 0$. See Figure 10.

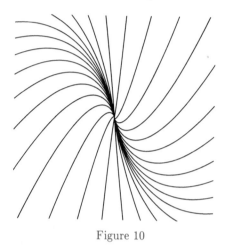

Figure 10

(iii) Singular node. We continue to assume $4q = p^2$ but we consider the special case $b = c = 0$, $a = d \neq 0$ that was excluded above. The linear system reduces to $x' = ax$, $y' = ay$, which can be solved by the methods of Chapter 1. The solution

$$x = c_1 e^{at}, \quad y = c_2 e^{at}$$

shows that y/x is constant and hence the orbits are radial lines through the origin; see Figure 11. In this case the origin is a *singular node*. It is stable if $a < 0$, unstable if $a > 0$. Plotting the orbits is trivial.

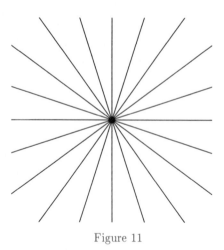

Figure 11

(iv) Vanishing determinant. Next we consider the case $q = 0$, which means that the determinant of the system

$$ax_0 + by_0 = 0, \quad cx_0 + dy_0 = 0$$

is 0. Hence there is a nontrivial solution (x_0, y_0). Since (kx_0, ky_0) is also a solution for any constant k, there is a whole line of critical points. Thus it makes no sense to talk about "the critical point" and analyze its character. It is shown in DETA, page 473, that the trajectories are parallel straight lines unless $a = b = c = d = 0$. In the latter case every point is a critical point and every solution is constant.

PROBLEMS

1. In the following examples the equation of the orbits can be found with ease. Find it, and verify that the orbits have the general character that one would expect from the classification given in the text. Some are degenerate:

(a) $x' = y, y' = 0$ (c) $x' = y, y' = -x$ (e) $x' = x, y' = 2y$

(b) $x' = x, y' = y$ (d) $x' = -x, y' = y$ (f) $x' = x, y' = x + y$

2. For $x' = x + y$, $y' = -x + y$ introduce polar coordinates (r, θ) and show that $r' = r$, $\theta' = -1$, hence $r = ke^{-\theta}$. Check by classifying the critical point.

3. Here are the coefficient matrices for some degenerate cases. Classify the critical point and sketch the orbits:

(a) $\begin{pmatrix} 1 & 2 \\ -3 & -1 \end{pmatrix}$ (d) $\begin{pmatrix} 1 & -1 \\ 2 & -1 \end{pmatrix}$ (g) $\begin{pmatrix} -3 & 2 \\ -2 & 1 \end{pmatrix}$ (j) $\begin{pmatrix} -3 & 0 \\ 0 & -3 \end{pmatrix}$

(b) $\begin{pmatrix} 1 & 1 \\ -1 & 3 \end{pmatrix}$ (e) $\begin{pmatrix} 1 & 2 \\ 2 & 4 \end{pmatrix}$ (h) $\begin{pmatrix} -1 & 1 \\ -2 & 2 \end{pmatrix}$ (k) $\begin{pmatrix} 5 & 1 \\ -1 & 3 \end{pmatrix}$

(c) $\begin{pmatrix} 2 & 0 \\ 0 & 2 \end{pmatrix}$ (f) $\begin{pmatrix} -2 & 3 \\ -3 & 2 \end{pmatrix}$ (i) $\begin{pmatrix} -5 & -4 \\ 7 & 5 \end{pmatrix}$ (l) $\begin{pmatrix} -5 & -2 \\ 2 & -1 \end{pmatrix}$

4. Let $u'' + 2bu' + a^2 u = 0$ where $a > 0$ and $b \geq 0$. In Chapter 7 we considered four cases: undamped, underdamped, critically damped, and overdamped. Show that these correspond respectively to a center, stable focus, stable degenerate node, and stable node in the phase plane.

---ANSWERS---

Further abbreviations: C= center, d= degenerate, ŝ= singular.

1. Partial answer: (a) $y = k$ (c) $x^2 + y^2 = k$ (e) $y = kx^2$

                 (b) $y = kx$ (d) $xy = k$ (f) $y = x \ln kx$

2. uF. In this special case the orbits are logarithmic spirals.

3. Partial answer: (ad) cC   (bk) udN   (c) uŝN   (eh) parallel lines

                 (f)   ccC   (gl) sdN   (i) ccC   (j)   sŝN

Chapter 20
A SURVEY OF NONLINEAR PROBLEMS

M ANY NONLINEAR systems can be approximated locally by linear systems to which the methods of the previous chapter apply. Three examples from mechanics and electrical engineering illustrate the important phenomenon of bifurcation, in which a minute change in a parameter can drastically alter the observed behavior. Further examples give insight into the behavior of interacting populations. After deriving criteria for coexistence and competitive exclusion in competitive systems, we turn to the predator-prey model of Lotka and Volterra. This model predicts a periodic behavior that arises, not from seasonal variation, but from the interaction. It also predicts, contrary to intuition, that the use of pesticides often increases the target population. We conclude with a brief discussion of self-limiting terms, which give another example of bifurcation.

Sections 4.3 and 5.1 are desirable and Chapter 19 is essential.

1. An autonomous system

Throughout this chapter, the variables and functions are real and primes denote differentiation with respect to t. In a system of the form

$$(1) \qquad x' = f(x,y), \qquad y' = g(x,y),$$

the right-hand side does not involve t explicitly, but only implicitly through the fact that x and y themselves depend on t. Such a system is called *autonomous*. The word "autonomous" means "self-governing" and implies that the slope of a solution-curve at (x,y) is determined solely by (x,y). Indeed, if one equation (1) is divided by the other the result is

$$(2) \qquad \frac{dy}{dx} = \frac{g(x,y)}{f(x,y)} \quad \text{or} \quad \frac{dx}{dy} = \frac{f(x,y)}{g(x,y)}$$

according as $f(x,y) \neq 0$ or $g(x,y) \neq 0$. The left-hand side is the slope of the solution-curve $y = \phi(x)$ or $x = \psi(y)$, as the case may be, and the right-hand side depends only on (x,y). Both formulations (2) are included in

$$(3) \qquad f(x,y)dy = g(x,y)dx.$$

If $x = \xi(t)$, $y = \eta(t)$ is a trajectory of (1), Equation (3) is a differential equation for the orbit or trace. The chain rule shows that

$$\bar{x} = \xi(t - t_0), \quad \bar{y} = \eta(t - t_0)$$

satisfies (1) on the corresponding interval for $\bar{t} = t - t_0$. In other words, the behavior of an orbit through a point is independent of the time t_0 at which that point is reached.

When $f(x,y)$ and $g(x,y)$ are defined and not both zero, at least one equation (2) is a consequence of (1) and also of (3). But if

$$f(x_0,\, y_0) = g(x_0,\, y_0) = 0$$

neither equation (2) is meaningful at (x_0, y_0), and (x_0, y_0) is called a *critical point*. Since the constant functions defined by $x(t) = x_0$, $y(t) = y_0$ satisfy (1) in this case, the critical point is also called a *stationary point*.

(a) General properties

Partial derivatives are denoted by subscripts as in Section 4.2 and the class of functions with continuous partial derivatives of order k is denoted by $C^{(k)}$. For example

$$f \in C^{(0)}, \quad g \in C^{(1)}, \quad (f,g) \in C^{(2)}$$

mean respectively that f is continuous, that g has continuous first partial derivatives, and that both f and g have continuous partial derivatives of order 2. If no region is specified, assume that the conditions hold in the whole (x, y) plane.

Under the hypothesis $(f,g) \in C^{(1)}$, the following hold:

(i) Equation (1) has one, and only one, solution through any given point.

(ii) If a solution is bounded for $t > 0$ as long as it exists, then it exists for all positive t.

(iii) If $\lim_{t \to \infty} (x, y) = (x_1, y_1)$ exists, then (x_1, y_1) is a critical point.

(iv) A critical point cannot lie on any trajectory $x = \xi(t)$, $y = \eta(t)$ with ξ or η nonconstant.

(v) In a region free of critical points, two orbits cannot cross nor can an orbit cross itself.

(vi) A periodic solution corresponds to a simple closed orbit.

For proof of (i)–(v) see DETA, pages 481–482. Here we discuss (vi), which requires explanation as well as proof. Although an orbit cannot cross itself, it can return to its starting point and start over. That is, we can have

(4) $$\xi(T) = \xi(0), \qquad \eta(T) = \eta(0), \qquad T > 0.$$

(7) $\xi(T) =$ $T > 0.$

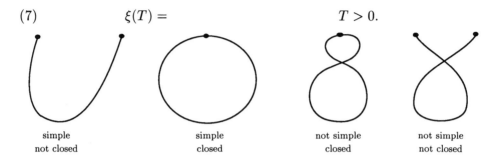

| simple | simple | not simple | not simple |
| not closed | closed | closed | not closed |

Figure 1

A nonconstant orbit with this property is said to be a *closed curve*, or, if it does not intersect itself, a *simple closed curve*. See Figure 1. When (4) holds, uniqueness yields

(5) $$\xi(T + t) = \xi(t), \quad \eta(T + t) = \eta(t).$$

This is true because the trajectories defined by

$$x = \xi(T + t), \quad y = \eta(T + t); \qquad x = \xi(t), \quad y = \eta(t)$$

both satisfy (1) and both go through the same point when $t = 0$. Equation (5) says that the functions ξ and η have period T. Conversely, setting $t = 0$ in (5) yields (4).

(b) Linear approximation

   An important special case of (1) is the linear system

(6) $$u' = au + bv, \quad v' = cu + dv \qquad ad \neq bc$$

where a and b are constant. In Chapter 19 it was seen that the trajectories of (6) fall into six classes as illustrated in Figure 2. The classification is determined by the signs of the three quantities

$$p = -(a + d), \quad q = ad - bc, \quad 4q - p^2.$$

Table 1 pertains to cases in which the behavior of the linear system is insensitive to small changes in the coefficients a, b, c, d. Cases that are sensitive are presented in Table 2. By Chapter 19, a focus or center is counterclockwise if $c > 0$, clockwise if $c < 0$. The degenerate node in Table 2 is obtained when $a \neq d$, the singular node when $a = d$:

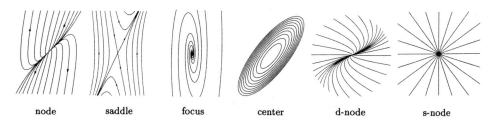

node   saddle   focus   center   d-node   s-node

Figure 2

Table 1

	stable		unstable		saddle	
p	+	+	−	−	±	0
q	+	+	+	+	−	−
$4q - p^2$	+	−	+	−	−	−
	focus	node	focus	node	saddle	saddle

Table 2

	stable	unstable	ambiguous
p	+	−	0
$4q - p^2$	0	0	+
	degenerate or singular node		center or focus

The following theorem may serve to explain why the tables can be applied to nonlinear systems:

THEOREM 1. *Let (x_0, y_0) be a critical point of (1) and let $(f, g) \in C^{(1)}$. Then (1) can be written in the form*

$$\frac{d}{dt} \begin{pmatrix} x - x_0 \\ y - y_0 \end{pmatrix} = \begin{pmatrix} A & B \\ C & D \end{pmatrix} \begin{pmatrix} x - x_0 \\ y - y_0 \end{pmatrix}$$

where A, B, C, D are functions of (x, y) satisfying

$$\lim_{(x,y) \to (x_0, y_0)} \begin{pmatrix} A & B \\ C & D \end{pmatrix} = \begin{pmatrix} f_x(x_0, y_0) & f_y(x_0, y_0) \\ g_x(x_0, y_0) & g_y(x_0, y_0) \end{pmatrix}.$$

A critical point is said to be *isolated* if there is some disk that contains this critical point and no other. Theorem 1 leads to:

THEOREM 2. *Let* (x_0, y_0) *be a critical point of the system*

$$x' = f(x, y), \quad y' = g(x, y),$$

let $(f, g) \in C^{(2)}$ *in a neighborhood of* (x_0, y_0), *and let*

$$\begin{pmatrix} a & b \\ c & d \end{pmatrix} = \begin{pmatrix} f_x(x_0, y_0) & f_y(x_0, y_0) \\ g_x(x_0, y_0) & g_y(x_0, y_0) \end{pmatrix}, \qquad ad \neq bc.$$

Then the critical point is isolated and its general character is given by Tables 1 and 2 with these values of a, b, c, d.

Under sufficiently strong continuity conditions, Theorem 1 can be deduced from the Taylor expansion

$$f(x, y) = f_x(x_0, y_0)(x - x_0) + f_y(x_0, y_0)(y - y_0) + \cdots$$

and from a similar expansion for g. Proof under the hypothesis $(f, g) \in C^{(1)}$ is in DETA, page 483.

The truth of Theorem 2 is suggested by the change of variable

$$u = x - x_0, \quad v = y - y_0$$

in Theorem 1. But for a complete analysis, and even for a mathematically precise statement, one must consult advanced treatises on differential equations.

The following comments give useful insight. In both theorems, the continuity hypothesis is needed in some small disk

$$(x - x_0)^2 + (y - y_0)^2 < \delta$$

and the conclusion pertains to the behavior in such a disk. It says nothing about what happens at points far away from (x_0, y_0).

The error introduced in replacing a nonlinear system by its linear approximation is similar to that introduced by a small change in the coefficients of the linear system. Table 2 is sensitive to such changes, and that is the reason why some of its entries are ambiguous. For example, if $p = 0$ and $q > 0$ a linear system always has a center. However a nonlinear system could have a center or focus.

The concept of *center* in nonlinear problems requires care as to its interpretation; see DETA, page 517. Aside from that, the trajectories of the nonlinear system near a critical point are a rather mild distortion of those in the linear case. The nature of the distortion is illustrated in Figure 3.

(c) An example

Theorem 2 is now used to analyze the system

(7) $$x' = x(x + y - 1), \quad y' = y(x - y + 3).$$

Setting $x' = y' = 0$ in (7) we are led to four cases

$$\begin{cases} x & = 0 \\ y & = 0 \end{cases} \quad \begin{cases} x & = 0 \\ x - y & = -3 \end{cases} \quad \begin{cases} x + y & = 1 \\ y & = 0 \end{cases} \quad \begin{cases} x + y & = 1 \\ x - y & = -3 \end{cases}$$

with corresponding stationary points $(0, 0)$, $(0, 3)$, $(1, 0)$, and $(-1, 2)$.
The functions f and g on the right of (7) satisfy

$$\begin{pmatrix} f_x & f_y \\ g_x & g_y \end{pmatrix} = \begin{pmatrix} 2x + y - 1 & x \\ y & x - 2y + 3 \end{pmatrix}.$$

At $(0, 0)$, $(0, 3)$, $(1, 0)$, and $(-1, 2)$ this equation yields the four matrices

(8) $$\begin{pmatrix} -1 & 0 \\ 0 & 3 \end{pmatrix}, \quad \begin{pmatrix} 2 & 0 \\ 3 & -3 \end{pmatrix}, \quad \begin{pmatrix} 1 & 1 \\ 0 & 4 \end{pmatrix}, \quad \begin{pmatrix} -1 & -1 \\ 2 & -2 \end{pmatrix}.$$

The pairs (p, q) for these matrices are respectively

$$(-2, -3), \quad (1, -6), \quad (-5, 4), \quad (3, 4).$$

Hence $4q - p^2 = -16, -25, -9, 7$.
According to the tables, the corresponding critical points for the linear
systems are:

saddle,   saddle,   node,   focus.

The node is unstable, the focus stable. By inspection of the matrix itself, the focus is counterclockwise.
The trajectories are shown in Figure
3. Near the critical points

   $(0, 0)$, $(0, 3)$, $(1, 0)$, $(-1, 2)$

the trajectories have the general behavior illustrated for the corresponding type of critical point in Figure 2.

Figure 3

(a) Undamped pendulum

(b) Damped pendulum

(c) van der Pol equation

Figure 4

(d) Bifurcation

We conclude with three additional examples, two from mechanics and one from electrical engineering. The discussion illustrates the important concept of bifurcation. In general, a system undergoes *bifurcation* when an arbitrarily small change in a parameter can drastically alter its behavior. A familiar example is given by the boundary-value problem

$$y'' + k^2 y = 0, \quad y(0) = y(1) = 0$$

which suddenly allows the nonzero solution $\sin \pi x$ when $k = \pi$.

EXAMPLE 1. By Section 9.1 the motion of an undamped pendulum with $x = \theta$ is described by

$$x' = y, \quad \ell y' = -g \sin x.$$

At the stationary points $(0, n\pi)$ the pendulum is hanging straight down when the integer n is even and it is balanced on its pivot when n is odd. In Problem 7 you are asked to get the linearized systems

$$x' = y, \quad \ell y' = -gx; \qquad x' = y, \quad \ell y' = gx$$

for these two cases, respectively. The first has a center and predicts oscillation, while the second has a saddle and predicts instability.

The nonclosed curves in Figure 4(a) correspond to an initial velocity so large that the pendulum makes complete revolutions about its pivot. The separatrix is obtained when the initial conditions are such that the pendulum tends to its unstable equilibrium position (exactly above the pivot) as $t \to \infty$.

EXAMPLE 2. The motion of a nonlinear damped pendulum is described by

$$x' = y, \quad \ell y' = -g \sin x - ky$$

where the damping constant $k > 0$. The stationary points are as in Example 1. In Problem 7 you are asked to verify that the linear approximation gives a stable focus, stable node, or saddle, in the three cases

$$n \text{ even}, \quad 4g > k^2; \qquad n \text{ even}, \quad 4g < k^2; \qquad n \text{ odd},$$

respectively. See Figure 4(b). If $k < 0$ (negatively damped pendulum) the sole change is that the node and focus are unstable.

As k changes from negative to positive, the pendulum goes from the stable regime of Figure 4(b) through Figure 4(a) to the unstable regime of Figure 4(b). Also the focus changes to a node when k^2 passes through the value $4g$. Both are examples of bifurcation.

EXAMPLE 3. In 1924 the Dutch applied mathematician and engineer Balthasar van der Pol investigated a series electrical circuit containing a nonlinear resistance. His analysis led to the *van der Pol equation*

$$x' = y, \quad y' = k(1 - x^2)y - x, \qquad k > 0 \text{ constant},$$

for which he proved existence of a nonconstant periodic solution.

In Problem 7 you are asked to check that the linear approximation near the origin has an unstable focus if $k < 2$, an unstable node if $k > 2$. The change from a focus to a node as k passes through the value 2 is another example of bifurcation. The closed curve in Figure 4(c) is a *limit cycle*; it yields the periodic solution that was discovered by van der Pol. Existence of a limit cycle cannot be deduced from the linear theory but is established in DETA, pages 513–514.

PROBLEMS

1. Check that if the quadratic terms in (7) are dropped, the resulting linear system has as its matrix the first matrix (8).

2. To get linear approximations in (7) near $(0,3)$, $(1,0)$, $(-1,2)$, we introduce the respective changes of variables

$$u = x, \quad v = y - 3; \qquad u = x - 1, \quad v = y; \qquad u = x + 1, \quad v = y - 2$$

Express (7) in terms of (u,v), drop quadratic terms, and verify that the matrices of the three linear systems so obtained are the second, third, and fourth matrices (8), respectively.

3. Find the stationary points and classify them by use of Theorem 2:

(a) $\begin{cases} x' = x(6 - 2x - 3y) \\ y' = y(4 - 3x - 4y) \end{cases}$
(e) $\begin{cases} x' = x(-18 - 2x + 2y) \\ y' = y(3 - x - 3y) \end{cases}$

(b) $\begin{cases} x' = x(3 - 3x - 4y) \\ y' = y(6 - 2x - 3y) \end{cases}$
(f) $\begin{cases} x' = x(6 - 3y) \\ y' = y(4 - 2x) \end{cases}$

(c) $\begin{cases} x' = x(6 - 2x + 2y) \\ y' = y(15 - x - 3y) \end{cases}$
(g) $\begin{cases} x' = 5x(1 - y) \\ y' = y(-12 + 3x) \end{cases}$

(d) $\begin{cases} x' = x(-6 - 2x + 2y) \\ y' = y(21 - x - 3y) \end{cases}$
(h) $\begin{cases} x' = x(-6 + 2x - 3y) \\ y' = y(12 - 4x + 6y) \end{cases}$

4P. On a full $8\frac{1}{2}$-by-11 inch sheet of graph paper, plot trajectories for the linear systems associated with the matrices (5) near the corresponding points $(0,0)$, $(0,3)$, $(1,0)$, $(-1,2)$. Then join these local charts by suitable curves to get a *global phase-plane portrait* as shown in Figure 3.

5. (abcdefg) This problem is time-consuming but instructive. After the manner of Problem 4, obtain global phase-plane portraits for the equations of Problem 3. Some of the characteristic equations have irrational roots.

6P. Make a tracing of Section 19.3, Figure 5. On your tracing put seven small disks to represent the seven possibilities for p, q, and $4q - p^2$ in Table 1. Also put three small circles to represent the three possibilities in Table 2. This figure gives insight into the reason for ambiguity when Table 2 is applied to nonlinear problems.

7. (abc) In Examples 1, 2, and 3, find the linear approximations and confirm what was said about them.

─────────────────────────ANSWERS─────────────────────────

2. (a) $u' = u(u + v + 2)$, $v' = (v + 3)(u - v)$
   (b) $u' = (1 + u)(u + v)$, $v' = v(u + v + 4)$
   (c) $u' = (u - 1)(u + v)$, $v' = (v + 2)(u - v)$

3. (a) $(0, 0)$uN, $(0, 1)$S, $(3, 0)$sN, $(-12, 10)$sF
   (b) $(0, 0)$uN, $(0, 2)$sN, $(1, 0)$S, $(-15, 12)$S
   (c) $(0, 0)$uN, $(0, 5)$S, $(3, 0)$S, $(6, 3)$sF
   (d) $(0, 0)$S, $(0, 7)$S, $(-3, 0)$uN, $(3, 6)$(sdN or sŝN)
   (e) $(0, 0)$S, $(0, 1)$sN, $(-9, 0)$uN, $(-6, 3)$S
   (f) $(0, 0)$uN, $(2, 2)$(cC or cF)   (g) $(0, 0)$S, $(4, 1)$(ccC or ccF)
   (h) The critical points are $(3\alpha, 2\alpha - 2)$ where α is arbitrary. These are not isolated. By (vi), or by direct calculation, $q = 0$ and Theorem 1 does not apply.

───

2. Problems in ecology

Let $x = \xi(t)$ and $y = \eta(t)$ denote the sizes, at time t, of two interacting populations: trees, bacteria, animals, or fish, to name four examples. These functions are nonnegative and a value 0 for either of them means that the corresponding population is extinct.

If $dx/dt = kx$, the constant k measures the rate of growth of x and is called the growth constant. A more general, and often a more realistic, model is obtained if k is replaced by a function F whose value depends on one or both populations. Upon making a similar generalization of $dy/dt = ky$, we are led to the following autonomous system for $t \geq 0$:

(1) $x' = xF(x, y)$, $y' = yG(x, y)$, $\xi(0) > 0$, $\eta(0) > 0$.

The initial conditions mean that both populations are present at the start. Special systems of the form (1) were analyzed by the American biologist Alfred James Lotka and the Italian mathematician Vito Volterra in the decade before 1930, and (1) is a *generalized Lotka-Volterra system*. It has a host of applications not only to ecology but to other disciplines.

The first quadrant is denoted by Q and its interior by Q_0; thus

$$Q = \{(x,y) : x \geq 0,\ y \geq 0\}, \qquad Q_0 = \{(x,y) : x > 0,\ y > 0\}.$$

If $(F,G) \in C^{(1)}$ in Q, solutions of (1) have the following properties:

(i) The trajectory remains in Q_0 as long as it exists.

(ii) If $F(0,0) > 0$ and $G(0,0) > 0$, the trajectory stays away from the origin.

(iii) For $(x,y) \in Q$ let $F(x,y) \leq 0$ and $G(x,y) \leq 0$ outside of some disk $r \leq r_0$. Then all solutions are bounded.

(iv) Near a critical point $(x_0, y_0) \in Q_0$ where $(F,G) \in C^{(1)}$ the matrix for the linear approximation is

$$\begin{pmatrix} x_0 F_x(x_0, y_0) & x_0 F_y(x_0, y_0) \\ y_0 G_x(x_0, y_0) & y_0 G_y(x_0, y_0) \end{pmatrix}.$$

The proof of (i),(ii), (iii) is in DETA, page 485. To get (iv) note that

$$\frac{\partial}{\partial x} x F(x,y) = x F_x(x,y) + F(x,y).$$

At a critical point (x_0, y_0) the second term is 0. When combined with a similar relation for $yG(x,y)$ this gives (iv).

(a) Competitive systems

Let x and y be the numbers of individuals in two competing populations; for example, two varieties of trees competing for sunlight, sheep and goats competing for a limited amount of forage, or two kinds of fish that do not prey on each other but do have a common source of food. Let us suppose that, in the absence of competition, growth would be governed by the logistic equations

$$x' = kx - ax^2, \quad y' = my - dy^2.$$

The effect of competition is assumed to be negative and jointly proportional to both populations. Adjoining a term $-bxy$ to the first equation and $-cxy$

to the second, we are led to the system

(2) $$x' = x(k - ax - by), \quad y' = y(m - cx - dy)$$

where all the coefficients are positive constants. This is a mathematical model for competition. Besides the origin, there are two stationary points $(k/a, 0)$, $(0, m/d)$ on the axes, and in the text we assume that there is a stationary point $(x_0, y_0) \in Q_0$. The case in which there is no such point is considered in the problems.

When (2) is written in the notation (1) we have

(3) $$F(0,0) = k > 0, \quad G(0,0) = m > 0$$

and if $x + y$ is sufficiently large both functions F and G are negative. By Properties (i), (ii), and (iii), the trajectories of (2) exist for $0 \le t < \infty$, they remain in a bounded region of Q_0, and they stay away from the origin. In DETA, pages 512-513 it is proved that no trajectory of (2) is a closed curve. From this it can be deduced that $\xi(t)$ and $\eta(t)$ have limits as $t \to \infty$, and the existence of these limits will be taken for granted here. Section 1, Item (iii) shows that the limit is a stationary point.

There is a simple graphical procedure for deciding whether the limit is (x_0, y_0) or is one of the stationary points on the axes. The decision is important, because it determines the long-time fate of the two populations. The first step is to plot the lines

$$ax + by = k, \quad cx + dy = m$$

on which $x' = 0$ and $y' = 0$, respectively. These lines are indicated as xx and yy in Figure 5. Their point of intersection is the stationary point

(a)

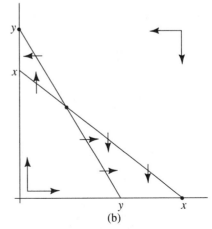
(b)

Figure 5

$(x_0, y_0) \in Q_0$ whose existence was postulated above. It is isolated if, as we now assume, $ad \neq bc$. On xx the velocity vector is vertical, on yy it is horizontal. To determine the sense of the vector (up or down, right or left) note that at points far from the axes both x' and y' are negative by (2), while at points near the origin they are both positive. This information is summarized by two pairs of orthogonal vectors in the figure. For want of a better term, such pairs are referred to as *indicators*.

When we move across the line xx from one side to the other, x' in (2) changes sign, going through the value 0 on the line itself. Similarly for y' and the line yy. If an indicator can be moved to one of these lines without crossing the other line, it gives the direction on that line immediately; no reversal of sign is needed. This feature of the analysis is exploited in Figure 5. Starting with the indicator at distant points, we move it to the upper part of the xx line. The horizontal component decreases during this motion until it becomes 0 when we reach the line. The remaining nonzero component gives the direction field on this part of the xx line. The same indicator also gives the direction on the lower part of the yy line.

In a like fashion, the indicator near the origin can be moved, without crossing either line, to the lower part of the xx line and to the upper part of the yy line. We replace the horizontal component of the indicator by 0 on the xx line, the vertical component by 0 on the yy line. Without the need for any computation, this gives arrows as shown in the figure.

The arrows can be interpreted as giving the direction of flow, and it is not hard to prove that a trajectory can never cross an xx or yy line in a direction contrary to the arrows. In Figure 5(a) the arrows suggest that the stationary point (x_0, y_0) is stable, and that every trajectory tends to this point for $t \to \infty$ no matter where it starts. Figure 5(b) suggests that the stationary point (x_0, y_0) is unstable and that, except for infinitely rare cases, one or the other population will go to extinction. These conclusions are correct, and are confirmed by the local behavior as discussed next.

Property (iv) above yields the following matrix and values p, q for the linear approximation near (x_0, y_0):

$$\begin{pmatrix} -ax_0 & -bx_0 \\ -cy_0 & -dy_0 \end{pmatrix}, \quad p = ax_0 + dy_0, \quad q = x_0 y_0 (ad - bc).$$

The difference between Figure 5(a) and 5(b) is that the xx line slants down more steeply than the yy line in Figure 5(a), less steeply in 5(b). Since the respective slopes are $-a/b$ and $-c/d$ we see that $ad > bc$ in the first case and $ad < bc$ in the second. In Figure 5(a) we have $q > 0$ and

$$4q - p^2 = -4x_0 y_0 bc - (ax_0 - dy_0)^2 < 0.$$

Hence (x_0, y_0) is a stable node. In Figure 5(b) we have $q < 0$ and (x_0, y_0) is a saddle. See Figure 6.

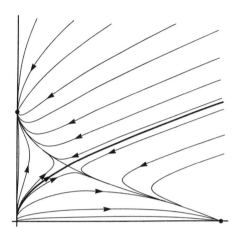

Figure 6

(b) The separatrix

In Figure 6(b) the only critical point allowing coexistence is unstable, and it is virtually certain that one of the populations will drive the other to extinction. The main ecological problem is to decide from the initial values whether the x-population or the y- population is the one that ultimately survives. The curve separating these two cases is called the *separatrix*. In Figure 6(b) the separatrix is the smooth curve that goes through the origin and through the stationary point (x_0, y_0).

The regions I and II bounded by the separatrix and by the axes are *basins of attraction* for $(k/a, 0)$ and $(0, m/d)$ respectively. If a trajectory starts in I it tends to $(k/a, 0)$ and only the x population survives. If it starts in II it tends to $(0, m/d)$ and the x population becomes extinct. Only when the initial point is exactly on the separatrix do both populations survive.

The discussion illustrates the important concept of competitive exclusion. The term *principle of competitive exclusion* is usually reserved, however, for the much simpler case of Problem 2, under the additional hypothesis that the two lines there considered are parallel.

(c) Predator-prey equations

The equations introduced next are best understood in a specific context. Let x be the number of foxes and y the number of rabbits in two populations. The foxes eat rabbits and the rabbits eat vegetation, of which there is an unlimited supply. In the absence of rabbits the number of foxes would decline exponentially, and in the absence of foxes the number of rabbits would increase exponentially. The situation is described by

(4) $$x' = -kx, \quad y' = my$$

where k measures the deathrate of foxes and m the birthrate of rabbits.

We assume next that the interaction of rabbits and foxes is jointly proportional to both populations. Growth of the fox population is helped by the rabbits and growth of the rabbit population is inhibited by the foxes. Hence we add a term bxy to x' and we subtract a term cxy from y' to get

(5) $$x' = x(-k + by), \quad y' = y(m - cx)$$

where k, m, b, c are positive constants. Equations (5) are the Lotka-Volterra *predator-prey equations*.

The lines $by = k$ and $cx = m$ on which $x' = 0$ and $y' = 0$ intersect at $(m/c, k/b)$ and this yields a stationary point $(x_0, y_0) \in Q_0$. Hence the populations can always coexist in a steady state. It is left for you to check that $p = 0$, $q = bcx_0y_0$ at (x_0, y_0) and hence the stationary point for the linear approximation is a center. This means that the phase-plane orbits are closed curves and the corresponding approximate solutions are periodic. It turns out that the nonlinear system has the same properties, and we now explain why this is so.

Leaving (5) for a moment, consider a system of the form

(6) $$x' = R(x,y)f(y), \quad y' = R(x,y)g(x)$$

where f and g are continuous. If F is an integral of f and G of g it is easily checked that

$$F(y) - G(x) = \text{const.}$$

Indeed, the derivative of the expression on the left is $f(y)y' - g(x)x'$ and this reduces to 0 by virtue of the differential equations.

Returning to (5), note that the equations can be written

$$x' = xy\left(\frac{-k}{y} + b\right), \quad y' = xy\left(\frac{m}{x} - c\right).$$

The functions f, g and their integrals F, G are found by inspection, with the result

$$-k \ln y + by - m \ln x + cx = \text{const.}$$

It is left for you to verify that exponentiation yields

(7) $$x^m y^k e^{-(cx+by)} = H$$

where H is constant. When interpreted in three dimensions, (7) describes the intersection of the surface

(8) $$z = x^m y^k e^{-(cx+by)}$$

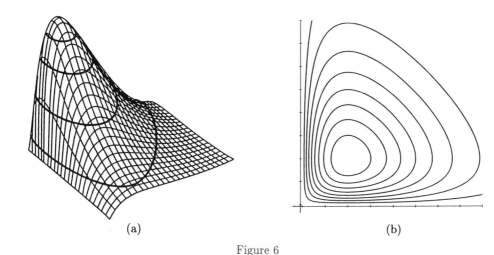

(a)   (b)

Figure 6

with the plane $z = H$. We regard the surface as a mountain in which the locus $z = H$ gives a contour line of constant height H above sea level. The elevation of the surface is positive in Q_0, is zero on the axes, and it tends to 0 on the line $cx + by = L$ as L tends to infinity. By a short calculation the equations

$$\frac{\partial z}{\partial x} = 0, \quad \frac{\partial z}{\partial y} = 0$$

yield $x = x_0$ and $y = y_0$, where (x_0, y_0) is the stationary point in Q_0 found above. Hence the maximum of z in Q is unique and is given by

$$H_0 = x_0{}^m y_0{}^k e^{-(cx_0 + by_0)}.$$

This means that the mountain has a single peak and its height is H_0.

A typical example of a surface defined by (8) is shown graphically in Figure 7(a) and contour lines $z = H$ are shown in Figure 7(b). Since these are closed curves, the corresponding solutions are periodic. The interior of the contour line for H is, by definition, the region in Q in which $z > H$. Hence if $H_1 > H_2$ the contour for H_1 is interior to that for H_2. Further information is obtained by considering the behavior as $H \to 0$ and as $H \to H_0$.

From the standpoint of biology, the periodicity is explained as follows: When the fox population is large, the rabbit population declines, and eventually the foxes do not have enough to eat. Then the fox population declines, and when this happens, the rabbit population goes up. The process of successive increase and decline leads to periodic variation in both populations. Such variation has been observed in nature, and formulation of an equation that predicts it was one of the first major successes of mathematical ecology.

(d) Self-limiting terms

When a growth law $y' = my$ is replaced by $y' = my - dy^2$ the extra term $-dy^2$ describes a kind of self-interaction that prevents the population from becoming infinite. It is called the *self-limiting term*. The presence or absence of self-limiting terms can have a profound effect on the behavior of an ecological system.

As an illustration, let us introduce self-limiting terms $-ax^2$ and $-dy^2$ for x and y, respectively, in the predator-prey equations (5). The result is the system

$$(9) \qquad x' = x(-k - ax + by), \quad y' = y(m - cx - dy)$$

where a, b, c, d, k, m are positive constants. The equations for a stationary point $(x_0, y_0) \in Q_0$ give

$$x_0 = \frac{bm - dk}{ad + bc}, \quad y_0 = \frac{am + ck}{ad + bc}.$$

Hence $y_0 > 0$ automatically but $x_0 > 0$ only if $bm > dk$. The fact that (9) need not have a stationary point in Q_0 is an important difference between (9) and (5).

A second difference is that a stationary point $(x_0, y_0) \in Q_0$ is a center in (5), but is not a center in (9) even when it exists. To see this, we write the matrix (iv) for (9) and the corresponding quantities p, q:

$$\begin{pmatrix} -ax_0 & bx_0 \\ -cy_0 & -dy_0 \end{pmatrix}, \quad p = x_0 a + y_0 d, \quad q = x_0 y_0 (ad + bc).$$

 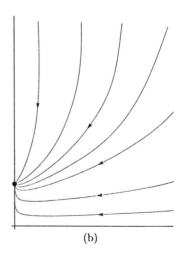

(a)   (b)

Figure 8

Since $p > 0$ and $q > 0$ the point (x_0, y_0) is a stable focus or stable node when $4q - p^2$ is positive or negative, respectively. If $4q = p^2$ the node is singular or degenerate, but still stable.

These results suggest that the solutions of (9) tend to a unique limit as $t \to \infty$, and a proof that they do is given in DETA, pages 500–501. If $bm > dk$ the limit is the stationary point $(x_0, y_0) \in Q_0$, and both populations survive. But if $bm \le dk$ there is no stationary point in Q_0 and the limit is $(0, m/d)$. Thus the predators become extinct and only the prey survive. See Figure 8.

PROBLEMS

1. Near the origin, linear approximations to the competition equations (2) and the general predator-prey equations (9) are respectively

$$x' = kx, \ y' = my; \qquad x' = -kx, \ y' = my$$

Solve, eliminate t, and sketch in four representative cases.

2. The principle of competitive exclusion. In the competition equations (2) suppose the lines xx and yy do not intersect in Q. Using indicators as in the text, predict the ultimate fate of the populations (a) when the xx line and (b) when the yy line is the farther from the origin.

3. Analyze the following competitive systems by indicators, as in the text:

(a) $\begin{cases} x' &= x(3 - 2x - y) \\ y' &= y(3 - x - 2y) \end{cases}$ 　　　 (d) $\begin{cases} x' &= x(2 - x - y) \\ y' &= y(5 - x - 4y) \end{cases}$

(b) $\begin{cases} x' &= x(3 - x - 2y) \\ y' &= y(3 - 2x - y) \end{cases}$ 　　　 (e) $\begin{cases} x' &= x(5 - x - 4y) \\ y' &= y(2 - x - y) \end{cases}$

(c) $\begin{cases} x' &= x(2 - x - 2y) \\ y' &= y(3 - x - y) \end{cases}$ 　　　 (f) $\begin{cases} x' &= x(6 - x - 2y) \\ y' &= y(3 - 3x - y) \end{cases}$

4. (abcdef) Confirm the results of Problem 3 by finding the character of the critical points.

5. Here are two predator-prey systems. Draw a figure like Figure 5 and use it to predict the qualitative behavior:

　(a) $x' = x(y - 1), \ y' = y(1 - x)$ 　　　 (b) $x' = 2x(2 - y), \ y' = 3y(x - 5)$

6. In the two cases $bm > kd$, $bm < kd$ make figures for (9) analogous to Figure 5.

7. Solve the linear approximation to the predator-prey equations (5) near the stationary point $(m/c, k/b)$ and show that the solutions have period $T = 2\pi/\sqrt{km}$. Show also that the orbits are ellipses with horizontal and vertical axes, and that they are circles if $b^2 m = c^2 k$.

8. Let $x = \xi(t)$, $y = \eta(t)$ be a solution of the predator-prey equations (5) of period T. The mean values of x, y, and xy are defined by the integrals on the left below. Show that they have the values on the right:

$$\frac{1}{T}\int_0^T \xi(t)\,dt = x_0 \qquad \frac{1}{T}\int_0^T \eta(t)\,dt = y_0 \qquad \frac{1}{T}\int_0^T \xi(t)\eta(t)\,dt = x_0 y_0$$

Hint: Write the first equation as $x'/x = -k + by$ and integrate from 0 to T. Since $\ln x$ has period T, the term x'/x integrates to 0 and you get an equation for the mean value of y.

9. Pesticides. In the predator-prey equations (5) suppose the prey is the target population for a pesticide that acts equally on both populations; that is, it subtracts hx from x' and hy from y'. Here h is a positive constant measuring the strength of the pesticide. Using the result of Problem 8, show that the mean value of the target population is increased by the pesticide.

10. Pesticides, continued. Let the basic equations in Problem 9 be the general predator-prey equations (9). Show that the pesticide never drives the prey to extinction, but drives the predator to extinction if

$$h > \frac{bm - dk}{b + d}$$

According to this model, environmental stress affects the predator more than it affects the prey. The same conclusion is suggested by the simpler model of Problem 9.

11. The case $k = m$ has special interest, because it means that both populations have the same intrinsic growth rates and any difference in survival depends wholly upon the interaction. Show that the competition equations (2) admit a solution of the form $y = \alpha x$ if, and only if, $k = m$, and in that case $\alpha = y_0/x_0$. The line $y = \alpha x$ so obtained goes through both the stationary point and the origin. If (x_0, y_0) is unstable, $y = \alpha x$ is the separatrix.

12. (abcdef) This problem is time-consuming but instructive. Plot the trajectories near each of the four critical points in Problem 3 and thus get a global phase-plane portrait in Q. Some of the characteristic equations have irrational roots.

13. Here are two predator-prey equations with self-limiting terms:

$$\text{(a)}\ \begin{cases} x' & = x(-1 - x + y) \\ y' & = y(3 - x - y) \end{cases} \qquad \text{(b)}\ \begin{cases} x' & = x(-3 - x + y) \\ y' & = y(1 - x - y) \end{cases}$$

Analyze by indicators as in Problems 5 and 6, sketch the behavior near the critical points in Q, and thus get a global phase-plane portrait in Q.

————————————————ANSWERS————————————————

1. Partial answer: For (2), $y^k = Cx^m$. When $t \to \infty$, the trajectories approach the origin tangent to the y axis if $k > m$, tangent to the x axis if $k < m$, and in any direction if $k = m$. For (9), $x^m y^k = C$. The trajectories near $(0,0)$ resemble the hyperbolas $xy = C$ with x and y axes as asymptotes.

2. (a) Only the x-population survives. (b) Only the y-population survives.

3. Partial answer: (d) is the basis for Figure 6(a) and (a) is like it. (e) is the basis for Figure 6(b) and (b) is like it. In (c) and (e) just one population survives. See also answers to Problem 4.

4. (a) $(1,1)$sN, $(3/2,0)$S, $(0,3/2)$S   (d) $(1,1)$sN, $(2,0)$S, $(0,5/4)$S
   (b) $(0,3)$sdN, $(3,0)$sdN, $(1,1)$S   (e) $(0,2)$sdN, $(5,0)$sdN, $(1,1)$S
   (c) $(0,3)$sN, $(2,0)$S   (f) $(8,0)$sN, $(0,3)$S
   The character of the critical point $(0,0)$ is given by Problem 1.

5. Partial answers: The line xx is vertical, yy is horizontal.

(a) $(1,1)$cC, $xye^{-(x+y)} = H$. See Figure 7.

(b) $(5,2)$ccC, $x^{15}y^4 e^{-3x-2y} = H$. Here x denotes prey.

13. Partial answer: (a) $(0,0)$S, $(0,3)$S, $(1,2)$scF
                        (b) $(0,0)$S, $(0,1)$sN. See Figure 8.

————————————————

Alexander, the king of the Macedonians, began like a wretch to learn geometry. Teach me, saith he, easy things. To whom his master said: These things be the same, and alike difficult unto all.

—Seneca

Chapter 21

THE LAPLACE TRANSFORM

W̲E̲ NOW PRESENT a systematic and elegant procedure that is widely used in circuit analysis and in the study of feedback and control. The theory takes its form from a symbolic method developed by the English engineer Oliver Heaviside. It enables one to solve many initial-value problems without going to the trouble of finding the general solution and then evaluating the arbitrary constants. The procedure applies to systems of equations, to partial differential equations, and to integral equations. It often yields results more readily than other techniques.

1. Basic concepts

The *Laplace transform* of the function f is defined by the equation

$$(1) \qquad F(s) = \int_0^\infty e^{-st} f(t)\, dt = \lim_{r \to \infty} \int_0^r e^{-st} f(t)\, dt$$

provided the limit exists for all sufficiently large s. This means that there is a number $s_0 = s_0(f)$, depending on f, such that the limit exists whenever $s > s_0$. If there is no such number s_0 then f has no Laplace transform. The role of the condition $s > s_0$ is illustrated in the following example.

EXAMPLE 1. If a is a real constant then

$$\int_0^r e^{-st} e^{at}\, dt = \int_0^r e^{t(a-s)}\, dt = \left. \frac{e^{t(a-s)}}{a-s} \right|_0^r.$$

The behavior as $r \to \infty$ depends on the sign of $a - s$. If $s < a$ then $a - s$ is positive and the expression tends to ∞ as $r \to \infty$. When $s = a$ the value of the integral is r, which also tends to ∞ as $r \to \infty$. In both cases the integral defining the Laplace transform is divergent. If $s > a$, however, $e^{r(a-s)}$ tends to 0 as $r \to \infty$. Hence

$$(2) \qquad \int_0^\infty e^{-st} e^{at}\, dt = \frac{1}{s-a}, \qquad s > a$$

and the integral is convergent. This shows that e^{at} has the Laplace transform $F(s) = 1/(s-a)$, and that the constant s_0 in the definition can be any number $\geq a$. If a is a complex constant, the same calculation shows that $F(s) = 1/(s-a)$ for $s > \operatorname{Re} a$.

Two notations for the Laplace transform will be used here. In the first notation, the connection between a function f and its transform F is indicated by lower case and capital letters. Thus

$$U(s) = \int_0^\infty e^{-st} u(t)\, dt, \qquad V(s) = \int_0^\infty e^{-st} v(t)\, dt$$

and so on. In the second notation the Laplace transform of f is denoted by Lf where L is the Laplace transform operator. Thus

$$Lu = \int_0^\infty e^{-st} u(t)\, dt, \qquad Lv = \int_0^\infty e^{-st} v(t)\, dt$$

and so on. The domain of L is the set of functions f for which Lf exists.

Let us show that L is linear. That is,

$$L(u + v) = Lu + Lv, \qquad L(cu) = cL(u)$$

for all functions u and v in the domain of L and for all constants c. The first of these equations follows by letting $r \to \infty$ in the equation

$$\int_0^r e^{-st}(u(t) + v(t))\, dt = \int_0^r e^{-st} u(t)\, dt + \int_0^r e^{-st} v(t)\, dt.$$

The calculation shows that if $U(s)$ exists for $s > s_0$ and $V(s)$ for $s > s_1$ then $L(u + v)$ exists for $s > \max(s_0, s_1)$. The proof that $L(cu) = cL(u)$ is similar and is left for you.

Since L is linear, all results valid for general linear operators apply to L. In particular, the principle of equating real parts holds provided, as we shall assume, s is real. That is,

$$L(f + ig) = F + iG \quad \Rightarrow \quad Lf = F \quad \text{and} \quad Lg = G.$$

This is true because Lf is real for real f, so that L is a real linear operator. We illustrate in an example.

EXAMPLE 2. If $a = ib$ in Example 1 the result is

$$L(e^{ibt}) = \frac{1}{s - ib} = \frac{s + ib}{s^2 + b^2}.$$

Writing $e^{ibt} = \cos bt + i \sin bt$ and equating real and imaginary parts, we get

(3) $$L(\cos bt) = \frac{s}{s^2 + b^2}, \qquad L(\sin bt) = \frac{b}{s^2 + b^2}.$$

The Laplace transform of a function can often be deduced by calculations that do not involve integration. Sometimes the justification requires a rather careful analysis and sometimes the justification is trivial. The following examples illustrate both cases.

EXAMPLE 3. Equation (2) can be written in the form

$$\int_0^\infty e^{-st} e^{at}\, dt = (s-a)^{-1}.$$

Proceeding without regard to rigor, let us differentiate both sides with respect to s. Since

$$\frac{d}{ds} e^{-st} = -te^{-st} \quad \text{and} \quad \frac{d}{ds}(s-a)^{-1} = -(s-a)^{-2}$$

the result is

$$\int_0^\infty e^{-st} te^{at}\, dt = (s-a)^{-2}.$$

Differentiating again we get a factor $-t^2$ on the left and $-2(s-a)^{-3}$ on the right. Differentiating n times leads to the formula

$$(4) \qquad\qquad L(t^n e^{at}) = \frac{n!}{(s-a)^{n+1}}, \qquad n = 0,\, 1,\, 2,\, \cdots$$

where by convention $0! = 1$. Although our argument is suggestive only, both the procedure and the result are in fact correct, as seen in DETA page 544.

EXAMPLE 4. The identity

$$\int_0^\infty e^{-st} e^{-ct} f(t)\, dt = \int_0^\infty e^{-(s+c)t} f(t)\, dt$$

shows that if $Lf(t) = F(s)$ then

$$(5) \qquad\qquad Le^{-ct} f(t) = F(s+c).$$

In contrast to (4), Equation (5) is an immediate consequence of the definitions. When applied to the result of Example 2, Equation (5) gives

$$(6) \qquad L(e^{-ct} \cos bt) = \frac{s+c}{(s+c)^2 + b^2}, \qquad L(e^{-ct} \sin bt) = \frac{b}{(s+c)^2 + b^2}.$$

The following definition describes a rather large class of functions that have Laplace transforms:

DEFINITION. A real- or complex-valued function $f(t)$ belongs to the class E *of functions of exponential type if the following two conditions hold:*

(a) *On any interval $[0, r]$ the function is defined and continuous except perhaps at finitely many points, and*

(b) *For constants A and B, depending on f, the inequality $|f(t)| \le Ae^{Bt}$ holds at all points of $[0, \infty)$ where $f(t)$ is defined.*

If f satisfies these two conditions and $s > B$, a calculation similar to that in Example 1 gives

$$\int_0^\infty e^{-st}|f(t)|\,dt \le \int_0^\infty e^{-st}Ae^{Bt}\,dt = \frac{A}{s-B}.$$

This suggests that the integral without absolute values exists for $s > B$, and a proof is given in DETA, page 577. Thus if $f \in$ E, then $L(f)$ exists. The class E is used in the following theorem:

THEOREM 1.   *Let f be continuous on $[0, \infty)$ and let $f' \in$ E. Then $f \in$ E and $Lf' = sLf - f(0)$.*

If f is defined only on $[0, \infty)$, continuity at $t = 0$ means that

$$f(0) = f(0+) = \lim_{t \to 0+} f(t).$$

This right-hand limit for $f(0)$ must be used in Theorem 1. If $f(t)$ is defined for $t < 0$ as well as for $t \ge 0$, but is discontinuous at $t = 0$, the left-hand limit $f(0-)$ would give an incorrect result.

We now prove Theorem 1. Integration by parts gives

(7) $$\int_0^r e^{-st}f'(t)\,dt = e^{-st}f(t)\Big|_0^r + s\int_0^r e^{-st}f(t)\,dt.$$

The fact that integration by parts is permissible under the hypothesis of Theorem 1 is established in DETA, Chapter G, where it is also shown that $f \in$ E. If

$$|f(t)| \le Ae^{Bt} \quad \text{and} \quad s \ge B + 1,$$

as we may assume, the integrated term in (7) at the upper limit r satisfies

$$|e^{-sr}f(r)| \le e^{-(B+1)r}Ae^{Br} = Ae^{-r}.$$

This tends to 0 as $r \to \infty$. Hence letting $r \to \infty$ in (7) gives Theorem 1.

The differential equations considered in this section have solutions y that are sums of functions of the form $ct^j e^{kt}$ with $j = 0, 1, 2, \cdots$. Therefore the solutions as well as their derivatives satisfy the hypothesis of Theorem 1. Using this fact, we take $f = y$ in Theorem 1 to get

(8) $$Ly' = sLy - y(0).$$

The choice $f = y'$ in Theorem 1 gives

$$Ly'' = sLy' - y'(0) = s(sLy - y(0)) - y'(0)$$

where the second equality follows from (8). Hence

(9) $$Ly'' = s^2 Ly - sy(0) - y'(0).$$

These relations enable us to transform first- or second-order initial-value problems into algebraic equations for Ly. Once Ly is found, the solution y is given by a table of Laplace transforms such as the table on the inside back cover of this book. The procedure is justified by the following theorem, which is established in DETA, Chapter G:

UNIQUENESS THEOREM. Let y and $\tilde{y}$ be functions of class E such that their Laplace transforms agree for all large s. Then $y(t) = \tilde{y}(t)$ at all points $t \geq 0$ where both these functions are continuous.

EXAMPLE 5. Solve the initial-value problem

$$y' - y = e^t, \qquad y(0) = 1.$$

Taking the Laplace transform gives

$$Ly' - Ly = Le^t \quad \text{or} \quad sLy - 1 - Ly = \frac{1}{s-1}$$

where the second equation follows from (8) with $y(0) = 1$ and from (2) with $a = 1$. Hence

$$(s-1)Ly = \frac{1}{s-1} + 1 \quad \text{or} \quad Ly = \frac{1}{(s-1)^2} + \frac{1}{s-1}.$$

By (4) with $a = 1$, $n = 1$, and $n = 0$ we see that

$$Ly = L(te^t + e^t), \quad \text{hence} \quad y = te^t + e^t.$$

EXAMPLE 6. Solve $y'' + 4y = 0$, $y(0) = 5$, $y'(0) = 6$. Taking the transform gives

$$Ly'' + 4Ly = 0 \quad \text{or} \quad s^2 Ly - 5s - 6 + 4Ly = 0$$

where the second equation follows from (9). Hence

$$Ly = \frac{5s + 6}{s^2 + 4} = 5\frac{s}{s^2 + 4} + 3\frac{2}{s^2 + 4}.$$

Using the two equations (3) with $b = 2$ we get

$$y = 5\cos 2t + 3\sin 2t.$$

Note the manner in which the two constants 5 and 3 were built into the analysis. The correctness of this procedure follows from the fact that L is linear.

The operation by which we recover $f(t)$ from $F(s)$ is called *inverse Laplace transformation* and is denoted by L^{-1}. Linearity of L^{-1} follows from linearity of L. According to the uniqueness theorem, $f(t) = L^{-1}F(s)$ is substantially determined for $t \geq 0$ even if $F(s)$ is known only for large s. The condition that s be large is just what is needed to ensure that the integral defining the transform converges. Hence, whenever we use a Laplace transform, it is assumed that s is real and large enough to justify the calculation.

PROBLEMS

1P. The purpose of this problem is to help you become familiar with the table of Laplace transforms on the inside back cover of this book. Only in Part (c) are you expected to write anything.

(a) Notice that entries ABCDEF follow from abcdef and (5), and that aA follow from bB with $0! = 1$.

(b) Section 6.4 gives

$$\cos ibt = \cosh bt, \qquad \frac{\sin ibt}{ib} = \frac{\sinh bt}{b}$$

Hence ef and EF follow from cd and CD with b replaced by ib.

(c) Replace b by ib in gG and thus get the two transform pairs

$$t\frac{\sinh bt}{b}, \qquad \frac{2s}{(s^2 - b^2)^2}; \qquad \frac{2b^2}{(s^2 - b^2)^2}, \qquad t\cosh bt - \frac{\sinh bt}{b}$$

2. Find $L1$ and Lt directly from the definition and check by the table.

3. Using the table find the transform $F(s)$ of the following functions $f(t)$:

(a) $7e^{2t}$ (b) $\cos 2t + \sin 2t$ (c) $e^t(t - \sin t)$ (d) $(t + e^t)^3$

4. Find $L\cos^2 t$ and $L\sin^2 t$. Hint: $2\cos^2 t = 1 + \cos 2t$.

5. Find functions $f(t)$ whose transforms are the following functions $F(s)$:

(a) $\dfrac{6}{s^2+1}$ (c) $\dfrac{5s-6}{s^2+4}+\dfrac{2}{s}$ (e) $\dfrac{7(s-3)+16}{(s-3)^2+4}$

(b) $\dfrac{s^2+4}{s^3}$ (d) $\dfrac{4s+3}{s^2-9}+\dfrac{6}{(s+1)^3}$ (f) $\dfrac{6s-1}{(s+4)^2-25}$

6P. Completing the square. The identity $s^2+2as=(s+a)^2-a^2$ will be used here. Check that

$$\frac{2s+6}{3s^2-18s+15}=\frac{2s+6}{3(s^2-6s+9)-12}=\frac{2(s-3)+12}{3((s-3)^2-4)}$$

Thus find a function $f(t)$ whose transform is the above function $F(s)$.

7. Find a function $f(t)$ whose transform is the given function $F(s)$:

(a) $\dfrac{s-2}{s^2-4s+5}$ (b) $\dfrac{2}{s^2+2s}$ (c) $\dfrac{s+8}{2s^2+8s+26}$ (d) $\dfrac{9s+3}{9s^2+6s+19}$

8P. The *rest solution* of a first-order equation satisfies $y(0)=0$. If y is the rest solution of the equation on the left below, check that Ly satisfies the other two equations. Thus find y, and verify that it satisfies both the differential equation and the condition $y(0)=0$:

$$y'+2y=1,\qquad sLy+2Ly=\frac{1}{s},\qquad Ly=\frac{1}{(s+1)^2-1}$$

9. By means of the Laplace transform, find the rest solution:

(a) $y'+4y=8$ (b) $y'-y=e^t$ (c) $y'+2y=e^{3t}$ (d) $y'=7y$

10. If $y'+2y=4$ and $y(0)=1$ check that $Ly=(s+4)/(s^2+2s)$. Thus get

$$y=e^{-t}(\cosh t+3\sinh t)=2-e^{-2t}$$

11. (abcd) By means of the Laplace transform, solve the differential equation subject to the initial condition below it:

$$\begin{array}{llll}
y'+4y=0 & y'-y=3e^{4t} & y'+y=e^t & y'+y=1 \\
y(0)=2 & y(0)=1 & y(0)=1 & y(0)=3
\end{array}$$

12. (abcd) Using the Laplace transform, solve subject to the given values of $y_0=y(0)$ and $y_1=y'(0)$:

$$\begin{array}{llll}
y''+y=0 & y''-4y=0 & y''+y=2y' & y''=y'+2y \\
y_0=1,\ y_1=1 & y_0=1,\ y_1=2 & y_0=0,\ y_1=1 & y_0=3,\ y_1=6
\end{array}$$

---------------------ANSWERS----------------------

3. (a) $7/(s-2)$ (c) $1/((s-1)^4 + (s-1)^2)$

    (b) $(s+2)/(s^2+4)$ (d) $6/s^4 + 6/(s-1)^3 + 3/(s-2)^2 + 1/(s-3)$

4. $(1/2)(1/s \pm s/(s^2+4))$

5. (a) $6\sin t$ (d) $4\cosh 3t + \sinh 3t + 3t^2 e^{-t}$

    (b) $1 + 2t^2$ (e) $e^{3t}(7\cos 2t + 8\sin 2t)$

    (c) $5\cos 2t - 3\sin 2t + 2$ (f) $e^{-4t}(6\cosh 5t - 5\sinh 5t) = (e^t + 11e^{-9t})/$:

6. $3f(t) = e^{3t}(2\cosh 2t + 6\sinh 2t) = 4e^{5t} - 2e^t$

7. (a) $e^{2t}\cos t$ (c) $e^{-2t}((1/2)\cos 3t + \sin 3t)$

    (b) $2e^{-t}\sinh t = 1 - e^{-2t}$ (d) $e^{-t/3}\cos \sqrt{2}t$

8. $y = e^{-t}\sinh t = (1/2)(1 - e^{-2t})$

9. (a) $2(1 - e^{-4t})$ (b) te^t (c) $5y = e^{3t} - e^{-2t}$ (d) 0

11. (a) $2e^{-4t}$ (b) e^{4t} (c) $\cosh t$ (d) $1 + 2e^{-t}$

12. (a) $\cos t + \sin t$ (b) e^{2t} (c) te^t (d) $3e^{2t}$

---

2. The transformed equation

We begin with the equation

(1) $$Ty = y^{(n)} + p_{n-1}y^{(n-1)} + \cdots + p_1 y' + p_0 y = f \in \mathrm{E},$$

where the p_j are constants and the linear operator T is defined by the first equality in (1). The characteristic polynomial of T is

$$P(s) = s^n + p_{n-1}s^{n-1} + \cdots + p_1 s + p_0.$$

When $f(t)$ is discontinuous, as it is in many applications, the word *solution* will be used in the sense of *generalized solution*. This means that $y^{(n-1)}$ is continuous but that existence of $y^{(n)}$ is allowed to fail at points where f is discontinuous. In DETA, Chapter G, it is shown that the condition $f \in \mathrm{E}$ ensures that all the derivatives in (1) also belong to E. This makes it easy to find Ly.

Calculation of Ly is especially easy for the *rest solution* of (1), which is the solution satisfying null initial conditions

$$y(0) = y'(0) = \cdots = y^{(n-1)}(0) = 0.$$

By Section 1 Theorem 1, these conditions give

$$Ly' = sLy, \qquad Ly'' = sLy' = s^2 Ly, \qquad Ly''' = sLy'' = s^3 Ly$$

and so on. In general, the rest solution satisfies

(2) $$\qquad\qquad Ly^{(j)} = s^j Ly, \qquad j = 0, 1, 2, \cdots, n.$$

Hence the transform of (1) gives

$$s^n Ly + p_{n-1} s^{n-1} Ly + \cdots + p_1 s Ly + p_0 Ly = Lf$$

when we use the fact that L is linear together with (2). Factoring out Ly gives the following:

THEOREM 1. *If y is the rest solution of (1) then $P(s)Ly = Lf$.*

EXAMPLE 1. Find Ly for the rest solution of

$$y^{iv} + 3y''' + 2y'' + 6y' = e^{-t} \sin t + t^2.$$

The Laplace transform of the function $f(t)$ on the right is given by the table and the characteristic polynomial is read off from the left side. Theorem 1 then produces the equation

$$(s^4 + 3s^3 + 2s^2 + 6s)Ly = \frac{1}{(s+1)^2 + 1} + \frac{2}{s^3}$$

and division by the polynomial on the left gives Ly.

If y is not the rest solution but satisfies arbitrary initial conditions

$$y(0) = y_0, \quad y'(0) = y_1, \quad y''(0) = y_2, \quad \cdots, \quad y^{(n-1)}(0) = y_{n-1}$$

the computation of Ly is only slightly more difficult. By results of the previous section

(3) $$\qquad\qquad Ly' = sLy - y_0, \qquad Ly'' = s^2 Ly - sy_0 - y_1.$$

Since $y''' = (y'')'$ the first equation (3) applied to y'' gives

$$Ly''' = sLy'' - y''(0) = sL''y - y_2.$$

Substituting Ly'' as given by the second equation (3), we get

$$Ly''' = s^3 Ly - s^2 y_0 - sy_1 - y_2.$$

The process continues in an obvious manner and leads to the following:

THEOREM 2.   *If $y^{(n)} \in E$ and $y^{(n-1)}$ is continuous then*

$$Ly^j = s^j Ly - (s^{j-1} y_0 + s^{j-2} y_1 + \cdots + s y_{j-2} + y_{j-1})$$

for $j = 1, 2, \cdots, n$, where $y_j = y^{(j)}(0)$ are the initial values. This holds in particular for the solutions of (1).

EXAMPLE 2.   Find Ly if $y(0) = 5$, $y'(0) = 6$, $y''(0) = 7$, and

$$y''' - 3y'' + 2y' = f \in \text{E}.$$

Taking the transform of both sides gives $Ly''' - 3Ly'' + 2Ly' = Lf$. By Theorem 2 with $y_0 = 5$, $y_1 = 6$, $y_2 = 7$ we get

$$(s^3 Ly - 5s^2 - 6s - 7) - 3(s^2 Ly - 5s - 6) + 2(sLy - 5) = Lf.$$

Moving the terms not involving Ly to the right gives

$$(s^3 - 3s^2 + 2s)Ly = Lf + (5s^2 - 9s - 1)$$

and Ly is readily found from this. Note that the coefficient of Ly is the characteristic polynomial $P(s)$, and that the two terms on the right are associated with the forcing function and with the initial conditions respectively.

If the Laplace transform of (1) is taken, the result is

(4) $$P(s)Ly = F(s) + P_0(s)$$

where $F = Lf$ and where $P_0(s)$ is a polynomial of degree $n-1$ depending on the initial conditions. A formula for $P_0(s)$ is readily obtained by Theorem 2, but it is better to use Theorem 2 directly as in the foregoing example.
Equation (4) gives

(5) $$Ly = \frac{F(s)}{P(s)} + \frac{P_0(s)}{P(s)}.$$

To interpret (5), consider two simpler problems. The first problem is

$$Tu = f(t), \quad u(0) = u'(0) = \cdots = u^{(n-1)}(0) = 0$$

which describes a forced response with null initial conditions. The second problem is

$$Tv = 0, \quad v(0) = y_0, \quad v'(0) = y_1, \quad \cdots, \quad v^{n-1}(0) = y_{n-1}$$

which describes an unforced response with arbitrary initial conditions. By (5) with $P_0 = 0$ or $f = 0$, as the case may be, the two functions u and v satisfy

(6)
$$Lu = \frac{F(s)}{P(s)}, \qquad Lv = \frac{P_0(s)}{P(s)}$$

respectively. Equation (5) indicates that $Ly = Lu + Lv$. Hence

$$Ly = L(u + v), \qquad y = u + v,$$

where the first of these equations follows from linearity and the second from the uniqueness theorem. Thus y is a superposition of the two simpler functions u and v. For dissipative systems u is a representative of the steady state as discussed in Section 8.3, and v represents a transient that depends on the initial conditions. The separation of steady-state and transient effects embodied in (5) is one of many advantages of the transform method.

The Laplace transform is very convenient for the solution of systems of differential equations. As an illustration, consider the system

(7)
$$x' = ax + by + f, \qquad x(0) = x_0$$
$$y' = cx + dy + g, \qquad y(0) = y_0$$

where a, b, c, d are constant and where f and g belong to the class E. Assuming that x and y satisfy the hypothesis of Theorem 2 with $n = 1$ we take the Laplace transform of each equation and obtain

$$sX - x_0 = aX + bY + F$$
$$sY - y_0 = cX + dY + G$$

where $X = Lx$, $Y = Ly$, $F = Lf$, and $G = Lg$. Moving the unknowns X, Y to the left and the other terms to the right gives

$$(s - a)X - bY = F + x_0$$
$$-cX + (s - d)Y = G + y_0$$

For large s the coefficient determinant

$$P(s) = \begin{vmatrix} s - a & -b \\ -c & s - d \end{vmatrix} = s^2 - (a + d)s + (ad - bc)$$

is not 0 and we can use Cramer's rule. The result is

$$P(s)X = \begin{vmatrix} F + x_0 & -b \\ G + y_0 & s - d \end{vmatrix} = \begin{vmatrix} F & -b \\ G & s - d \end{vmatrix} + \begin{vmatrix} x_0 & -b \\ y_0 & s - d \end{vmatrix}$$

where $P(s)$ is the *characteristic polynomial* of the system. It is left for you to get a similar equation for $P(s)Y$. Note that the effects of the forcing function and of the initial condition are separated, much as in the case $Ty = f$ discussed above.

EXAMPLE 3. Solve the following system by means of the Laplace transform:

$$x' = 6x + y, \qquad x(0) = 2; \qquad y' = 4x + 3y, \qquad y(0) = 7.$$

Taking the transform of each equation, we get

$$sX - 2 = 6X + Y \qquad \text{or} \qquad (s - 6)X - Y = 2$$
$$sY - 7 = 4X + 3Y \qquad \text{or} \qquad -4X + (s - 3)Y = 7.$$

By Cramer's rule applied to the system on the right

$$\begin{vmatrix} s - 6 & -1 \\ -4 & s - 3 \end{vmatrix} X = \begin{vmatrix} 2 & -1 \\ 7 & s - 3 \end{vmatrix} \qquad \text{or} \quad (s^2 - 9s + 14)X = 2s + 1.$$

Thus

$$X = \frac{2s + 1}{s^2 - 9s + 14} = \frac{3}{s - 7} - \frac{1}{s - 2}$$

where the second form is obtained by the same method of partial fractions that is commonly used for integrating rational functions. (A systematic discussion is given in the following section.) By the table of transforms

$$x = 3e^{7t} - e^{2t}.$$

A similar method gives $y = 3e^{7t} + 4e^{2t}$, or y can be found from $y = x' - 6x$.

The transform method applies to linear systems of far greater generality than (7), and the key to such applications is given by Theorems 1 and 2. As an illustration, consider a system of the form

(8)
$$T_{11}x + T_{12}y = f$$
$$T_{21}x + T_{22}y = g$$

where each T_{ij} is an operator like T in (1) with the characteristic polynomial $P_{ij}(s)$. Assuming $f, g \in E$ the rest solution of (8) satisfies

$$P_{11}(s)X + P_{12}(s)Y = F$$
$$P_{21}(s)X + P_{22}(s)Y = G$$

If the *characteristic polynomial*

$$P(s) = \begin{vmatrix} P_{11}(s) & P_{12}(s) \\ P_{21}(s) & P_{22}(s) \end{vmatrix}$$

does not vanish for large s we can solve for X and Y by Cramer's rule. General initial conditions are introduced by Theorem 2.

Since Cramer's rule applies to n-by-n linear systems, there is no difficulty of principle in extending the foregoing methods to systems

$$\sum_{j=1}^{n} T_{ij} x_j = f_i, \qquad i = 1, 2, \cdots, n$$

with the characteristic polynomial $P(s) = \det P_{ij}(s)$. The Laplace transform thus gives an alternative approach to the method of eigenvectors as presented in Chapter 16.

PROBLEMS

In all problems, assume that the functions whose transforms you need belong to E and that there is enough continuity for the use of Theorem 2.

1. By inspection, find Ly for the rest solution:

(a) $y''' + y'' - 2y' = f$ (b) $y''' + 2y'' + y' = f$ (c) $y^{iv} + y = f$

2. (abc) Instead of considering the rest solution in Problem 1, assume appropriate initial conditions of the form $y^{(j)}(0) = y_j$, and again find Ly.

3. (abcd) Find $X = Lx$ and $Y = Ly$ for the rest solution:

$$
\begin{aligned}
x' &= x - y + f & x' &= 2x + 3y + f & x' + y &= x & x' + f &= y \\
y' &= x - y + g & y' &= 3x + 2y + g & y' - x &= g & y' - y &= x
\end{aligned}
$$

4. (abcd) In the following systems let $x(0) = k$ and $y(0) = 0$. Express $X = Lx$ in a form that shows the separate roles of f and k:

$$
\begin{aligned}
x' + y &= 0 & x' + x &= y & x' &= 3x + 2y + f & x' + f &= y + x \\
y' + x &= f & y' - x &= f & y' &= 4x + 2y & y' &= x + y
\end{aligned}
$$

5. Let x and y satisfy the second-order system

$$
\begin{aligned}
x'' &= ax + by + f, & x(0) &= x_0, & x'(0) &= x_1 \\
y'' &= cx + dy + g, & y(0) &= y_0, & y'(0) &= y_1
\end{aligned}
$$

where a, b, c, d are constant. Show that

$$
\begin{vmatrix} s^2 - a & -b \\ -c & s^2 - d \end{vmatrix} Lx = \begin{vmatrix} F & -b \\ G & s^2 - d \end{vmatrix} + \begin{vmatrix} sx_0 + x_1 & -b \\ sy_0 + y_1 & s^2 - d \end{vmatrix}
$$

and obtain a similar formula for Ly. No answer is given.

6. Find Lx and Ly if $x(0) = 1$, $x'(0) = 2$, $y(0) = 3$, $y'(0) = 4$ and

$$x'' = 3x - 4y, \qquad y'' = 2x - 3y.$$

7. Using the Laplace transform, solve $y^{(5)} = 6!t^2$ subject to the initial conditions $y^{(j)}(0) = j!$, $j = 0, 1, 2, 3, 4$.

———————————————————ANSWERS———————————————————

1. (a) $(s^3 + s^2 - 2s)Ly = F$　(b) $(s^3 + 2s^2 + s)Ly = F$　(c) $(s^4 + 1)Ly = F$
2. Add the following terms to F in the above solutions for Problem 1:
      (a) $y_0(s^2 + s - 2) + y_1(s + 1) + y_2$　(c) $y_0 s^3 + y_1 s^2 + y_2 s + y_3$
      (b) $y_0(s + 1)^2 + y_1(s + 2) + y_2$
3. (a) $s^2 X = (s + 1)F - G$, $s^2 Y = (s - 1)G + F$
      (b) $(s^2 - 4s - 5)X = (s - 2)F + 3G$, $(s^2 - 4s - 5)Y = (s - 2)G + 3F$
      (c) $(s^2 - s + 1)X = -G$, $(s^2 - s + 1)Y = (s - 1)G$
      (d) $(s^2 - s - 1)X = (1 - s)F$, $(s^2 - s - 1)Y = -F$
4. (a) $(s^2 - 1)X = ks - F$
      (b) $(s^2 + s - 1)X = ks + F$
      (c) $(s^2 - 5s - 2)X = (ks - 2k) + (Fs - 2F)$
      (d) $(s^2 - 2s)X = (ks - k) + (F - Fs)$
6. $(s^4 - 1)X = s^3 + 2s^2 - 9s - 10$, $(s^4 - 1)Y = 3s^3 + 4s^2 - 7s - 8$
7. $(2/7)t^7 + 1 + t + t^2 + t^3 + t^4$

---

3. Techniques of computation

We continue to denote by T the linear operator introduced in the previous section. Suppose the forcing function $f(t)$ is a finite sum of functions of the form

$$t^m e^{at}(A \cos bt + B \sin bt)$$

where $m \geq 0$ is an integer and a, b, A, B are constant. It follows from Sections 8.3 and 11.2 that every solution of $Ty = f$ is a finite sum of the same general form, though the constants in it may be different. Hence Ly is a rational function, that is, a quotient $Q(s)/P(s)$ of two polynomials. This can be seen from the formula

$$Lt^m e^{kt} = \frac{m!}{(s - k)^{m+1}}$$

which holds when $k = a + ib$ is any complex constant.

(a) Partial fractions

When $F(s)$ is rational, the basic method of recovering $f(t)$ from $F(s)$ is by expansion into partial fractions, and a brief review is given now. Let a_j denote unequal real or complex numbers and let

$$P(s) = (s - a_1)(s - a_2) \cdots (s - a_n).$$

If $Q(s)$ is any polynomial of degree less than n, there are constants A_j such that the partial-fraction expansion

$$\frac{Q(s)}{P(s)} = \frac{A_1}{s - a_1} + \frac{A_2}{s - a_2} + \cdots + \frac{A_n}{s - a_n}, \qquad s \neq a_j$$

holds. (For proof see DETA page 534). The constants A_j can be determined by clearing of fractions and equating coefficients of corresponding powers of s. But a more efficient method is to multiply by $s - a_j$ and let $s \to a_j$. This gives

$$A_j = \lim_{s \to a_j} \frac{Q(s)(s - a_j)}{P(s)} = Q(a_j) \lim_{s \to a_j} \frac{s - a_j}{P(s)}.$$

By the definition of derivative

$$\lim_{s \to a_j} \frac{P(s)}{s - a_j} = \lim_{s \to a_j} \frac{P(s) - P(a_j)}{s - a_j} = P'(a_j)$$

where the first equality follows from the fact that $P(a_j) = 0$. Thus we get the following theorem:

THEOREM 1. *Let $P(s)$ be a polynomial of degree $n \geq 1$ with n distinct real or complex zeros a_j and let $Q(s)$ be any polynomial of degree $< n$. Then for $s \neq a_j$*

$$\frac{Q(s)}{P(s)} = \sum_{j=1}^{n} \frac{A_j}{s - a_j} \quad \text{where} \quad A_j = \frac{Q(a_j)}{P'(a_j)}.$$

If a_j is a root of multiplicity m, it is shown in books on algebra that the term $A_j/(s - a_j)$ in the partial-fraction expansion must be replaced by

$$(1) \qquad \frac{A_{j1}}{s - a_j} + \frac{A_{j2}}{(s - a_j)^2} + \cdots + \frac{A_{jm}}{(s - a_j)^m}$$

where the A_{jk} are constant. A similar procedure applies to quadratic factors. For example, a factor $(s^2 + as + b)^3$ leads to

$$\frac{A_{11} + B_{11}s}{s^2 + as + b} + \frac{A_{12} + B_{12}s}{(s^2 + as + b)^2} + \frac{A_{13} + B_{13}s}{(s^2 + as + b)^3}.$$

The following theorem is an aid in checking whether a partial-fraction expansion is in the correct form. For proof see DETA, page 535:

THEOREM 2. *Let P be a polynomial of degree $n \geq 1$ and Q a polynomial of degree $< n$. Then the number of arbitrary constants in the partial-fraction expansion of $Q(s)/P(s)$ equals the degree n of the denominator. This is the same as the number of constants needed to specify the numerator Q.*

For example, let $Q(s)$ be a polynomial of degree ≤ 8 and let

$$F(s) = \frac{Q(s)}{s^3(s^2+4)^2(s-3)(s+5)}.$$

The appropriate form for the partial fraction expansion is

$$\frac{A}{s^3} + \frac{B}{s^2} + \frac{C}{s} + \frac{Ds+E}{(s^2+4)^2} + \frac{Fs+G}{s^2+4} + \frac{H}{s-3} + \frac{I}{s+5}$$

which has nine constants $A, B, \cdots, I$.

(b) Properties of the transform

The scope of the transform method is increased by a number of general properties, some of which have already been used. For example we showed in Section 1 that L is linear and we derived the *first shift theorem*

$$Le^{-ct}f(t) = F(s+c).$$

This holds under the sole hypothesis that Lf exists, and the constant c can be complex. Further properties are obtained now. However, some of these have a restriction on f and all require that c be real.

Corresponding to the first shift theorem is *second shift theorem*

$$Lf(t-c) = e^{-sc}F(s) \quad \text{if} \quad c \geq 0 \quad \text{and} \quad f(t) = 0 \quad \text{for} \quad t < 0.$$

For proof, let $u = t - c$ and $t = c + u$ to get

$$\int_0^\infty e^{-st}f(t-c)\,dt = \int_{-c}^\infty e^{-s(c+u)}f(u)\,du = e^{-sc}\int_{-c}^\infty e^{-su}f(u)\,du.$$

The limits $(-c, \infty)$ can be changed to $(0, \infty)$ since $f(u) = 0$ for $u < 0$, and the result follows.

To emphasize the condition $f(t) = 0$ for $t < 0$, the function $f(t)$ is often replaced by $h(t)f(t)$. Here h is the *Heaviside unit function*,

$$h(t) = 0 \text{ for } t < 0, \qquad h(t) = 1 \text{ for } t \geq 0.$$

This is used in Entry h of the table and in Section 22.1.

If $c > 0$ the following *scaling theorems* hold:

$$Lf(ct) = \frac{1}{c} F\left(\frac{s}{c}\right), \quad F(cs) = L\frac{1}{c} f\left(\frac{t}{c}\right).$$

The first is obtained by the change of variable $u = ct$. Replacing c by $1/c$ gives the second.

The *first differentiation theorem* and *first integration theorem* are respectively

$$Lf'(t) = sF(s) - f(0), \quad L\int_0^t f(u)\, du = \frac{F(s)}{s}.$$

The left-hand equation was derived in Section 1 under the hypothesis that $f' \in E$ and f is continuous. If $f \in E$, we can apply this equation to the integral of f. This gives the equation on the right.

The *second differentiation theorem* and *second integration theorem* are, respectively,

$$F'(s) = -Ltf(t), \quad \int_s^\infty F(s)\, ds = L\frac{f(t)}{t}.$$

The equation on the left holds if $f \in E$, that on the right if $f(t)/t \in E$. It is easily checked that these results agree with those obtained by formal differentiation or integration of the Laplace transform. A proof using uniform convergence is in DETA, pages 544–545.

The following result shows why the numerator of a rational transform $F(s)$ always has lower degree than the denominator:

$$(2) \qquad\qquad f \in E \Rightarrow \lim_{s\to\infty} F(s) = 0.$$

The truth of this is obvious because the hypothesis $f \in E$ gives

$$|F(s)| \leq \int_0^\infty e^{-st} Ae^{Bt}\, dt = \frac{A}{s - B}, \qquad s > B$$

where A and B are constant. If $f' \in E$ and f is continuous, we can apply (2) to the equation $Lf' = sLf - f(0)$ and get

$$\lim_{s\to\infty} sF(s) = f(0).$$

PROBLEMS

In all problems, $Q(s)$ introduced without explanation denotes a polynomial of degree lower than that of the denominator.

1P. We want to find constants A, B, C such that

$$\frac{s^2 - 10s + 13}{(s-1)(s-2)(s-3)} = \frac{A}{s-1} + \frac{B}{s-2} + \frac{C}{s-3}, \qquad s \neq 1, 2, 3$$

(a) Multiply by $s - 1$ to obtain

$$\frac{s^2 - 10s + 13}{(\cancel{s+1})(s-2)(s-3)} = \frac{A}{\cancel{s+1}} + \frac{B}{s-2}(s-1) + \frac{C}{s-3}(s-1)$$

where the crossed-out factors can be covered by your finger or by a small coin such as a dime. Let $s \to 1$ and get $A = (1 - 10 + 13)/(-1)(-2) = 2$.

(b) By the same technique find $B = 3$ and $C = -4$, and show that these values of A, B, C lead to an identity when you clear of fractions.

2. As in Problem 1, find A, B, C, D:

$$\frac{Q(s)}{(s-1)(s-2)(s+1)(s-3)} = \frac{A}{s-1} + \frac{B}{s-2} + \frac{C}{s+1} + \frac{D}{s-3}$$

3. (a) Multiply by $(s-1)^3$ and let $s \to 1$ to get A. Also find D and E:

$$\frac{Q(s)}{(s-1)^3(s+2)(s-3)} = \frac{A}{(s-1)^3} + \frac{B}{(s-1)^2} + \frac{C}{s-1} + \frac{D}{s+2} + \frac{E}{s-3}$$

(b) Multiply by s and let $s \to \infty$ to get $C + D + E = b$ where b is the leading coefficient of $Q(s) = bs^4 + \cdots$. This gives C.

(c) Set $s = 0$ to get $Q(0)/6 = -A + B - C + D/2 - E/3$, which gives B.

The gist of this important method is that the values $s = 1, -2, 3$ were used to find A, D, E and any other values, including ∞, can be used to find the remaining coefficients B, C. Letting $s \to \infty$ in the equation as it stands produces the equation $0 = 0$, which gives no information. That is why we multiply by s before letting $s \to \infty$.

4. (a) Find A, B, C as in Problem 1:

$$\frac{Q(s)}{(s+1)(s+3)(s-2)(s^2 + 2s + 2)} = \frac{A}{s+1} + \frac{B}{s+3} + \frac{C}{s-2} + \frac{Ds + E}{s^2 + 2s + 2}$$

(b) Multiply by s and let $s \to \infty$ to get $A + B + C + D = b$ where b is the leading coefficient of $Q(s) = bs^4 + \cdots$. This gives D.

(c) Check that setting $s = 0$ gives an equation from which E can be found.

5. We assume Q real here because the calculation simplifies in that case:

$$\frac{Q(s)}{(s^2 + 1)^2(s^2 + 4)} = \frac{As + B}{(s^2 + 1)^2} + \frac{Cs + D}{s^2 + 1} + \frac{Es + F}{s^2 + 4}$$

(a) Multiply by $(s^2 + 1)^2$ and let $s \to i$ to get $Ai + B = Q(i)/3$. Equating real and imaginary parts gives both A and B.

(b) Similarly get E and F.

(c) Check that multiplying by s and letting $s \to \infty$ gives C, and the choice $s = 0$ gives D.

6. Expand in partial fractions:

(a) $\dfrac{4s^2 + 5s - 12}{(s - 1)(s - 2)(s^2 + s + 1)}$
(b) $\dfrac{4(s^4 - s^3 + 1)}{(s - 2)(s + 1)^2(s - 1)^2}$

7. Find the inverse transform $f = L^{-1}F$:

(a) $\dfrac{2s^3 + 6s^2 + 21s + 52}{s(s + 2)(s^2 + 4s + 13)}$
(b) $\dfrac{s^2 - 18s + 5}{(s - 1)^3(s - 2)(s - 3)}$

8. Using the Laplace transform, solve $y' + y = 2\sin t$, $y(0) = 1$.

9. For the initial-value problem

$$y''' - y' = 4\sin t, \qquad y(0) = 8, \qquad y'(0) = 0, \qquad y''(0) = 4$$

check that

$$Ly = \frac{8s^3 + 4s}{(s - 1)(s + 1)(s^2 + 1)}, \qquad y = 6\cosh t + 2\cos t$$

10. Solve by the Laplace transform:

(a) $y'' - 3y' + 2y = 4$, $y(0) = 2$, $y'(0) = 3$
(b) $y'' + 4y = 6\sin t$, $y(0) = 6$, $y'(0) = 0$
(c) $y''' + y'' = 6(e^t + t + 1)$, $y(0) = y'(0) = y''(0) = 0$
(d) $y''' - y' = 2$, $y(0) = y'(0) = y''(0) = 4$
(e) $y''' - y' = 6 - 3t^2$, $y(0) = y'(0) = y''(0) = 1$

11. Find the rest solution of $(D+1)^3(D+2)y = 6$.

12. Solve subject to the initial conditions $x(0) = 0$, $y(0) = 25$:

$$x' + 2y' + x - y = 25, \qquad 2x' + y = 25e^t$$

13. (abcd) Solve subject to the initial conditions $x(0) = 1$, $y(0) = 0$:

$$x' + 5x = 3y \qquad x' + 5x = y \qquad x' + 6x = 2y \qquad x' = x + 2y$$
$$y' + 4x = 2y \qquad y' + 3y = -x \qquad y' + 7y = 3x \qquad y' + 2x = -3y$$

14. Deduce from the second and first integration theorems that

$$L\frac{\sin t}{t} = \frac{\pi}{2} - \tan^{-1} s, \qquad L\left(\int_0^t \frac{\sin u}{u}\,du\right) = \frac{\pi}{2s} - \frac{\tan^{-1} s}{s}$$

———————————————————ANSWERS———————————————————

2. $A = Q(1)/4$, $B = -Q(2)/3$, $C = -Q(-1)/24$, $D = Q(3)/8$
3. $A = -Q(1)/6$, $D = Q(-2)/135$, $E = Q(3)/40$
4. $A = -Q(-1)/6$, $B = Q(-3)/50$, $C = Q(2)/150$
5. Partial answer: $2Ei + F = Q(2i)/9$
6. Partial answer: (a) The sum of the coefficients is -4; their product is 24.
   (b) The sum of the coefficients is 2; their product is -4.
7. (a) $2 - e^{-2t} + e^{-2t}(\cos 3t - 2\sin 3t)$ 8. $2e^{-t} - \cos t + \sin t$
   (b) $e^t(-3t^2 - 17t - 22) + 27e^{2t} - 5e^{3t}$
10. (a) $2 - 3e^t + 3e^{2t}$ (c) $t^3 - 6t + 6\sinh t$ (e) $t^3 + e^t$
    (b) $2\sin t + 6\cos 2t - \sin 2t$ (d) $-e^{-t} + 5e^t - 2t$
11. $y = -3t^2 e^{-t} - 6e^{-t} + 3e^{-2t} + 3$
12. $x = 25 - 9e^t + 5te^t - 16e^{-t/4}$, $y = 25e^t - 2x'$
13. Partial answers: (a) $y = 4e^{-2t} - 4e^{-t}$ (c) $5y = 3(e^{-4t} - e^{-9t})$
                      (b) $y = -te^{-4t}$ (d) $y = -2te^{-t}$

I do not refuse my dinner, just because I don't understand
the process of digestion.

—Oliver Heaviside

Chapter 22

APPLICATIONS

THE LAPLACE transform is useful in dealing with discontinuous inputs (closing of a switch) and with periodic functions (sawtooth waves, rectified sine waves). Analysis of the effect of such inputs proceeds most smoothly in the frequency domain, that is, in the domain of the transform-variable s. However, an integral expression known as the convolution allows an easy passage from the frequency domain to the time domain and leads to explicit solutions even in the case of a general forcing function $f(t)$. The chapter concludes with a brief introduction to feedback, the Dirac function, and the generalized factorial.

1. The frequency domain

For suitable functions $f(t)$ we have defined the Laplace transform by

$$Lf = \int_0^\infty e^{-st} f(t)\, dt = F(s).$$

This formula represents a superposition of exponential functions, e^{-st}, where the superposition is over t and s is a parameter. By contrast, an expression such as

$$c_1 e^{s_1 t} + c_2 e^{s_2 t} + \cdots + c_n e^{s_n t}$$

describes a superposition over s with t as a parameter.

In most applications t is the time, and the domain of the real variable t is the *time domain*. But the product st in e^{-st} is necessarily a pure number. One way to see this is by the Taylor series

$$e^{-st} = 1 - st + \frac{(st)^2}{2!} - \frac{(st)^3}{3!} + \cdots.$$

If st had any units, such as centimeters or seconds, the terms in the series would all have different units and the summation would make no sense. Since st is a pure number, s must have the units of $1/t$. Hence it represents the number of something dimensionless (such as cycles) per second, and has the units of frequency. This interpretation agrees with the familiar fact that ω in $e^{i\omega t}$ represents the angular frequency of the corresponding harmonic oscillations, $\cos \omega t$ and $\sin \omega t$.

It is often advantageous to let s be complex. In that case s is referred to as the *complex frequency* and the domain of the complex variable s is the *frequency domain*. Unless the contrary is indicated by context, however, we shall continue to regard s as real.

In those scientific fields that make robust uses of the Laplace transform there is a growing tendency to work in the frequency domain rather than in the time domain. It is an important feature of the frequency domain that many transcendental operations on $f(t)$ correspond to algebraic operations on $F(s)$. For example, the formula

$$L \int_0^t f(t)\,dt = \frac{F(s)}{s}$$

shows that integration corresponds to division by s.

(a) Discontinuous inputs

The function $f(t)$ is "admissible on a finite interval $[a, b]$" if it is defined and continuous except at finitely many points of this interval and is bounded on the set of points where it is defined. The notation $f \in A$ means that f is admissible on $[0, a]$ for every a, and we refer to such functions as *admissible*. The class E of functions introduced in Section 21.1 is the class of admissible functions that are at most of exponential growth. The condition $f \in A$ ensures that f is Riemann integrable on every finite interval $[0, a]$, and $f \in E$ ensures existence of the Laplace transform. That is why these two classes are appropriate.

A *simple discontinuity* c is a discontinuity at which the left- and right-hand limits

$$\lim_{t \to c-} f(t) = f(c-), \qquad \lim_{t \to c+} f(t) = f(c+)$$

exist and are not equal. A function f is said to be *piecewise continuous* if it is admissible and all of its discontinuities are simple. In the theory of Chapter 21 it did not matter whether the discontinuities were simple or not. We want the discontinuities to be simple here, however, for the following theorem:

THEOREM 1. *Let $f'(t) \in E$ and let $f(t)$ be continuous except at a single value $t = c > 0$ where $f(t)$ has a simple discontinuity. Then*

$$Lf'(t) = sLf(t) - f(0+) - J(c)e^{-sc}$$

where $J(c) = f(c+) - f(c-)$ is the jump of $f(t)$ at c.

This follows by addition of the two equations

$$\int_0^c e^{-st} f'(t)\,dt = e^{-st} f(t) \Big|_0^c + s \int_0^c f(t)e^{-st}\,dt$$

$$\int_c^\infty e^{-st} f'(t)\,dt = e^{-st} f(t) \Big|_c^\infty + s \int_c^\infty f(t)e^{-st}\,dt$$

which are obtained by partial integration. As in the more detailed discussion given in Section 21.1, the integrated term at ∞ disappears when s is large. To get the one-sided limits replace c by $c - \delta$ in the first equation, by $c + \delta$ in the second, and let $\delta \to 0+$.

If $f(t)$ has finitely many discontinuities, all simple, there is a term $J(c)e^{-sc}$ for each of them. Applying Theorem 1 to $f^{(n)}$ instead of f and solving for $Lf^{(n)}$, we get

$$(1) \qquad Lf^{(n)} = \frac{Lf^{(n+1)}}{s} + \frac{J_n(0) + e^{-sc_1}J_n(c_1) + e^{-sc_2}J_n(c_2) + \cdots}{s}$$

where $J_n(c_j)$ is the jump of $f^{(n)}$ at c_j. For uniformity of notation we have considered that $f(t) = 0$ for $t < 0$ and hence that

$$J_n(0) = f^{(n)}(0+).$$

The form of Equation (1) alerts you to the fact that a term $J_n(c_j)$ is needed at each discontinuity.

(b) Piecewise polynomial inputs

We shall find Lf if $f(t) = 0$ when $t > 2$ and

$$f(t) = t^3, \qquad 0 \le t \le 1; \qquad f(t) = (2-t)^3, \qquad 1 \le t \le 2.$$

The graph of $f(t)$ is shown in Figure 1(a). The function f is defined by a polynomial on each of finitely many intervals and is said to be *piecewise polynomial*; the word "polynomial" here is an adjective not a noun. Equation (1) is especially useful for such functions, because sufficiently high derivatives of a polynomial are 0.

We begin by differentiating $f(t)$ on the intervals $(0, 1)$, $(1, 2)$, and $(2, \infty)$. The first row in the following table is obtained by copying the definition of $f(t)$ and the other rows follow by inspection:

Table 1

interval	$(0, 1)$	$(1, 2)$	$(2, \infty)$
$f^{(0)}(t)$	t^3	$(2-t)^3$	0
$f^{(1)}(t)$	$3t^2$	$-3(2-t)^2$	0
$f^{(2)}(t)$	$6t$	$6(2-t)$	0
$f^{(3)}(t)$	6	-6	0

The next table is found by evaluating these derivatives at 0,1,2. The limits at 0+ and 1− are obtained from values on (0,1), hence from the first

column of Table 1. The limits at 1+ and 2− are obtained from the second
column, and the limit 0 at 2+ is obtained from the last column:

Table 2

n	$f^{(n)}(0+)$	$f^{(n)}(1-)$	$f^{(n)}(1+)$	$f^{(n)}(2-)$	$f^{(n)}(2+)$
0	0	1	1	0	0
1	0	3	−3	0	0
2	0	6	6	0	0
3	6	6	−6	−6	0

The values of $J_n(0)$ are given by the first column of Table 2, $J_n(1)$ is
given by the third column minus the second, and $J_n(2)$ is given by the last
column minus the fourth. Thus we get a third table:

Table 3

n	$J_n(0)$	$J_n(1)$	$J_n(2)$
0	0	0	0
1	0	−6	0
2	0	0	0
3	6	−12	6

The first row has only zeros because $f(t)$ is continuous as seen by Figure
1(a). For continuous functions f, the jump is 0 at every value c. We have
included this third table only for clarity of exposition. In practice, one could
insert the three columns of Table 3 between appropriate columns of Table 2.

Since $f^{(4)}(t) = 0$ except at 0,1,2 where it is undefined, Equation (1)
with $n = 3$ gives

$$Lf^{(3)} = \frac{6 - 12e^{-s} + 6e^{-2s}}{s} = 6\frac{(1 - e^{-s})^2}{s}.$$

Here we used the last line of Table 3. The third line of the table and (1)
show that $Lf^{(2)}$ is obtained when the above formula is divided by s. The
second line of the table then gives

$$Lf' = 6\frac{(1 - e^{-s})^2}{s^3} - 6\frac{e^{-s}}{s}$$

and, referring to the first line of the table, we see that division by s gives Lf.
Direct calculation of Lf from its definition involves elementary integrals, but
the above method is easier. It applies to all piecewise polynomial functions.

In the course of the computation we obtained the Laplace transforms of f', f'', f''' automatically. The graphs of these functions are shown in Figure 1(bcd). It does not matter how the functions are defined, or even whether they are defined, at points of discontinuity. This fact is indicated by vertical lines at the jumps in Figures 1 and 3. Graphs with the jumps filled in are called *maximal*. They are used for mathematical reasons in some parts of the theory of differential equations, but are introduced for visual reasons here.

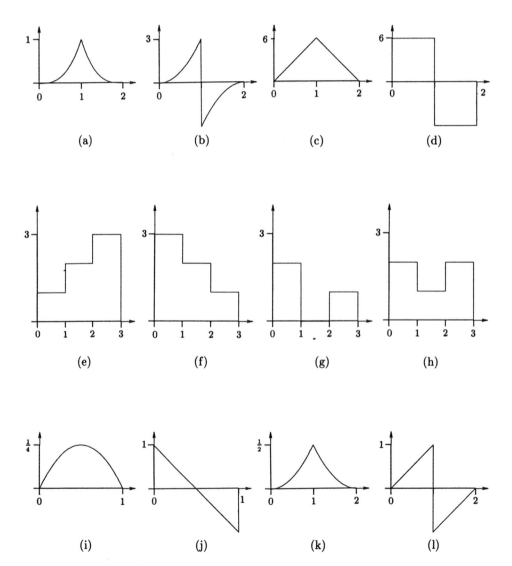

Figure 1

(c) The Heaviside unit function

The *Heaviside unit function* is defined by the equations

$$h(t) = 0, \qquad t < 0; \qquad h(t) = 1, \qquad t \geq 0.$$

Clearly $Lh = L1 = 1/s$, a fact used below.

One of the main uses of the Heaviside function is in connection with the second shift theorem,

$$Lf(t - c) = e^{-sc}Lf(t), \qquad c \geq 0,$$

which requires the side condition $f(t) = 0$ for $t < 0$. If $f(t)$ is defined for $t < 0$ (this is essential) then $h(t)f(t)$ agrees with $f(t)$ for $t \geq 0$ and is 0 for $t < 0$. Hence, it satisfies the side condition. If $0 \leq a < b$ the first of the functions

$$h(t - a)f(t), \qquad h(t - b)f(t)$$

agrees with $f(t)$ for $t \geq a$ and is 0 for $t < a$. The second of these functions agrees with $f(t)$ for $t \geq b$ and is 0 for $t < b$. A moment's thought shows then that the difference

$$h(t - a)f(t) - h(t - b)f(t)$$

agrees with $f(t)$ on the interval $[a, b)$ and is 0 elsewhere. This function describes the effect of closing a switch at time $t = a$ and opening it at $t = b$. Since $Lh(t) = 1/s$ the second shift theorem gives

(2) $$\qquad\qquad\qquad Lh(t - c) = \frac{e^{-cs}}{s}, \qquad c \geq 0.$$

EXAMPLE 1.   The response y to an input f satisfies $Ty = f$ where T is a linear differential operator with constant coefficients. Find the frequency-domain response corresponding to the rest solution, if a switch to a unit input is closed at time $t = a \geq 0$ and opened at time $t = b > a$.

The desired frequency-domain response is $Y = Ly$ where y is the rest solution of

$$Ty = h(t - a) - h(t - b).$$

There is no factor $f(t)$ because the term *unit input* means $f = 1$. Equation (2) with $c = a$ and $c = b$ yields

$$P(s)Y = \frac{e^{-as} - e^{-bs}}{s}$$

where $P(s)$ is the characteristic polynomial of T. Division by $P(s)$ gives Y.

EXAMPLE 2. If $f_0(t)$ represents one arch of a sine curve, so that $f_0(t) = \sin t$ for $0 \le t < \pi$ and $f_0(t) = 0$ elsewhere, find $F_0 = Lf_0$. Since

$$f_0(t) = h(t)\sin t - h(t - \pi)\sin t = h(t)\sin t + h(t - \pi)\sin(t - \pi)$$

the second shift theorem with $c = 0$ and with $c = \pi$ gives

$$F_0(s) = (1 + e^{-s\pi})L\sin t = \frac{1 + e^{-s\pi}}{1 + s^2}.$$

As a check, you can equate imaginary parts in

$$\int_0^\pi e^{-st}e^{it}\,dt = \frac{e^{(i-s)t}}{i - s}\Big|_0^\pi = \frac{1 + e^{-s\pi}}{s - i}.$$

(d) Periodic functions

Let f be a periodic function of period p, so that $p > 0$ and

$$f(t + p) = f(t)$$

for all t. We assume $f \in E$, which holds if $f(t)$ is admissible on an interval covering a period, such as $0 \le t \le p$.

The function

(3) $$f_0(t) = h(t)f(t) - h(t - p)f(t)$$

agrees with $f(t)$ on $0 \le t < p$ and is 0 elsewhere. Hence f_0 represents f on a single period, and f will be called the *periodic extension* of f_0. Since f is periodic we can replace $f(t)$ by $f(t - p)$ in the second term of (3). Taking the Laplace transform then gives

$$Lf_0 = Lf - e^{-ps}Lf$$

and, solving for Lf,

(4) $$Lf = \frac{Lf_0}{1 - e^{-ps}} \quad \text{where} \quad Lf_0 = \int_0^p e^{-st}f(t)\,dt.$$

This is the basic formula for the Laplace transform of a periodic function.

EXAMPLE 3. A *full rectified wave* is defined by

$$f(t) = |\sin t|$$

and has the graph shown in Figure 2(a), while for a half-rectified wave every other arch is omitted as shown in Figure 2(b). For the full wave $f_0(t) = \sin t$ on the interval $0 \le t < \pi$ whereas the half wave has

$$f_0(t) = \sin t, \qquad 0 \le t < \pi; \qquad f_0(t) = 0, \qquad \pi \le t < 2\pi.$$

(a)

Figure 2

(b)

In both cases Lf_0 is given by $F_0(s)$ from Example 2. Hence by (4) the respective Laplace transforms are

$$\frac{F_0(s)}{1 - e^{-\pi s}}, \qquad \frac{F_0(s)}{1 - e^{-2\pi s}}.$$

We have taken $p = \pi$ in the first case and $p = 2\pi$ in the second. Substituting $F_0(s)$ as given by Example 2 yields

$$\frac{1}{1 + s^2} \frac{1 + e^{-\pi s}}{1 - e^{-\pi s}}, \qquad \frac{1}{1 + s^2} \frac{1}{1 - e^{-\pi s}}$$

where the second expression was simplified by cancellation of a common factor. These two functions give the Laplace transform of the full-rectified and half-rectified waves, respectively.

PROBLEMS

1. (efgh) A *step function* is a piecewise polynomial function in which each polynomial is constant. Since the derivative is 0 except at points of discontinuity, the choice $n = 0$ in (1) gives

$$sF(s) = J(0) + e^{-sc_1} J(c_1) + e^{-sc_2} J(c_2) + \cdots$$

for step functions $f(t)$, where $J(c) = J_0(c)$ is the jump of $f(t)$ at $t = c$. Use this to find $sF(s)$ for step functions that are defined on (0,3) by the graphs in Figure 1(efgh) and are 0 elsewhere.

Hint for (e): By inspection, $J(0) = J(1) = J(2) = 1$, $J(3) = -3$.

2. (efgh) Solve Problem 1 by use of $sLh(t - c) = e^{-sc}$, $c \geq 0$

Hint for (e): $f(t) = h(t) + h(t - 1) + h(t - 2) - 3h(t - 3)$.

3. What function must the Laplace transforms of f, f', f'', f''' in Part (b) be divided by to get the transforms of the periodic functions shown in Figure 3(abcd)? The period is $p = 2$ in each case.

4. What function must the transforms $F(s)$ in Problems 1 and 2 be divided by to get the transforms of the periodic functions shown in Figure 3(efgh)? The period is $p = 4$ in each case.

5. Let $f(t) = t(1 - t)$ for $0 \leq t \leq 1$ and $f(t) = 0$ elsewhere. The graphs of f and f' are shown in Figures 1(i) and 1(j).

(a) Starting at the left-hand edge of your paper, make a two-column table showing f, f', f'' on $(0,1)$ versus $n = 0, 1, 2$.

(b) To the above two columns add two more giving the values of $f^{(n)}$ at $0+$ and $1-$. The values at $1+$ are all 0.

(c) To your four columns add two more giving $J_n(0)$ and $J_n(1)$.

(d) Thus find the Laplace transforms of f, f' and of the periodic functions whose graphs are shown in Figure 3(ij).

6. Let $f(t) = 2t$ on $(0,1)$, $f(t) = 3 - t$ on $(1,3)$, and $f(t) = 0$ elsewhere.

(a) Sketch the graph of $f(t)$ and note that, by continuity, $J_0(c) = 0$ at each c.

(b) Find $J_1(c)$ at $c = 0, 1, 3$ by inspection of $f'(t)$.

(c) Thus find the transform of f and of its periodic extension of period 3.

(d) Sketch the graph of the periodic extension.

7. Let $f(t) = t^2/2$ for $0 \leq t \leq 1$, $f(t) = (t - 2)^2/2$ for $1 \leq t \leq 2$, and $f(t) = 0$ elsewhere. The graphs of f and f' are shown in Figures 1(k) and 1(ℓ). Proceeding as in Example 1 or as in the foregoing problems, show that the Laplace transforms of f'', f', and f are respectively

$$\frac{1 - e^{-2s}}{s}, \qquad \frac{1 - e^{-2s}}{s^2} - \frac{2e^{-s}}{s}, \qquad \frac{1 - e^{-2s}}{s^3} - \frac{2e^{-s}}{s^2}$$

8. If the three functions $F(s)$ of Problem 7 are divided by $1 - e^{-2s}$ you get the Laplace transforms of three periodic functions $f(t)$. Sketch the graphs of these periodic functions.

(a)

(b)

(c)

(d)

(e)

(f)

(g)

(h)

(i)

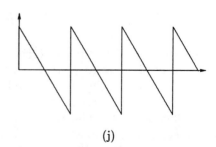

(j)

Figure 3

9. This problem requires knowledge of the geometric series for $1/(1-r)$. Setting $r = e^{-cs}, c > 0$ and proceeding without regard to rigor, give reasons for believing that, in general,

$$\frac{F(s)}{1 - e^{-sc}} = L((h(t)f(t) + h(t-c)f(t-c) + h(t-2c)f(t-2c) + \cdots)$$

If $c = p$ and if $f(t) = f_0(t)$ is 0 outside the interval $(0, c)$, this result is equivalent to the formula for the transform of a periodic function. The method in the text avoids questions about whether the series can be integrated term by term.

———————————————————————ANSWERS———————————————————————

1. (e) $1 + e^{-s} + e^{-2s} - 3e^{-3s}$ (g) $2 - 2e^{-s} + e^{-2s} - e^{-3s}$
   (f) $3 - e^{-s} - e^{-2s} - e^{-3s}$ (h) $2 - e^{-s} + e^{-2s} - 2e^{-3s}$
3. $1 - e^{-2s}$ 4. $1 - e^{-4s}$
5. Partial answer, Figure 3(i): $(1 - e^{-s})Lf = 2s^{-3}(e^{-s} - 1) + s^{-2}(e^{-s} + 1)$
6. Partial answer (c): $s^2(1 - e^{-3s})Lf = 2 - 3e^{-s} + e^{-3s}$ for periodic extension

———

2. The convolution

Consider a linear system in which the effect at the present time t of a stimulus $f(t_1)\, dt_1$ at the past time t_1 is proportional to the stimulus. On physical grounds we assume that the proportionality constant depends only on the elapsed time $t - t_1$ and hence has the form $g(t - t_1)$. The effect at the present time t is therefore

$$f(t_1)g(t - t_1)dt_1.$$

Since the system is linear, the response to the whole past history can be obtained by adding these separate effects and we are led to the integral

(1) $$w(t) = \int_0^t f(t_1)g(t - t_1)\, dt_1.$$

The lower limit is 0 because the process is assumed to have started at time $t = 0$; in other words, $f(t_1) = 0$ for $t_1 < 0$.

The expression (1) is called the *convolution* of f and g. It gives the response at the present time t as a weighted superposition over the inputs at times $t_1 \leq t$. The weighting factor $g(t - t_1)$ characterizes the system and $f(t_1)$ characterizes the past history of the input. Because of this physical interpretation, the convolution is also called the *superposition integral.*

To ensure existence of the integral, we assume from now on that $f \in A$ and $g \in A$, where A is the class of admissible functions introduced in Section 1.

The function w in (1) is denoted by $f * g$, so that $(f * g)(t) = w(t)$ is the value of $f * g$ at the argument t. Naturally the meaning of $*$ does not depend on the letters f and g. For example if $a \in A$ and $b \in A$ then

$$(a * b)(t) = \int_0^t a(t_1)b(t - t_1)dt_1.$$

The $*$ operation acts like ordinary multiplication in that, if a, b, c are admissible,

$$a * (b + c) = a * b + a * c, \qquad a * b = b * a, \qquad a * (b * c) = (a * b) * c.$$

The following theorem shows that convolution in the time domain really does correspond to multiplication in the frequency domain:

CONVOLUTION THEOREM. If $f \in E$ and $g \in E$ then $f * g \in E$ and $L(f * g) = (Lf)(Lg)$.

For proof of these results, see DETA, pages 560–562.

The importance of the convolution could hardly be overestimated. It plays a prominent role in the study of heat conduction, wave motion, plastic flow and creep, and in other branches of mathematical physics. It is also encountered in those areas of sociology, economics, ecology, and genetics in which an effect at a given time t_1 induces a delayed response at a later time t. The convolution theorem gives a bridge between the time and frequency domains, as illustrated in the following example.

EXAMPLE 1. Consider the problem

$$y'' + \omega^2 y = \omega^2 f(t), \qquad y(0) = y'(0) = 0$$

where ω is constant and $f \in E$. Taking the transform yields

(2) $$(s^2 + \omega^2)Ly = \omega^2 Lf \quad \text{or} \quad Ly = (Lf)\frac{\omega^2}{s^2 + \omega^2}.$$

By the table of transforms

$$\frac{\omega^2}{s^2 + \omega^2} = Lg \quad \text{where} \quad g(t) = \omega \sin \omega t.$$

Hence Equation (2) can be written

$$Ly = (Lf)(Lg) = L(f * g)$$

where the second equality follows from the convolution theorem. By uniqueness $y = f * g$ or, more explicitly,

$$y(t) = \omega \int_0^t f(t_1) \sin \omega(t - t_1)\, dt_1.$$

Here we have a formula for the rest solution corresponding to the arbitrary forcing function f.

The method of Example 1 applies to the general equation $Ty = f(t)$ where $f \in E$ and where T is the usual linear operator with constant coefficients,

$$Ty = y^{(n)} + p_{n-1}y^{(n-1)} + \cdots + p_1 y' + p_0 y.$$

Namely, let $P(s)$ be the characteristic polynomial of T and let g be a function such that

(3) $$Lg = \frac{1}{P(s)}.$$

Then the rest solution of $Ty = f(t)$ satisfies $P(s)Ly = Lf$ or

(4) $$Ly = (Lf)\frac{1}{P(s)} = (Lf)(Lg) = L(f * g).$$

Therefore $y = f * g$ by uniqueness. The two equations

$$Y(s) = \frac{F(s)}{P(s)}, \qquad y = f * g$$

describe the solution in the frequency and time domains, respectively.

Suppose now that $P(s)$ has only simple zeros s_j. By Section 21.3, Theorem 1, the partial-fraction expansion of $1/P$ is

$$\frac{1}{P(s)} = \sum_{j=1}^{n} \frac{A_j}{s - s_j} \quad \text{where} \quad A_j = \frac{1}{P'(s_j)}.$$

The equation $Le^{s_j t} = 1/(s - s_j)$ yields

$$\frac{1}{P(s)} = Lg \quad \text{where} \quad g(t) = \sum_{j=1}^{n} A_j e^{s_j t}.$$

If $f \in A$ the method of Part (a) gives $y = f * g$ for the rest solution of $Ty = f$ or, substituting the above expression for g,

$$y(t) = \int_0^t f(t_1) \sum_{j=1}^n A_j e^{s_j(t-t_1)} dt_1.$$

This is the *Heaviside expansion theorem*. It gives the rest solution of $Ty = f$.

Although the above argument requires $f \in E$ the final result $y = f * g$ is valid for $f \in A$, which is a much larger class. For example, it holds with

$$f(t) = e^{e^t} \quad \text{or} \quad f(t) = e^{t^2} \sin e^t$$

neither of which is in E. See DETA, page 560.

EXAMPLE 2. For the equation $y'' + 9y' + 20y = f(t) \in A$ the characteristic polynomial satisfies

$$P(s) = s^2 + 9s + 20, \qquad P'(s) = 2s + 9, \qquad P(s) = (s+4)(s+5).$$

Hence $s_1 = -4$, $s_2 = -5$, $A_1 = 1/P'(-4) = 1$, $A_2 = 1/P'(-5) = -1$ and, by the Heaviside expansion theorem, the rest solution is

$$y(t) = e^{-4t} \int_0^t f(t_1)e^{4t_1} dt_1 - e^{-5t} \int_0^t f(t_1)e^{5t_1} dt_1.$$

For greater clarity, the factors e^{-4t} and e^{-5t} have been moved outside the integral signs. As a check, it is readily verified that the formula gives the correct result when $f(t) = 1$, namely,

$$y = \frac{1}{20} - \frac{1}{4}e^{-4t} + \frac{1}{5}e^{-5t}.$$

PROBLEMS

1P. We are going to find the rest solution of $(D+1)(D+2)y = f(t)$. Check that the characteristic polynomial satisfies

$$\frac{1}{P(s)} = \frac{1}{s+1} - \frac{1}{s+2} = Lg, \quad \text{where} \quad g(t) = e^{-t} - e^{-2t}.$$

Deduce that $Ly = (Lf)(Lg)$ and, from this, that

$$y = f * g = e^{-t} \int_0^t f(t_1)e^{t_1} dt_1 - e^{-2t} \int_0^t f(t_1)e^{2t_1} dt_1$$

2. (abc) Taking $f(t) = h(t)$, $20e^{3t}$ and $4t$ evaluate the solution in Problem 1 and verify that it satisfies both the initial conditions and the differential equation.

3. Find the rest solution:

(a) $y' + y = f(t)$ (b) $y'' = f(t)$ (c) $y'' - y = f(t)$ (d) $y'' + 2y' + 2y = f(t)$

4. (abcd) Evaluate your general formulas in Problem 3 when $f(t) = 1$ and verify that the solutions so obtained satisfy both the initial conditions and the differential equation.   ·

5. Let $y'' + (a+b)y' + aby = f(t)$ where a and b are constant. Find the rest solution (a) when $a \neq b$ and (b) when $a = b$.

6. If ω and ω_1 are constant find the rest solution of $y'' + \omega^2 y = \omega^2 \sin \omega_1 t$ by the general formula obtained in Example 1, and also by the method of undetermined coefficients. Carry out the calculation (a) when $\omega_1 \neq \omega$ and (b) when $\omega_1 = \omega$. Since you are asked to check, no answer is given. The point is that, for problems to which they apply, the elementary techniques of Chapters 6 and 8 are more efficient than general methods, and are not made obsolete by the latter.

6. An *integral equation of Volterra type* is an equation of the form

$$y(t) = f(t) + \int_0^t g(t_1)y(t - t_1)\,dt_1$$

where f and g are known and y is to be found. If the functions belong to E check that the convolution theorem gives $Ly = Lf/(1 - Lg)$.

—————————————————ANSWERS—————————————————

2. (a) $2y = 1 - 2e^{-t} + e^{-2t}$ (b) $e^{3t} + 4e^{-2t} - 5e^{-t}$ (c) $-3 + 2t + 4e^{-t} - e^{-2t}$

3. (a) $e^{-t} \int_0^t f(t_1)e^{t_1}\,dt_1$ (c) $\int_0^t f(t_1)\sinh(t - t_1)\,dt_1$

   (b) $\int_0^t f(t_1)(t - t_1)\,dt_1$ (d) $e^{-t} \int_0^t f(t_1)e^{t_1}\sin(t - t_1)\,dt_1$

5. (a) $(b - a)y = \int_0^t f(t_1)(e^{-a(t-t_1)} - e^{-b(t-t_1)})\,dt_1$

   (b) $\int_0^t f(t_1)(t - t_1)e^{-a(t-t_1)}\,dt_1$

3. Supplementary topics

Three topics taken up here give a very brief introduction to three major areas of engineering and mathematics. We begin with the theory of the transfer function, which brings us back to the frequency domain of Section 1. Next we give a brief discussion of the Dirac delta function $\delta(t)$, and finally we use the Laplace transform to define $p!$ for nonintegral p.

(a) The transfer function and feedback

Throughout Parts (a) and (b) the letter T denotes a linear differential operator with constant coefficients and with the characteristic polynomial $P(s)$. Before taking up the Laplace transform, we found that the equation

$$Tx = Ae^{st}$$

could be solved, in general, by the trial solution $x = Be^{st}$. Under the assumption that $P(s) \neq 0$ the result was $P(s)Be^{st} = Ae^{st}$ or

$$(1) \qquad B = G(s)A \quad \text{where} \quad G(s) = \frac{1}{P(s)}.$$

The function $G(s) = 1/P(s)$ that transforms the input amplitude A into the output amplitude B is called the *transfer function*. Its main property is given by the equation

$$(2) \qquad (\text{output}) = (\text{transfer function})(\text{input}).$$

Let us now characterize the input f and output x by their transforms, F and X. If $f \in E$ the rest solution of the equation $Tx = f$ satisfies

$$(3) \qquad P(s)Lx = Lf \quad \text{or} \quad X = G(s)F$$

where $G(s) = 1/P(s)$ is the transfer function as defined above. The difference between this equation and (1) is that (1) applies to exponential inputs only, while (3) allows the arbitrary input f. This increased scope of the transfer function is one of the advantages of working in the frequency domain.

The transfer function is especially useful when two or more systems are connected together, so that the output of one system becomes the input of another. A simple case of this kind is shown in Figure 4. For uniformity of notation we denote the input by X_0 rather than F. The coupled system is then described by

$$X_1 = G_1(s)X_0, \quad X_2 = G_2(s)X_1, \quad X_2 = G_2(s)G_1(s)X_0.$$

Thus the lower system in the figure is equivalent to the upper one. By repetition a similar result is obtained for any number of systems connected in this fashion. The transfer function for the whole system is the product of the individual transfer functions.

A more elaborate interconnection is shown in Figure 5. Here the basic system is G_1 and the other system G_2 provides a *feedback*. What is meant by this term is that the output is fed back through G_2 and used to modify the input. The encircled plus sign means that the feedback output is added to the input. Hence the total input to G_1 is

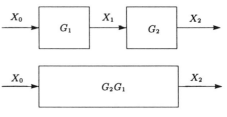

Figure 4

$$X_1 = X_0 + G_2X_2.$$

Using the input-output relation (2) as applied to G_1, we get

$$X_2 = G_1(X_0 + G_2X_2).$$

Therefore the transfer function for the feedback system is

(4) $$G = (1 - G_1G_2)^{-1}G_1$$

Figure 5

as indicated in the lower figure.

In the theory of control, G_2 is replaced by kG_2, so that

$$G = (1 - kG_1G_2)^{-1}G_1$$

k is a real constant. By changing k one can change the transfer function of the system, hence the term *control*.

(b) The Dirac distribution

Here we consider the response of a system to an impulse that acts for a very short time but produces a large effect. The physical situation is exemplified by a lightning strike on a transmission line or a hammer blow on a mechanical system.

To formulate the idea of an impulse, let a be a small positive constant and let $\delta_a(t)$ be the function whose graph is shown in Figure 6. That is,

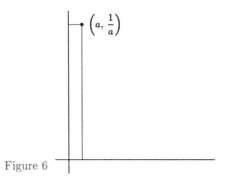

Figure 6

$$\delta_a(t) = \frac{1}{a}, \qquad 0 \leq t < a; \qquad \delta_a(t) = 0 \quad \text{elsewhere.}$$

Although the function is nonzero only on a small interval, the area under the curve is the area of the rectangle, which is 1. The Laplace transform is

$$L\delta_a(t) = \int_0^a \frac{1}{a} e^{-st} dt = \frac{1 - e^{-sa}}{sa}$$

and this tends to 1 as $a \to 0$.

It is a conceptual aid to introduce an expression $\delta(t)$ that describes the effect of $\delta_a(t)$ as $a \to 0$ and to say that $L\delta(t) = 1$. The symbol $\delta(t)$ is called the *Dirac delta function* or the unit impulse. It acts as if it were a function with the two properties

$$\delta(t) = 0 \quad \text{for} \quad t \neq 0, \qquad \int_0^1 \delta(t)dt = 1.$$

For example, by definition,

$$L\delta(t - c) = e^{-sc}, \quad c \geq 0; \qquad L\delta(t - c) = 0, \quad c < 0.$$

These matters are discussed in DETA, pages 578–586, where it is seen that we do not (and cannot) define $\delta(t)$ as a function of t, but rather we define the action $\delta(\phi)$ of δ on a class of test functions ϕ. Instead of developing theory here, we give an example.

Let y and k be rest solutions of the two equations

$$Ty = f(t) \in E, \qquad Tk = \delta(t).$$

Clearly $P(s)Ly = Lf$, $P(s)Lk = 1$. Hence

$$Ly = \frac{1}{P(s)} Lf = (Lk)(Lf) = L(k * f)$$

and by uniqueness of the transform, $y = k * f$. That is,

$$y(t) = \int_0^t k(t_1)f(t - t_1)\, dt_1.$$

The equation gives the rest solution of $Ty = f$ in terms of the rest solution of $Tk = \delta$. Thus we have established the remarkable fact that the response to $\delta(t)$ determines the response to any input. This is quite typical, and is one of the reasons why the unit impulse is important.

(c) The gamma function

Entry (b) of the transform table shows that $L(t^n)$ reduces to $n!$ when n is a positive integer and $s = 1$. That fact is now used to define $p!$ even if p is not an integer. In other words, by definition,

$$p! = \int_0^\infty e^{-t} t^p dt, \qquad p > -1.$$

The condition $p > -1$ ensures that the integral converges.

The key relation that gives a connection between this formula and the familiar factorial function is

$$p(p-1)! = p!, \qquad p > 0.$$

For proof, apply the first differentiation theorem with $f(t) = t^p$ and set $s = 1$ in the resulting equation

$$pLt^{p-1} = sLt^p - 0^p = sLt^p.$$

This theory is due to Leonhard Euler, who used the notation $\Gamma(p) = (p-1)!$. The symbol Γ is the Greek letter "gamma" and the function $\Gamma(p)$ is called the *gamma function*. For a more detailed development along the same lines, see DETA, Section 22.4.

PROBLEMS————————————————————————

1. One of the reasons for introducing negative feedback is to keep the system characteristics more or less constant when some of the components change. For a numerical illustration consider

$$y = \frac{G_1}{1 + G_1 G_2} x$$

where the G_j, instead of being general transfer functions, are merely constants. The characteristics of the overall system are measured by the ratio y/x of output to input. Let us suppose that G_1 jumps from its design value of 100 to a new value 10,000. Without feedback this would multiply y/x by 100. If $G_2 = 0.1$, so there is a small amount of negative feedback, check that the above malfunction in G_1 changes y/x from about 9.1 to about 10.

2. Show that an attempt to get the rest solution of $y'' + y = \delta(t)$ leads to $y = h(t) \sin t$. Sketch the graph. The unit impulse produces a jump of magnitude 1 in $y'(t)$ at $t = 0$, and the graph of y versus t has a corner there.

3. A weightless beam clamped at one end, free at the other, bears a concentrated load at its center as shown in Figure 7(a). The equation and boundary conditions are

$$y^{iv}(x) = 6\delta(x-1), \qquad y(0) = y'(0) = y''(2) = y'''(2) = 0$$

With $y''(0) = 2a$, $y'''(0) = 6b$ find Ly and

$$y = ax^2 + bx^3 + (x-1)^3 h(x-1)$$

Determine a and b from the boundary conditions at $x = 2$ and thus find the maximum deflection.

4. If the beam of Problem 3 is supported by a pivot at both ends as shown in Figure 7(b) the boundary conditions become

$$y(0) = y''(0) = y(2) = y''(2) = 0$$

By the Laplace transform find a two-parameter family of solutions y of the differential equation, depending on $y'(0) = a$, $y'''(0) = 6b$. Determine a and b to satisfy the boundary conditions at $x = 2$ and check that the maximum deflection is $y(1) = 1$.

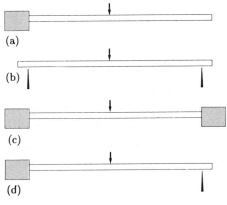

(a)

(b)

(c)

(d)

Figure 7

5. If the beam of Problem 3 is clamped at both ends as shown in Figure 7(c) the boundary conditions are

$$y(0) = y'(0) = y(2) = y'(2) = 0$$

Find y and check that the maximum deflection is $1/4$. Hence the maximum deflections in Figures 7(abc) are in the ratio 20:4:1, respectively.

6. Find the maximum deflection in Problem 3 when one end of the beam is clamped and the other pivoted as shown in Figure 7(d).

7. By setting $t = x^2$ and using the known value $\sqrt{\pi}/2$ for the integral, check that

$$\left(-\frac{1}{2}\right)! = 2 \int_0^\infty e^{-x^2}\, dx = \sqrt{\pi}$$

--------------------------ANSWERS--------------------------

3. (b) $a = 3$, $b = -1$, $y(2) = 5$
4. (a) $y = ax + bx^3 + (x-1)^3 h(x-1)$ (b) $a = 3/2$, $b = -1/2$.
5. $4y(x) = 3x^2 - 2x^3 = 4y(2-x)$, $0 \le x \le 1$
6. $16y(x) = 18x^2 - 11x^3 + 16h(x-1)(x-1)^3$, $y_{\max} = \sqrt{5}/5$.

Chapter 23

NUMERICAL METHODS

WHETHER for qualitative results such as a graph or for quantitative results such as the window of safety of a returning spacecraft, numerical analysis of differential equations has become an indispensable tool of great power. After a brief discussion of truncation and roundoff errors, we take up the subject of Richardson extrapolation. This is a technique by which two calculations can be combined to get a more accurate result than is given by either one alone. Ten algorithms for approximate solution of $y' = f(x, y)$ are compared empirically in a single table, and two more are mentioned in connection with the important concept of "stiff problem." All twelve are tested against differential equations that can be solved with ease. But this aspect of the exposition should not obscure the ultimate objective, which is to get accurate numerical answers when no elementary solution exists.

Prerequisites: Facility in the use of Taylor series is assumed; see Chapter B. Some of the problems require knowledge of computer programming.

1. Introduction

The goal of this chapter is to obtain numerical solutions to differential equations by use of a computer. A program of computation usually involves a large number m of separate steps. As m increases the theoretical error goes down but the computational errors go up. If the latter overwhelm the results before sufficient accuracy is reached, the problem is intractable with the method being used. Thus the science of computation involves a compromise between the demands of theoretical and numerical precision. It has subtle features that are easily overlooked on first acquaintance and it leads us to think about familiar topics in a new way.

(a) Algorithms

An algorithm is a definite, fully-defined procedure for solving a given class of problems. Even if two algorithms are mathematically equivalent, in the sense that they give the same theoretical result, they may be entirely different from the point of view of numerical computation. For example, when evaluating a polynomial

$$P(x) = a_0 x^m + a_1 x^{m-1} + \cdots + a_{m-1} x + a_m$$

one should arrange the calculations as follows:

$$u_1 = a_0 x, \qquad u_2 = (u_1 + a_1) x, \qquad \cdots, \qquad u_{n+1} = (u_n + a_n) x, \qquad \cdots.$$

Then $u_m + a_m = P(x)$. As an illustration, if $m = 5$

$$P(x) = ((((a_0 x + a_1)x + a_2)x + a_3)x + a_4)x + a_5.$$

This method is commonly named after Horner (1819) but is actually due to Newton. It requires only m multiplications for a polynomial of degree m and is easily programmed on a computer. The number of multiplications in a computation gives a rough measure of its cost.

The above algorithm is iterative; that is, it involves a passage from n to $n + 1$ in a single formula. Nearly all algorithms for the numerical solution of differential equations are iterative. In the early days of computation it was a common practice to make a list of intermediate results for use in subsequent steps. But this pencil-and-paper technique has long been obsolete. The procedure currently used, which is much faster, is to substitute $n + 1$ for n repeatedly in a single basic formula. Values computed at Step n are automatically stored for use in Step $n + 1$.

The choice of algorithm can have a profound effect on the accuracy of a calculation. For example, let us use the quadratic formula to get the root $x = 1$ of

$$1.0000x^2 - 88,888x + 88,887 = 0.$$

Assuming that the coefficients are correct to 5 significant digits and rounding each step of the calculation to that accuracy, we find the approximate root

$$44,444 - \sqrt{1,975,200,000} \approx 0.8.$$

This is barely correct to one digit. To be sure, rounding off the answer 0.8 yields 1, but all digits after the first remain in doubt.

It might be thought that the root $x = 1$ depends critically on the coefficients, and that if the latter are given only to 5 digits then no more than one correct digit can be expected in the answer. But this is not the case. For example, if the equation is changed to

$$1.0000x^2 - 88,888x + 90,000 = 0$$

the smaller root, correct to 5 digits, is 1.0125.

The trouble does not come from any inherent instability in the problem. It comes, rather, from the use of an inappropriate algorithm. A better way to compute the smaller root is to get the larger root by the quadratic formula and then use the fact that the product of the roots is the constant term. You can easily convince yourself that this procedure does not lead to the degradation of accuracy seen above.

In conclusion, we mention that such words as *method, procedure,* and *formula* are sometimes used instead of *algorithm.* The object is to avoid excessive repetition and no mathematical distinction is intended.

(b) Roundoff error

Since a computer can store only a finite number of decimal digits at any given time, it is necessary to round off the results of decimal calculation. This operation introduces an error, called roundoff error, that was already illustrated above. Namely, we found that roundoff in a single use of the quadratic formula led to a loss of four significant digits. It is not surprising that similar difficulty arises, in aggravated form, when a problem involves hundreds or thousands of steps.

There is, however, one favorable aspect. If at each step the average error introduced by roundoff is ϵ, in n steps the probable roundoff error is of the order $\epsilon\sqrt{n}$, and not ϵn as one might expect. This is so because the individual errors act like independent random variables with mean value zero, and positive and negative errors tend to cancel. It is as if the errors perform a random walk, with positive errors representing a step to the right, negative errors a step to the left. (Such desirable behavior is not found, however, if the computer merely drops decimals beyond a certain point, instead of rounding.)

Roundoff error is sometimes estimated by use of a quantity called the *machine epsilon*. This is the smallest $\epsilon > 0$ such that the computer can distinguish $1 + \epsilon$ from 1. For example if a computer is programmed to keep 8 significant digits it will act as if

$$1.00000005 = 1.0000001 > 1, \qquad 1.00000004 = 1.0000000 = 1.$$

Two conventions that are often used in numerical work will be followed here. First, if the result of a numerical calculation is stated with a given number of significant digits, its truth is asserted only to within that accuracy. For example, "$\pi = 3.14$" means "To three significant digits, $\pi = 3.14$." Second, we ignore roundoff errors in most of the theoretical development and assume that our various approximate formulas are computed exactly.

(c) Truncation error

A formula that is valid in the limit as $h \to 0$ is often used as if it were true for a small positive value of the constant h. The resulting error is an example of truncation error. In general, *truncation error* is introduced whenever a formula in a numerical calculation is only approximate (and would therefore give a wrong answer even if computed exactly.) For example, it is a familiar fact that

$$e = \lim_{h \to 0} (1 + h)^{1/h}.$$

If we use this with $h = 0.01$ instead of passing to the limit, the error in the resulting equation

$$e \approx (1.01)^{100} = 2.705$$

is due to truncation.

A convenient notation for describing truncation errors was introduced by the German mathematician Edmund Landau in connection with the analytic theory of numbers. If f and g are functions defined for small positive values of a parameter h, Landau's notation $f = O(g)$ means

$$|f(h)| \leq (\text{const.})|g(h)| \quad \text{as} \quad h \to 0+ .$$

The equation $f = O(g)$ is read aloud thus: "f is oh of g." Both f and g can be real, complex, or vector valued. Here, however, we shall have $g(h) = h^p$ where p is a nonnegative integer. The symbol $O(h^p)$ by itself is thought to mean "terms of order h^p" and an error of magnitude $O(h^p)$ is said to be of order p. Here are some illustrations:

$$9h^2 + h^3 = O(h^2), \quad h^2 = O(h), \quad \sin h = h + O(h^3), \quad e^h = 1 + h + O(h^2).$$

The last two follow from the Taylor series for $\sin h$ and e^h, respectively. More generally, we get a truncation error of order p whenever a convergent Taylor series is cut off, or *truncated*, just after its pth term. This fact is used below.

An equation such as $f = O(h)$ does not denote equality in the sense of logical identity, but signifies merely that f belongs to a certain class of functions. Algebraic properties suggested by the notation are interpreted by going back to the definition. For example, the implication

$$|\phi(h)| \leq Jh^p, \quad |\psi(h)| \leq Kh^q \quad \Rightarrow \quad |\phi(h)\psi(h)| \leq JKh^{p+q}$$

yields first an interpretation, then a proof, of the equation

$$O(h^p)O(h^q) = O(h^{p+q}).$$

As another example, let us recall that the function f is *Lipschitzian* if there is a constant K such that, for all relevant values of the variables,

$$|f(t, y + z) - f(t, y)| \leq K|z|.$$

Taking $z = O(h^p)$ we get

$$f(t, y + O(h^p)) = f(t, y) + O(h^p).$$

This is a concise way of saying that a change in the second argument of a Lipschitzian function f produces a change of at most the same order of magnitude in f itself.

(d) Richardson extrapolation

Suppose a numerical algorithm produces an approximation $y(x, h)$ to the desired exact answer, $y(x)$. In the applications we have in mind y will be the solution to a differential equation and h will be a parameter called the step size. The smaller the step size, the more steps are needed to calculate $y(x, h)$. By definition, an algorithm of this kind is *convergent* if

$$\lim_{h \to 0+} y(x, h) = y(x).$$

The rate of convergence is generally measured by an estimate of the form

$$(1) \qquad y(x) - y(x, h) = \mathrm{O}(h^p), \qquad p = \text{const.}$$

If $y(x) - y(x, h)$ admits a Taylor expansion

$$y(x) - y(x, h) = b_0(x) + b_1(x)h + b_2(x)h^2 + \cdots + b_p(x)h^p + \mathrm{O}(h^{p+1})$$

it is easily checked that (1) implies $b_j(x) = 0$ for $j < p$. Hence

$$(2) \qquad y(x) - y(x, h) = c(x)h^p + \mathrm{O}(h^{p+1})$$

where $c(x) = b_p(x)$. This is a sharpened form of (1).

Equation (2) forms the basis for a simple but powerful procedure known as Richardson extrapolation. Let us write (2) with step-size h and step-size $2h$, thus:

$$y(x) = y(x, h) + c(x)h^p + \mathrm{O}(h^{p+1}), \quad y(x) = y(x, 2h) + c(x)(2h)^p + \mathrm{O}(h^{p+1}).$$

If the first of these equations is multiplied by 2^p and the second is then subtracted, the result is

$$(3) \qquad (2^p - 1)y(x) = 2^p y(x, h) - y(x, 2h) + \mathrm{O}(h^{p+1}).$$

This gives $y(x)$ in terms of computed quantities within $\mathrm{O}(h^{p+1})$ even though, by (1), the basic method was accurate only within $\mathrm{O}(h^p)$. Since one should use at least two different step sizes anyway, as a check, Richardson's procedure is very effective in applications.

To make free use of results such as these, we assume whenever convenient (and without further comment) that functions such as $y(t)$ or $y(t, h)$ have convergent Taylor series expansions. This assumption simplifies the exposition but does not follow automatically from the hypotheses in Part (e) below.

(e) The basic differential equation

Having given a brief discussion of numerical methods in general, we turn to the specific problem

$$(4) \qquad y' = f(t, y), \qquad t \geq 0; \qquad y(0) = y_0$$

with which this chapter is chiefly concerned. It is assumed that f and f_y are continuous in a rectangle sufficiently large to contain all values (t, y) that arise in the analysis. By Chapters 14 and A, the solution $y(t)$ exists and is unique. The functions and variables are assumed real. However, the main results extend to complex vector-valued functions; see DETA, Sections 14.4 and 24.5. This gives a corresponding extension to higher-order equations and to systems. Since $t - t_0$ could be taken as a new variable, prescribing $y(0)$ rather than $y(t_0)$ involves no loss of generality.

Our problem is to approximate the solution of (4) numerically at any given point $t = x$. To this end the interval $[0, x]$ is divided into m equal parts by points

$$t_0 = 0, \quad t_1 = h, \quad t_2 = 2h, \quad t_3 = 3h, \quad \cdots, \quad t_m = mh = x.$$

The positive constant $h = x/m$ is the step size. A suitable function $y_n = y(t_n, h)$ is computed successively at each point t_n, starting with y_0 at $t_0 = 0$, and $y(t_m, h) = y(x, h)$ yields an approximation to $y(x)$.

Most algorithms of this kind are based upon approximate identities that are obtained by neglecting certain error terms. If the term neglected is $O(h^{p+1})$ the identity is said to be accurate to order p. For example, the approximate identities obtained by dropping error terms in Problem 1 are accurate to orders 1, 2, 1, 2, and 4 respectively. That they are not accurate to a higher order is seen in Problem 2.

In addition to the identity error mentioned above, another quantity of importance is the local error. This measures the difference between what the algorithm actually produces in a single step and what would be obtained by following a trajectory of the differential equation. It turns out that the local error usually has the same order of magnitude as the identity error in the approximate identity that gave rise to the algorithm.

Most important is the global error $y(x) - y(x, h)$. If the local error is $O(h^{p+1})$ at each step, one would naturally expect the error after $m = x/h$ steps to have the order of magnitude $(x/h)O(h^{p+1}) = O(h^p)$. In that case it would be correct to say, "A local error of order $p + 1$ yields a global error of order p." Despite its plausibility, this statement is not always true. A principal goal of the theory is to characterize those algorithms for which it holds.

PROBLEMS

The letter c or C accompanying a problem means that it requires a calculator or a programmable computer, respectively.

1. Given the convergent Taylor series

$$y(t + h) = y(t) + hy'(t) + \frac{h^2}{2}y''(t) + \frac{h^3}{6}y'''(t) + \cdots$$

and the corresponding series with y' in place of y, check that:

(a) $y(t + h) = y(t) + hy'(t) + O(h^2)$

(b) $y(t + h) - y(t - h) = 2hy'(t) + O(h^3)$

(c) $y(t + h) - y(t) = hy'(t + h) + O(h^2)$

(d) $y(t + h) - y(t) = \frac{h}{2}(y'(t) + y'(t + h)) + O(h^3)$

(e) $y(t + h) - y(t - h) = \frac{h}{3}(y'(t - h) + 4y'(t) + y'(t + h)) + O(h^5)$

Hint for (e): Write the series for $y(t + h) - y(t - h)$ and $y'(t + h) + y'(t - h)$ through terms in h^4.

2. (abcde) In most cases of interest here, an approximate identity is accurate to order p if, and only if, it is an exact identity for all polynomials of degree $\leq p$. By trying $y = t^p$ for suitable p verify this behavior in Problem 1, and at the same time show that the exponent in the error terms cannot be increased. For instance, in Problem 1(a) try $y = 1, t, t^2$.

3P. (abcde) In Problem 1 replace $y(t - h)$, $y(t)$, $y(t + h)$ by y_{n-1}, y_n, y_{n+1}, and $y'(t - h)$, $y'(t)$, $y'(t + h)$ by f_{n-1}, f_n, f_{n+1}, respectively, and drop the O term. This produces all but one of the algorithms discussed in Section 2.

4. Let q be a positive integer and let each ϕ_n be a real-valued function defined for all real values of the variables. Prove by mathematical induction that the recurrence relation

$$y_{n+1} = \phi_n(y_n, y_{n-1} \cdots, y_{n-q+1}), \qquad n \geq q - 1$$

has one and only one solution $\{y_n\}$ satisfying prescribed initial conditions

$$y_0 = Y_0, \qquad y_1 = Y_1, \qquad \cdots, \qquad y_{q-1} = Y_{q-1}$$

5. The *Fibonacci sequence* is defined by $y_{n+1} = y_n + y_{n-1}$ with the initial conditions $y_0 = y_1 = 1$.

(a) Find the first seven y_n and verify that $y_7 = 21$.

(b) Check that $\{y_n\} = \{s^n\}$ satisfies $y_{n+1} = y_n + y_{n-1}$ if, and only if, s is one of the roots s_j of $s^2 = s + 1$.

(c) With s_j as in (b) verify that $z_n = c_1 s_1^n + c_2 s_2^n$ satisfies $z_{n+1} = z_n + z_{n-1}$ for every choice of the constants c_j.

(d) Choose c_j in (c) so that $\{z_n\}$ has the same inital values as $\{y_n\}$, in which case Problem 4 gives $y_n = z_n$. Thus get the formula

$$ y_n = \frac{1}{\sqrt{5}} \left(\frac{1 + \sqrt{5}}{2} \right)^{n+1} - \frac{1}{\sqrt{5}} \left(\frac{1 - \sqrt{5}}{2} \right)^{n+1} $$

6C. Working to as many digits as you can, find y_{100} in Problem 5 by use of the recurrence relation and compare with the formula of Part (d).

7P. (a) Solve Richardson's formula (3) for $y(x)$ when $p = 1, 2, 3, 4$.

(b) If the step size is changed from κh to h, where $\kappa > 1$ is constant, show that (3) holds with κ in place of 2.

8. (a) Show that $h \sin(1/h) = O(h)$ and that nevertheless $h \sin(1/h)$ does not have the form $ch + O(h^2)$ where c is constant.

(b) Interpret and prove: $O(h^p) + O(h^p) = O(h^p)$.

(c) Briefly discuss $O(h^4) + O(h^5) = O(h^4)$ and state a generalization that contains both this result and the result (b). No answer is provided.

9. If we required $f(t, y)$ to be globally Lipschitzian, that is, Lipschitzian for $-\infty < y < \infty$, show that the equation $y' = y^2$ would be excluded on every interval $0 \le t \le b$ no matter how short.

10C. If $P(x)$ denotes the sum of the first 101 terms of the series for e^x, set up a program for computing $P(x)$ by Horner's algorithm or by direct summation of terms as you prefer. Using this, evaluate $e^x / P(x) \approx 1$ for $x = 20, 100, -20$ and explain.

11c. Apply the method of Richardson extrapolation to the Taylor series for $\sin h$ and thus show that

$$ 3^5 \sin h = \sin 3h + 240h - 36h^3 + O(h^7) $$

Taking $h = \pi/18$ find $\sin 10°$ and compare with the exact value.

————————————————ANSWERS————————————————

The c problems were done on an HP-32S scientific calculator, the others on a NEXT with machine epsilon $2^{-52} \approx 2 \cdot 10^{-16}$. Different equipment may give somewhat different results.

6. $y_{100} = 573147844013817084101$

7. (a) Partial answer: $p = 3$ gives $y(x) \approx (8/7)y(x, h) - (1/7)y(x, 2h)$

10. 1.0000000000000, 1.89911087960. When $x = -20$ we got about .01 for direct sum, about 0.3 for Horner. The main source of error is truncation when $x = 100$ and roundoff when $x = -20$. It is for you to explain why.

11. $\sin 10° \approx 0.17364816987$

2. Five algorithms

Here we discuss five specific algorithms, each of which has both historical and practical importance. As in the last section $y(t)$ denotes a solution of the initial-value problem

$$y' = f(t, y), \qquad y(0) = y_0.$$

The successive values of the approximation $y(t, h)$ are denoted by y_n and corresponding values of f by f_n. Thus

$$y_n = y(t_n, h), \qquad f_n = f(t_n, y_n), \qquad t_n = nh.$$

The passage from approximate identities for $y(t)$ to numerical algorithms for $y(t, h)$ is facilitated by the equations

$$y(t - h, h) = y_{n-1}, \qquad y(t, h) = y_n, \qquad y(t + h, h) = y_{n+1}, \qquad t = t_n.$$

Each of the five algorithms has global accuracy of order p, where $p \geq 1$ is the order of accuracy of the approximate identity giving rise to the algorithm. Except in one case, this follows from a general theory of *linear multistep formulas* that was developed by the contemporary Swedish mathematician Germund Dahlquist in 1956. (Heun's formula is of a different type, but the conclusion still holds.) A development of the Dahlquist theory is in DETA, Section 24.3 and will not be repeated in this book. However, we use the fact that p as given by Section 1, Problem 1 is the correct value for Richardson extrapolation.

(a) Euler's method

Euler's method was introduced in Section 14.2 and is presented again now. The definition of derivative gives

$$\frac{y(t+h) - y(t)}{h} \approx y'(t).$$

Ordinarily this is used as a means of approximating $y'(t)$. Here, however, we use it to approximate $y(t+h)$:

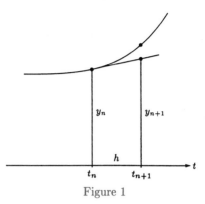

Figure 1

$$y(t+h) \approx y(t) + hy'(t) = y(t) + hf(t, y(t)).$$

Replacing $\approx$ by $=$ and $y(t)$ by $y(t, h)$ we are led to the formula (Euler 1768)

$$(1) \qquad\qquad y_{n+1} = y_n + hf_n.$$

A geometric interpretation is given in Figure 1. By Section 1, Problem 1(a) the underlying identity is accurate to order 1.

EXAMPLE 1. Let us apply this method to the *test problem*

$$(2) \qquad\qquad y' = 2ty, \qquad y(0) = 1, \qquad y(1) = ?$$

Here $f(t, y) = 2ty$, $f_n = 2nhy_n$ and the exact solution is

$$(3) \qquad\qquad y(t) = e^{t^2}, \qquad y(1) = 2.718281828459045\cdots.$$

Euler's formula gives $y_{n+1} = y_n(1 + 2nh^2)$. The sequence

$$y_0 = y_1 = 1, \quad y_2 = 1 + 2h^2, \quad y_3 = (1 + 4h^2)y_2, \quad y_4 = (1 + 6h^2)y_3, \quad \cdots$$

is easily computed, with the stopping-value $n = 4$ or $n = 9$ for $h = .2$ or $h = .1$. This makes the last computed value y_5 or y_{10} in the two cases, respectively, and yields

$$y(1, .2) = 2.0505830, \qquad y(1, .1) = 2.3346334.$$

By Richardson extrapolation with $p = 1$ we get

$$y(1) \approx 2y(1, .1) - y(1, .2) = 2(2.3346334) - 2.0505830 = 2.62.$$

Here, as elsewhere, we keep only so many significant figures as seem warranted by the accuracy of the calculation.

(b) The midpoint formula

A modification of Euler's method is illustrated in Figure 2. Here we go from (t_{n-1}, y_{n-1}) to (t_{n+1}, y_{n+1}) by following a line whose slope agrees with the slope f_n at the midpoint. The figure leads to the *midpoint formula*

$$y_{n+1} = y_{n-1} + 2hf_n$$

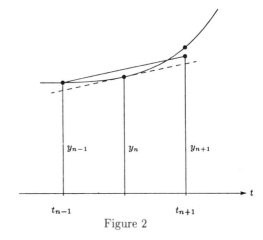

Figure 2

(Nyström 1925). This is a two-step method because we have to know both y_0 and y_1 to get started. In general, a q-step method gives y_{n+1} in terms of q consecutive earlier values y_j. Since the midpoint formula is based on the identity of Section 1, Problem 1(b), it is accurate to order $p = 2$.

EXAMPLE 2. For the test problem (2) the midpoint formula becomes

$$y_{n+1} = y_{n-1} + 4nh^2 y_n.$$

If we did not know the exact solution, y_1 would be computed by a one-step method based on a subdivision of the interval $[0, h]$. Here, however, we use our knowledge of the exact solution to get $y_1 = e^{h^2}$, hence

$$y_1 = 1.04081108, \qquad h = .2; \qquad y_1 = 1.0100502, \qquad h = .1.$$

The final results are recorded in the table at the end of this section.

(c) The trapezoid rule

A general method for constructing algorithms can be based on integration. Let $y' = f(t, y)$ and suppose for the moment that

$$y_n = y(t_n), \qquad y_{n+1} = y(t_{n+1}).$$

Integrating $y'(t)$ from t_n to t_{n+1} yields

$$(4) \quad y_{n+1} = y_n + \int_{t_n}^{t_{n+1}} f(t, y(t)) \, dt.$$

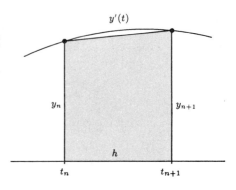

Figure 3

Since we are following the integral curve from t_n to t_{n+1}, the local error is zero. Such an algorithm at each step, starting from $t = 0$, would yield $y(t)$ exactly. It cannot be implemented because the integrand $f(t, y(t))$ depends on the unknown function y. But if the integral is approximated by a suitable quadrature formula, Equation (4) provides a practical means of computation.

For example, the approximation illustrated in Figure 3 yields the *trapezoid rule*

$$(5) \qquad y_{n+1} = y_n + \frac{h}{2} \left(f(t_n, y_n) + f(t_{n+1}, y_{n+1}) \right).$$

This is based on the approximate identity

$$(6) \qquad y(t + h) \approx y(t) + \frac{h}{2} \left(y'(t) + y'(t + h) \right)$$

for which Section 1, Problem 1(d) gives $p = 2$.

Since y_{n+1} occurs in f on the right of (5), the equation (5) determines y_{n+1} only implicitly. In general, a method is *implicit* if it does not give y_{n+1} directly in terms of preceding values but only gives an equation from which, in principle, y_{n+1} can be found. It is a universal characteristic of implicit methods that h must be sufficiently small to ensure that the implicit equations have a solution. For the test problem Equation (5) is easily solved for small h, as seen in Problem 1(c). Numerical results are given in the table.

(d) Heun's formula

One way to go from the implicit equation (5) to an explicit equation is to replace y_{n+1} on the right by an approximate value. For example the choice

$$y_{n+1} \approx y_n + h f_n$$

leads to the formula (Heun 1900)

$$y_{n+1} = y_n + \frac{h}{2} \left(f(t_n, y_n) + f(t_{n+1}, y_n + h f_n) \right).$$

Results for the test problem are given in Problem 1(d) and in the table.

The equation $y_{n+1} = y_n + h f_n$ that was used to predict the value of y_{n+1} is called the predictor and the final formula for y_{n+1} so obtained, in this case Heun's formula, is called the corrector. In general, a *predictor-corrector* method is a method in which a predicted value of y_{n+1} is used to simplify a more accurate, but usually implicit, formula for y_{n+1}.

If $y'(t) = f(t, y(t))$ then $y'(t + h) = f(t + h, y(t + h))$. Using this in (6) and then replacing $y(t + h)$ by $y(t) + y'(t)h + O(h^2)$, we get

$$(7) \qquad y(t + h) = y(t) + \frac{h}{2} \left(y'(t) + f(t + h, \, y(t) + h y'(t)) \right) + O(h^3).$$

Equation (7) without the O term is the approximate identity that underlies Heun's method. It is accurate to order 2, hence $p = 2$.

(e) The Milne-Simpson formula

The equation corresponding to (4) for passage from t_{n-1} to t_{n+1} is

$$y_{n+1} = y_{n-1} + \int_{t_{n-1}}^{t_{n+1}} f(t, y(t)) \, dt.$$

If the integral is approximated by Simpson's rule we get the formula

$$y_{n+1} = y_{n-1} + \frac{h}{3}(f_{n-1} + 4f_n + f_{n+1})$$

(Milne 1926). This is both two-step and implicit. Since it is based on the approximate identity

$$y(t + h) - y(t - h) \approx \frac{h}{3}(y'(t - h) + 4y'(t) + y'(t + h))$$

Section 1, Problem 2(e) shows that $p = 4$. For the test problem, see Problem 1(e) and the table.

	algorithm	$h = 0.2$	$h = 0.1$	$h = 0.05$	p
a	Euler	2.05058	2.334633	2.5106623	1
b	midpoint	2.59509	2.683557	2.7093196	2
c	trapezoid	2.79397	2.736598	2.7228243	2
d	Heun	2.68138	2.709057	2.7159898	2
e	Milne-Simpson	2.72060	2.718437	2.7182917	4
f	Adams-Bashforth	2.68588	2.713345	2.7178396	4
g	Moulton	2.72415	2.718814	2.7183215	4
h	Adams-Moulton	2.71827	2.718486	2.7183079	4
i	Milne predictor	2.69481	2.716304	2.7181465	4
j	Runge-Kutta	2.71811	2.718270	2.7182811	4

PROBLEMS————————————————————————————

The letters (abcde) in Problems 1–9 correspond to those in the table. Implicit algorithms are applied only to equations that allow explicit solution.

1. The following recursion formulas for the test problem were not given in the text. Verify them:

$$\text{(c)} \quad y_{n+1} = \frac{1 + h^2 n}{1 - h^2(n+1)} y_n, \qquad hx < 1$$

$$\text{(d)} \quad y_{n+1} = (1 + (2n+1)h^2 + 2n(n+1)h^4)y_n$$

$$\text{(e)} \quad y_{n+1} = \frac{8nh^2 y_n + (3 + 2h^2(n-1))y_{n-1}}{3 - 2h^2(n+1)}, \qquad hx < \frac{3}{2}$$

2c. (abcde) For the test problem find $y(.4, .2)$.

3C. (abcde) Verify the numerical answers in the table. Here and in analogous cases elsewhere, the preceding c problem is an aid in checking your program.

4c. (abcde) Repeated Richardson extrapolation. Get an improved value X from the entries for $h = .2$ and $h = .1$ in the table. Also get an improved value Y from the entries for $h = .1$ and $h = .05$. Finally, using $p+1$ instead of p, get an improved value Z from your X and Y.

5c. (abcde) Suppose the quantity $E(h) = y(x) - y(x, h)$ is of order $O(h^{p+1})$ and has a Taylor series. Then $E(h)/h^p$ tends to a finite limit as $h \to 0+$. For the test problem find three values of $E(h)/h^p$ from the table.

6c. (abcde) Check that $(1 + t^2)y' = 4$, $y(0) = 0$ yields $y(1) = \pi$. Then approximate π by using $h = 1/4$ in the algorithms of the text.

7C. (abde) In Problem 6 take $h = .1$, $h = .05$, and extrapolate. We omit (c) because (c)=(d) when f is independent of y. See answer to Problem 6.

8c. (acd) Given $y' = -20y$, $y(0) = 1$ approximate $y(1)$ by the algorithms of the text, taking $h = .2, .1, .05, .01, .0025$. This can be done as a c problem because y_n has the form α^n where $\alpha = \alpha(h)$.

9C. (be) Do Problem 8 for (be) by actually programming the algorithms. The strange behavior you will encounter is explained by the discussion of *stiff problems* in Section 3.

10P. Backward Euler. Estimating the integral (4) by the area of a suitable rectangle, get the respective equations

$$y_{n+1} = y_n + hf_n, \qquad y_{n+1} = y_n + hf_{n+1}$$

The second of these is the implicit or *backward* Euler formula. Check that the average of these two equations yields the trapezoid formula.

11c. Apply the backward Euler formula of Problem 10 to Problem 8, using h as prescribed there.

────────────────────ANSWERS────────────────────

2. 1.0800   1.1665   1.1775   1.1714   1.1736

4. $X = 2.618684$ 2.713046   2.717474   2.718283   2.718293

   $Z = 2.7093604$ 2.7186016   2.7183415   2.7183033   2.7182817

5. Partial answer: $E(.05)/(.05)^p = 4.152$ 3.585 -1.817 0.917 -1.579

6. 3.3811765   3.1623529   3.1311765   3.1311765   3.1415686

7. Last answer: 3.1424259850   3.1415926920   3.1415926530   3.1415926556

8. Partial answer: $10^9\, y(1, .0025) = 1.2286894$ 2.0525801   2.079064

9. (b) -449 41907   741088   223932   4756

   (e)   2.07 -2.28 $-.548$ $-.00059$ $6.56 \cdot 10^{-7}$

11. Partial answer: $y(1, .0025) = 3.3441086 \cdot 10^{-9}$

──

3. Supplementary topics

In scientific work the algorithms of Section 2 are often replaced by others that are more elaborate but have better convergence properties. Some of the most popular are described next. As in Sections 1 and 2, we set

$$t_j = jh, \qquad y(t_j, h) = y_j, \qquad f(t_j, y_j) = f_j.$$

The explicit four-step algorithm

$$(1) \qquad y_{n+1} = y_n + \frac{h}{24}(55 f_n - 59 f_{n-1} + 37 f_{n-2} - 9 f_{n-3})$$

was proposed in a book by the British mathematical scientists Adams and Bashforth in 1883, and in 1926 the American astronomer Moulton introduced the implicit three-step formula

$$(2) \qquad y_{n+1} = y_n + \frac{h}{24}(9 f_{n+1} + 19 f_n - 5 f_{n-1} + f_{n-2}).$$

To get an explicit algorithm he replaced the term y_{n+1} in f_{n+1} on the right by (1). Thus the *Adams-Bashforth formula* (1) is used as a predictor and the

Moulton formula (2) as a corrector. The final result, known as the *Adams-Moulton formula*, is thought by many to be the best of the three.

A fourth formula, called the *Milne predictor*, is

$$y_{n+1} = y_{n-3} + \frac{4h}{3}(2f_n - f_{n-1} + 2f_{n-2}).$$

The use of such formulas in specific problems is much the same as for the simpler methods of Section 2 and is taken up in the problems.

(a) The classical Runge-Kutta formula

An important method was introduced by the German mathematician Carl David Runge in 1895 and was developed further by another German, Wilhelm Kutta, in 1901. The basic idea is to introduce intermediate values

$$t_{n+c}, \qquad y_{n+c}, \qquad f_{n+c} = f(t_{n+c}, y_{n+c})$$

as an aid in getting from y_n to y_{n+1}. The process involves just one step but several stages. This means that a single use of the algorithm takes us from y_n to y_{n+1} through a number of intermediate values y_{n+c}.

Runge-Kutta algorithms are described by an array of coefficients called a *tableau*, plural tableaux. Here is the tableau for the classical Runge-Kutta formula of 1901, with equations involving intermediate values f_{n+c_i} on the right:

0	0	0	0	0	$k_1 = f(t_n, y_n)$
$\frac{1}{2}$	$\frac{1}{2}$	0	0	0	$k_2 = f\left(t_n + \frac{h}{2}, y_n + \frac{h}{2}k_1\right)$
$\frac{1}{2}$	0	$\frac{1}{2}$	0	0	$k_3 = f\left(t_n + \frac{h}{2}, y_n + \frac{h}{2}k_2\right)$
1	0	0	1	0	$k_4 = f(t_n + h, y_n + hk_3)$
	$\frac{1}{6}$	$\frac{1}{3}$	$\frac{1}{3}$	$\frac{1}{6}$	

After the k_i are thus determined, the step from n to $n+1$ is made by the formula

$$(3) \qquad\qquad y_{n+1} = y_n + \frac{h}{6}(k_1 + 2k_2 + 2k_3 + k_4).$$

This is a four-stage method because there are four intermediate values. The connection of the tableau with the equations is explained in DETA, Section 24.4, where it is also shown that the order of the method is $p = 4$. Suffice it to say here that the bottom row of the tableau gives the coefficients in (3) and the remaining entries describe the equations at the right of the tableau.

EXAMPLE 1.   In the test equation of Section 2 the step from $t = 0$ to $t = h$ is accomplished as follows. Using $f(t, y) = 2ty$, $t_n = 0$ and $y(0) = 1$, compute

$$k_1 = 0, \qquad k_2 = h, \qquad 2k_3 = 2h + h^3, \qquad k_4 = 2h + 2h^3 + h^5.$$

Then use (3) to get

$$y_1 = 1 + h^2 + \frac{1}{2}h^4 + \frac{1}{6}h^6.$$

Since the exact value is $y(h) = e^{h^2}$, the error is $O(h^8)$.

(b) Stiff problems

Some interesting conclusions are suggested by the equation

$$(4) \qquad\qquad y' = -100(y - \cos t) - \sin t.$$

This has an obvious solution $y = \cos t$ and the general solution is

$$(5) \qquad\qquad y(t) = \cos t + ce^{-100t} \quad \text{where} \quad c = y(0) - 1.$$

For moderately large t, say $t = 1$ or 2, Equation (5) shows that $y(t)$ is very insensitive to errors in the initial value. Since a roundoff or truncation error at the nth step is much like an initial-value error at t_n, one might naturally expect a high degree of numerical stability in the calculation of y.

To investigate this point let us try the following algorithms:

$$(6) \qquad 2y_{n+1} = 2y_n + 3hf_n - hf_{n-1}, \qquad 3y_{n+1} = 4y_n - y_{n-1} + 2hf_{n+1}.$$

The results for (4) with $y(0) = 1$ are in the accompanying table, where the algorithms are designated as AB2 and BD2 respectively. Although it can be shown that both have order 2, the first method, which is explicit, yields hopelessly bad approximations until $h \leq .01$ while the second, implicit method is much better:

	$h = 0.2$	$h = 0.1$	$h = 0.05$	$h = 0.02$	$h = .01$
AB2	14.4	$-6 \cdot 10^4$	$-2 \cdot 10^7$	$-6 \cdot 10^{10}$	.540302
BD2	.5404	.540330	.540309	.5403034	.5403026

Equation (4), and others that exhibit similar behavior, are said to be *stiff*. A formal definition is not easily given, but here are some of the main features:

(i) For stiff problems, many popular algorithms exhibit an extreme numerical instability that is not connected with any actual instability of the problem.

(ii) Because of (i), it is necessary to use a much smaller step-size h than one would expect from the local error.

Item (i) is worded so as to distinguish instability of numerical algorithms from true instability of the underlying problem. If the coefficient -100 on the right of (4) is changed to 100 the exponent $-100t$ in (5) becomes $100t$ and the problem itself is unstable. A corresponding numerical instability then reflects a genuine aspect of the situation.

To see why (4) causes difficulty, let z denote the difference of two solutions. One of these is thought to be the desired solution while the other has an incorrect initial value. Linearity gives $z' = -100z$ and the first algorithm (6) yields

$$z_{n+1} = z_n - 50h(3z_n - z_{n-1}).$$

It can be shown that the complete solution of this difference equation is on the left below where s_j are the two roots, supposed unequal, of the quadratic on the right:

$$z_n = c_1 s_1^n + c_2 s_2^n, \qquad s^2 = s(1 - 150h) + 50h.$$

We number the roots so that $|s_2| \geq |s_1|$. Since their product is $-50h$, clearly $|s_2| > 1$ if $h > .02$. In this case $\{z_n\}$ diverges as $n \to \infty$, (unless c_2 happens to be 0), even though $z(t)$ tends rapidly to 0 as $t \to \infty$. Closer analysis shows that the dividing line is $h = .01$. That is, if $h > .01$ then $|s_2| > 1$ and if $h \leq .01$ both roots satisfy $|s_j| \leq 1$. This behavior is reflected in the first row of the table.

When $|s_2| > 1$ the term $c_2 s_2^n$ is a *parasitic term* that comes from the algorithm and has nothing to do with the differential equation. For the second algorithm, on the contrary, Problem 4 shows that both roots satisfy $|s_j| < 1$. This explains the superior numerical results. An introduction to the general theory is in DETA, Section 14.5.

(c) The direction field

Before undertaking an elaborate calculation, or even choosing an algorithm, one should get whatever qualitative information can be easily obtained. This preliminary step may reveal unsuspected sources of difficulty and it also gives an idea of what to expect when the solution is computed numerically.

For a first-order scalar equation, the method of isoclines discussed in Section 19.3 is a useful source of qualitative information. It yields short segments, called *line elements*, that are tangent to the solution at points of the isocline. For example if

$$\frac{dy}{dx} = |y - x|^{1/3}$$

the line $y - x = 8$ gives an isocline at all points of which the slope of the solutions is 2. By use of a parallel ruler one can draw many line elements without the aid of a computer.

A computer allows us to construct a line element at every point of a grid; for example at the points $(.05m, .05n)$ where m and n range over a set of integers. One advantage of this approach is that the density of the line elements is uniform. Such a field for the above equation is shown in Figure 4(a) and some of the trajectories are shown in Figure 4(b).

A corresponding technique for certain types of second-order equations and 2-by-2 systems leads to the concept of the phase plane. This was introduced in Chapter 7, carried further in Chapter 19, and applied to nonlinear problems in Chapter 20. The main idea is to linearize by expansion in Taylor series near the stationary points, thus obtaining a local approximation. Further examples of local approximation can be found in the problems of Section 4.2 and in Chapters 12, 13, and B.

(a)   (b)

Figure 4

PROBLEMS

1c. In the test problem of Example 1 approximate $y(1/2)$ by one use of the classical Runge-Kutta formula and verify that it agrees with the result given there. Then repeat to get $y_2 = 2.7131450$.

2C. Confirm the entries (j) in the table at the end of Section 2.

3C (fghij) Apply the last five methods of the table (Section 2) to find $y(1)$ if $y' = -20y$, $y(0) = 1$. Take $h = .2, .1, .05, .01$. The answer is given for $h = .01$ only.

4. For the second algorithm (6), write the difference equation corresponding to $f(t, y) = -100y$ and show that the roots s of the associated quadratic satisfy $(s - 1)(3s - 1) = -200hs^2 < 0$. It follows that $1/3 < s < 1$.

5. (abc) Using direction fields, plot the trajectories:

$$\text{(a) } y' = \sin(xy) \qquad \text{(b) } y' = x^{1/3}y^3 \qquad \text{(c) } y' = (xy)^{1/5}$$

6. Show by any method that the equation $|x|(y - |x|)\, dy = x(y - 2|x|)\, dx$ leads to the figure at the bottom of this page.

ANSWERS

  3. $10^9\, y(1, .01) = 2.09186 \quad 2.05915 \quad 2.05654 \quad 5.89791 \cdot 10^9 \quad 2.06180$

  5. See Section 4.2, Figure 1. Figure 1(a) is correct but misleading, in that the trajectories omitted do not behave like the ones shown.

> I see well I was much deceived, and numbering is a more cunning thing than I took it to be.
>
>   —Robert Recorde

Chapter A

UNIQUENESS

In this chapter we give a simple, unified approach to a variety of significant uniqueness theorems. The differential equations and inequalities are assumed to hold for $t \in I$, where I is an open interval containing the point t_0. Primes denote differentiation with respect to t and the letters M, K denote positive constants. Except in Example 5, all functions are real valued.

THEOREM 1. *Let $|w'(t)| \leq Mw(t)$ and $w(t_0) = 0$. Then $w = 0$.*

For proof, suppose, contrary to the conclusion, that $w(t_1) > 0$ at some point $t_1 \in I$. We take $t_1 > t_0$; the case $t_1 < t_0$ is similar. We have

$$\frac{d}{dt} e^{-Mt} w(t) = e^{-Mt}(w'(t) - Mw(t)) \leq 0.$$

Hence $e^{-Mt} w(t)$ is decreasing and, since $t_1 > t_0$,

$$e^{-Mt_1} w(t_1) \leq e^{-Mt_0} w(t_0) = 0.$$

Multiplication by e^{Mt_1} gives $w(t_1) \leq 0$, contrary to our initial assumption. This completes the proof.

EXAMPLE 1. The function f is *Lipschitzian* if, for all relevant values of the variables,

$$|f(t, u) - f(t, v)| \leq K|u - v|.$$

Under this condition the initial-value problem

$$y' = f(t, y), \qquad y(t_0) = y_0$$

has at most one solution. To see why, let u and v be two solutions. Subtraction gives $u(t_0) - v(t_0) = 0$ and also

$$|u' - v'| = |f(t, u) - f(t, v)| \leq K|u - v|.$$

Theorem 1 with $K = M$ and $w = u - v$ implies $w = 0$, hence $u = v$.

In Section 14.2 it was seen that the Lipschitz condition holds in a rectangular region if f_y is bounded. Hence this example yields the uniqueness theorem for first-order equations as stated without proof in Section 4.2.

EXAMPLE 2. Assuming that the coefficients a, b, c, d are bounded, we shall show that the system

$$x' = a(t)x + b(t)y + f(t), \quad y' = c(t)x + d(t)y + g(t)$$

has at most one solution subject to the initial conditions

$$x(t_0) = x_0, \quad y(t_0) = y_0.$$

For proof, let (x_1, y_1) and (x_2, y_2) be two solutions and define

$$u = x_1 - x_2, \quad v = y_1 - y_2.$$

Then $u(t_0) = v(t_0) = 0$ and, with an obvious abbreviation,

$$u' = au + bv, \quad v' = cu + dv.$$

The function $w = u^2 + v^2$ satisfies

(1) $$w' = 2uu' + 2vv' = 2u(au + bv) + 2v(cu + dv).$$

Also (since squares are nonnegative) $w \geq u^2$, $w \geq v^2$. Hence

$$|u| \leq \sqrt{w}, \quad |v| \leq \sqrt{w}.$$

If $|a|, |b|, |c|, |d|$ are all bounded by K we conclude from (1) that

$$|w'| \leq 2K(u^2 + uv + vu + v^2) \leq 8Kw.$$

Theorem 1 with $M = 8K$ gives $w = 0$, which completes the proof. This example provides a mathematical foundation for some of the results in Chapters 9, 19, and 20.

EXAMPLE 3. If p and q are bounded, the problem

$$x'' + p(t)x' + q(t)x = f(t), \quad x(t_0) = x_0, \quad x'(t_0) = x_1$$

has at most one solution. For proof, write the equations as a system in x and $y = x'$ and use Example 2. This provides a mathematical foundation for the main results in Section 10.1.

EXAMPLE 4. If the functions a_{ij} are bounded, the linear system

$$u_i' = \sum_{j=1}^{n} a_{ij}(t)u_j, \quad u_i(t_0) = 0, \qquad i = 1, 2, \cdots, n$$

has $u = 0$ as sole solution. For proof, apply Theorem 1 with

$$w = u_1^2 + u_2^2 + \cdots + u_n^2.$$

As before, $|u_i| \leq \sqrt{w}$. If $|a_{ij}| \leq K$ the differential equation for u_i gives

$$|u_i'| \leq K(|u_1| + |u_2| + \cdots + |u_n|) \leq nK\sqrt{w}$$

and hence $|u_i u_i'| \leq nKw$. It follows that $|w'| \leq 2n^2 Kw$. Since $w(t_0) = 0$ by the initial conditions, Theorem 1 gives $w = 0$.

EXAMPLE 5. If the elements of the square matrix A are bounded, the inhomogeneous vectorial equation

$$y' = A(t)y + f(t), \quad y(t_0) = y_0$$

has at most one solution. For proof let y and z be two solutions and apply Example 4 to $u = y - z$. This example provides a mathematical foundation for some of the main results in Chapters 15,16,17.

Example 5 gives uniqueness for the linear nth-order initial-value problem when the latter is converted to a system as in Example 3. However, the next theorem applies to nth-order equations directly:

THEOREM 2. *If u satisfies the following two conditions, then $u = 0$:*

$$|u^{(n)}(t)| \leq K(|u(t)| + |u'(t)| + \cdots + |u^{(n-1)}(t)|),$$

$$u(t_0) = u'(t_0) = \cdots = u^{(n-1)}(t_0) = 0.$$

Again the proof is simple. Differentiate the equation

$$w = (u)^2 + (u')^2 + \cdots + (u^{(n-1)})^2$$

to get

(2) $$w' = 2uu' + 2u'u'' + \cdots + 2u^{(n-1)}u^{(n)}.$$

Since $|u^{(j)}| \leq \sqrt{w}$ for $j = 0, 1, \cdots, n-1$, the hypothesis of Theorem 2 gives

$$|u^{(n)}| \leq K(\sqrt{w} + \sqrt{w} + \cdots + \sqrt{w}) = Kn\sqrt{w}.$$

In (2) this yields $|w'| \leq 2(n-1)w + 2Knw$ and $w = 0$ by Theorem 1. The result is extended to complex-valued functions in DETA, page 248.

For readers familiar with the notion of a Lipschitz condition, Theorem 2 has an interesting bearing on the problem

$$y^{(n)} = f(t, y, y', \cdots, y^{(n-1)})$$

with the initial conditions

$$y(t_0) = y_0, \quad y'(t_0) = y_1, \cdots, y^{(n-1)}(t_0) = y_{n-1}.$$

If y and z are two solutions, and if f admits the Lipschitz constant K, then $u = y - z$ satisfies the hypothesis of Theorem 2. Hence $u = 0$ and $y = z$. This is a classical uniqueness theorem for nth-order nonlinear equations. It is not hard to show that the Lipschitz condition holds in a rectangular region if the partial derivatives of f with respect to its last n arguments are bounded. Hence the case $n = 2$ implies the uniqueness theorem for second-order equations that was stated without proof in Section 4.2.

Instead of assuming that the coefficients are bounded, we could assume that they are continuous on I. By a theorem of advanced calculus, functions continuous on I are automatically bounded on every closed subinterval J contained in I. The above results give $u(t) = 0$ on J and, since J is arbitrary, this implies $u(t) = 0$ on I. Thus the hypothesis of boundedness is both more general and more elementary than the hypothesis of continuity.

Under sufficiently strong continuity conditions, some of the foregoing results can be obtained by extending the Picard procedure of Chapter 14 to the vector case. But if the possibility of this extension is actually demonstrated, and not merely assumed, the Picard procedure is more difficult than the elementary method used here. For details see DETA, Section 14.4.

The public presentation of a proof usually begins with a statement of particular hypotheses and conditions, of unknown origin, and proceeds to the desired result as neatly as peeling a banana.

—Gordon Raisbeck

Mathematics has the inhuman quality of starlight, brilliant and sharp but cold.

—Hermann Weyl

A *power series* (about $x = 0$) is an expression of the form

(1)
$$\sum_{j=0}^{\infty} a_j x^j = a_0 + a_1 x + a_2 x^2 + \cdots + a_n x^n + \cdots$$

where the a_j are real or complex numbers. Unless the contrary is said we assume that x is real. The extension to complex values causes no trouble, because the theory of limits for complex sequences is virtually identical to that for real sequences. Indeed, if

$$s_n = u_n + iv_n, \qquad s = u + iv$$

then, by definition,

$$\lim_{n \to \infty} s_n = s \quad \text{means} \quad \lim_{n \to \infty} u_n = u \quad \text{and} \quad \lim_{n \to \infty} v_n = v.$$

The symbol $\sum$, read "sigma," is a Greek capital S and denotes "sum." The phrase *sigma notation* distinguishes the condensed style of writing on the left of (1) from the expanded form on the right. The term a_0 does not involve x and is called the constant term. In order to make the constant term on the right agree with $a_0 x^0$ on the left we set

$$a_0 0^0 = \lim_{x \to 0} a_0 x^0 = a_0.$$

Some such convention is needed because 0^0 is otherwise undefined. In this book we make the further convention that

$$0 \cdot 0^j = \lim_{x \to 0} 0 \cdot x^j = 0 \quad \text{for all } j.$$

The algebra of power series has two aspects: the formal theory, in which no numerical value is attached to the series, and the analytic theory, in which the series has a definite numerical value called its sum. Both aspects are needed when power series are used to solve differential equations. Since the formal theory is easy (it is much like the algebra of polynomials) it is presented first.

(a) The formal theory

Two power series are said to be equal if corresponding coefficients agree. That is,

$$\sum_{j=0}^{\infty} a_j x^j = \sum_{j=0}^{\infty} b_j x^j \quad \text{means} \quad a_j = b_j, \qquad j = 0, 1, 2, \cdots.$$

A power series is multiplied by a constant k according to the rule

$$k \sum_{j=0}^{\infty} a_j x^j = \sum_{j=0}^{\infty} k a_j x^j$$

and two power series are added by adding corresponding coefficients:

$$\sum_{j=0}^{\infty} a_j x^j + \sum_{j=0}^{\infty} b_j x^j = \sum_{j=0}^{\infty} (a_j + b_j) x^j.$$

The product is defined in the same way as for polynomials. Namely,

$$(a_0 + a_1 x + a_2 x^2 + a_3 x^3 + \cdots)(b_0 + b_1 x + b_2 x^2 + b_3 x^3 + \cdots)$$

$$= a_0 b_0 + (a_0 b_1 + a_1 b_0) x + (a_0 b_2 + a_1 b_1 + a_2 b_0) x^2 + \cdots$$

or, in sigma notation,

$$\left(\sum_{j=0}^{\infty} a_j x^j \right) \left(\sum_{j=0}^{\infty} b_j x^j \right) = \sum_{j=0}^{\infty} c_j x^j$$

where

$$c_j = a_0 b_j + a_1 b_{j-1} + a_2 b_{j-2} + \cdots + a_j b_0.$$

This is called the *Cauchy product* after the French mathematician Augustin Louis Cauchy. When formulas such as this are used, any missing coefficients in the power series are considered to have the value 0.

If the constant term b_0 of the denominator is not 0, the quotient $C = A/B$ can be defined by long division or, equivalently, by setting $CB = A$ and equating coefficients. Another method using geometric series is given in DETA, page 662.

The derived series, or the formal derivative, is defined by

$$(a_0 + a_1 x + a_2 x^2 + a_3 x^3 + \cdots)' = a_1 + 2a_2 x + 3a_3 x^2 + \cdots.$$

In sigma notation the equation can be written

$$\left(\sum_{j=0}^{\infty} a_j x^j \right)' = \sum_{j=0}^{\infty} j a_j x^{j-1}$$

since the term with $j = 0$ in the sum on the right is 0. Higher derivatives are obtained by repetition of the same procedure. A power series has formal derivatives of all orders, even though the function represented by the series may be defined only at $x = 0$, hence not even once differentiable anywhere. The formal derivative behaves much like the ordinary derivative. For example, if u, v are formal power series, the following holds in the sense of formal equality:

$$(uv)' = uv' + u'v.$$

(b) Convergence

The formal theory does not require that a power series have a numerical value at any given value of x. But a series of interest in differential equations generally does have a numerical value, which is obtained by considering the sequence $\{s_n\}$ of partial sums

$$a_0, \ a_0 + a_1 x, \ a_0 + a_1 x + a_2 x^2, \ a_0 + a_1 x + a_2 x^2 + a_3 x^3, \ \cdots .$$

The nth partial sum is

$$\sum_{j=0}^{n} a_j x^j = a_0 + a_1 x + a_2 x^2 + \cdots + a_n x^n.$$

If the nth partial sum tends to a limit as $n \to \infty$ the series is said to converge and its sum is the value of the limit. Thus, by definition,

$$\sum_{j=0}^{\infty} a_j x^j = \lim_{n \to \infty} \sum_{j=0}^{n} a_j x^j$$

provided the limit exists. If the limit does not exist the series is said to diverge and no value is assigned to the sum. Convergence expresses numerical equality, in contrast to the formal equality discussed above.

One might expect the set of points x where a power series converges to be a complicated set on the real axis. The following surprising theorem shows, however, that the set contains only the single point $x = 0$, or else it is an interval centered at the origin:

RADIUS OF CONVERGENCE. For any power series, one of the following holds: The series converges for $x = 0$ only; or the series converges for all x; or there is a positive number r such that the series converges when $|x| < r$ and diverges when $|x| > r$.

The number r is called the *radius of convergence* and the interval $|x| < r$ is called the *interval of convergence*. This terminology is extended by the convention that $r = 0$ if the series converges only for $x = 0$, and $r = \infty$ if the series converges for all x. The proof is outlined in DETA, pages 281–282.

The simplest and often the most effective way to determine the radius of convergence is by means of the following theorem, which applies whether the c_j are real or complex:

RATIO TEST. The series with general term c_j converges if

$$\lim_{n \to \infty} \left| \frac{c_{n+1}}{c_n} \right| < 1.$$

If the limit is greater than 1, the series diverges. If the limit is 1 or does not exist, the ratio test gives no information. For proof see DETA, pages 665–666.

When the ratio test is applied to power series, we make such choices as

$$c_n = a_n x^n, \qquad c_n = a_{2n} x^{2n}, \qquad c_n = a_{2n+1} x^{2n+1}$$

depending on the structure of the series. Here is an example.

EXAMPLE 1. To get the radius of convergence of

$$2x^3 + 8x^6 + \cdots + n2^n x^{3n} + \cdots,$$

we cannot take $c_n = a_n x^n$ because $a_n = 0$ for infinitely many values of n. Instead we set

$$c_n = n2^n x^{3n}, \qquad x \neq 0$$

and obtain

$$\left| \frac{c_{n+1}}{c_n} \right| = \left| \frac{(n+1)2^{n+1} x^{3(n+1)}}{n2^n x^{3n}} \right| = \frac{n+1}{n} 2|x|^3.$$

The limit as $n \to \infty$ is $2|x|^3$. Hence the series converges if $2|x|^3 < 1$, diverges if $2|x|^3 > 1$, and the radius of convergence is given by $2r^3 = 1$. Whether mentioned or not, the condition $x \neq 0$ is always understood when the ratio test is applied to power series.

(c) Properties of convergent series

The same symbol is used to denote both a power series and, when the series is convergent, the value of its sum. This ambiguity causes no trouble because, if x is in the interval of convergence, the formal theory and the analytic theory are in agreement. With $r > 0$, the facts are as follows:

(i) Equality. If two power series converge to the same sum for $|x| < r$ then corresponding coefficients agree.

(ii) Sum, difference, and product. If two power series converge to $A(x)$ and $B(x)$ respectively for $|x| < r$ then the formal sum, difference, and product of the series converge, respectively, to $A(x)+B(x)$, $A(x)-B(x)$, and $A(x)B(x)$ for $|x| < r$.

(iii) Quotient. If in addition to the above hypothesis (ii) we have $B(0) \neq 0$, the formal quotient converges to $A(x)/B(x)$ on some interval $|x| < r_1$. Here r_1 is positive but may be smaller than r.

(iv) Differentiation. A power series and its formal derivative always have the same radius of convergence. If the common radius of convergence is r then for $|x| < r$ the series converges to a differentiable function $A(x)$ and the derived series converges to $A'(x)$. A similar result holds for integration.

(v) Rearrangement. If a power series converges for $|x| < r$ then for $|x| < r$ its terms can be rearranged in any manner without changing its sum.

(vi) Composite functions. If $f(u)$ has a convergent power series expansion for $|u| < \infty$ and $g(x)$ has an expansion for $|x| < r$ then $f(g(x))$ has a convergent expansion at least for $|x| < r$. The coefficients can be obtained by substituting $u = g(x)$ in the series for $f(u)$ and collecting terms.

For a concise discussion of the proofs see DETA, page 664. Property (iv) pertaining to termwise differentiation and integration is established in DETA, page 711.

(d) Translation

Power series are often written in the form

$$(2) \qquad \sum_{j=0}^{\infty} a_j(z - c)^j$$

where c is a real constant. It is said that the expansion is *about the point* c. When $c = 0$, as has been the case up to now, the expansion is about the origin. The equation $x = z - c$ that converts the one case into the other represents a translation.

The general case is easily reduced to the case $c = 0$. Suppose, for example, that $r > 0$ and that, in the sense of convergence,

$$f(z) = \sum_{j=0}^{\infty} a_j (z - c)^j, \qquad |z - c| < r.$$

Setting $x = z - c$ we get a series about $x = 0$, namely

$$f(x + c) = \sum_{j=0}^{\infty} a_j x^j, \qquad |x| < r.$$

Differentiation yields

$$\frac{d}{dx} f(x + c) = f'(x + c) = \sum_{j=0}^{\infty} j a_j x^{j-1}$$

or, since $x + c = z$,

$$f'(z) = \sum_{j=0}^{\infty} j a_j (z - c)^{j-1}.$$

Hence the series (2) can be differentiated term by term just as it could be in the case $c = 0$. Other results extend to the case $c \neq 0$ with equal ease. The parameter c will be retained in the rest of this chapter, but we take $c = 0$ in Chapter 12.

(e) Taylor series

Let us suppose that

$$f(x) = a_0 + a_1(x - c) + \cdots + a_n(x - c)^n + \cdots$$

for $|x - c| < r$, where $r > 0$. Differentiating n times yields

$$f^{(n)}(x) = n! a_n + \cdots$$

where the terms not written all have $(x - c)$ as a factor. When $x = c$ these terms vanish and we get

$$f^{(n)}(c) = n! a_n, \quad \text{hence} \quad a_n = \frac{f^{(n)}(c)}{n!}.$$

This is *Taylor's formula*, named after the English mathematician Brook Taylor who stated it in 1715. The special case $c = 0$ is often called "MacLaurin's series" though Taylor's work preceded MacLaurin's.

When the coefficients are given by Taylor's formula the resulting series, whether convergent or not, is called the *Taylor series* for f. The following test gives conditions under which the series actually converges to the function that generated it:

CONVERGENCE TEST FOR TAYLOR SERIES. Suppose there are positive constants M, K, and r such that

$$|f^{(n)}(t)| \le MK^n, \qquad |t - c| < r, \quad n = 0, 1, 2, \cdots .$$

Then the Taylor series for f converges to $f(x)$ for $|x - c| < r$.

This is a consequence of Lagrange's formula, which asserts that the difference between $f(x)$ and the first n terms of its Taylor series satisfies

$$(3) \qquad\qquad f(x) - \sum_{j=0}^{n-1} a_j (x - c)^j = \frac{f^{(n)}(t)}{n!}$$

for some t between c and x. Lagrange's formula holds if

(i) $f(s)$ is real and continuous on a closed interval containing c and x,

(ii) $f^{(n)}(s)$ exists at every interior point of the interval, and

(iii) $a_j = f^j(c)/j!$, $j = 0, 1, \cdots, n - 1$.

In view of Problem 7, the convergence test follows when (3) is applied to the real and imaginary parts of the series. For proof of (3) see DETA, page 668.

EXAMPLE 2.   If xy is small, the equation on the left below is approximated by the linear equation in the center and by the Bernoulli equation on the right:

$$\frac{dy}{dx} = \sin xy, \qquad \frac{dy}{dx} = xy, \qquad \frac{dy}{dx} - xy = -\frac{x^3}{6} y^3 .$$

This follows when we use Problem 2(a) to get the Taylor expansion of $\sin xy$ in powers of xy. The linear equation gives the approximation

$$y \approx c e^{x^2/2} \approx c \left(1 + \frac{x^2}{2} \right)$$

where $c = y(0)$ is the initial value. On the right we used the Taylor expansion of e^x with x replaced by $x^2/2$. See Problem 2(b).

EXAMPLE 3. Another way to obtain a Taylor series is to differentiate the differential equation by the chain rule. In Example 2 this gives

$$y' = \sin xy, \quad y'' = (\cos xy)(xy' + y)$$

and so on. The next step is

$$y''' = (\cos xy)(xy'' + 2y') - (xy' + y)^2 \sin xy.$$

Setting $x = 0$ and using $y(0) = c$ we get, in succession,

$$y(0) = c, \quad y'(0) = 0, \quad y''(0) = c, \quad y'''(0) = 0.$$

In agreement with the result of Example 2, the Taylor series is

$$y = c\left(1 + \frac{x^2}{2} + 0x^3 + \cdots\right).$$

PROBLEMS

1. Check that, as a formal power series,

$$\sum_{j=0}^{\infty}(2^{j+1} - (j+1)^2)x^j = 1 + 0x - x^2 + 0x^3 + 7x^4 + \cdots$$

This example shows that there is a certain ambiguity when an infinite series is described by giving only the first few terms. That is, from knowledge of the first four terms in the displayed series one could hardly guess that the fifth term is $7x^4$. Nevertheless we shall often describe infinite series by giving just a few terms, because in cases when we do this there is no difficulty in surmising what general term is intended.

2. (abcdef) Without considering convergence, derive the following Taylor expansions:

(a) $\sin x = x - \dfrac{x^3}{3!} + \dfrac{x^5}{5!} - \cdots$

(d) $\sinh x = x + \dfrac{x^3}{3!} + \dfrac{x^5}{5!} + \cdots$

(b) $e^x = 1 + x + \dfrac{x^2}{2!} + \dfrac{x^3}{3!} + \cdots$

(e) $\cos x = 1 - \dfrac{x^2}{2!} + \dfrac{x^4}{4!} - \cdots$

(c) $\cosh x = 1 + \dfrac{x^2}{2!} + \dfrac{x^4}{4!} + \cdots$

(f) $\ln(1 + x) = x - \dfrac{x^2}{2} + \dfrac{x^3}{3} - \cdots$

3. (abcdef) Determine the radius of convergence of the series in Problem 2.

4. (abcde) Show that the Taylor expansions in Problem 2 (abcde) converge to the functions on the left for every value of x.

5P. (abcdefg) It is important to be able to pick out the coefficient of x^n in expressions of the form $x^m y^{(k)}$ where

$$y = \sum_{j=0}^{\infty} a_j x^j$$

This is best done by looking only at the general term, not at the whole series. For example, to find the coefficient of x^n in the formal expansions of y, y', y'', and xy'' we make the following table:

y	y'	y''	xy''
$a_j x^j$	$j a_j x^{j-1}$	$j(j-1)a_j x^{j-2}$	$j(j-1)a_j x^{j-1}$
a_n	$(n+1)a_{n+1}$	$(n+2)(n+1)a_{n+2}$	$(n+1)n a_{n+1}$

The second line gives the general term in the expansion of the first line and the third line gives the coefficient of x^n. The procedure is to take $j = n$, $j - 1 = n$, $j - 2 = n$, $j - 1 = n$ in the four entries of the second line to get the desired exponent n on x. Find the coefficient of x^n:

$$y, \quad xy', \quad x^2 y'', \quad x^{-1}y, \quad xy' + x^2 y'', \quad y' + xy'', \quad xy'' - y' + y + x^{-1}y$$

6. If you let $n \to \infty$ in the following identities, you get three theorems about convergent infinite series. What theorems? No answer is provided:

$$c_n = \sum_{j=0}^{n} c_j - \sum_{j=0}^{n-1} c_j, \qquad \sum_{j=0}^{n} k c_j = k \sum_{j=0}^{n} c_j, \qquad \sum_{j=0}^{n}(a_j + b_j) = \sum_{j=0}^{n} a_j + \sum_{j=0}^{n} b_j$$

7. Let M and K be positive constants. Show that the series with general term $MK^n/n!$ converges and thus deduce $\lim_{n \to \infty} MK^n/n! = 0$.

8. Geometric series. By inspection the two equations

$$s_n = 1 + r + r^2 + \cdots + r^n, \qquad r s_n = r + r^2 + \cdots + r^n + r^{n+1}$$

yield $s_n - r s_n = 1 - r^{n+1}$. Solve for s_n and let $n \to \infty$ to get

$$1 + r + r^2 + r^3 + \cdots + r^n + \cdots = \frac{1}{1-r}, \qquad |r| < 1.$$

9. Set $x = it$ in three of the series of Problem 2 to conclude that

     (a) $e^{it} = \cos t + i \sin t$ (b) $\cos it = \cosh t$ (c) $\sin it = i \sinh t$

10. Suppose two power series

$$u(x) = a_0 + a_1(x-c) + a_2(x-c)^2 + \cdots, \quad v(x) = b_0 + b_1(x-c) + b_2(x-c)^2 + \cdots$$

both converge for $|x - c| < r$ where $r > 0$. Show that the Wronskian W satisfies $W(c) = a_0 b_1 - a_1 b_0$. Hence if $a_0 b_1 \neq a_1 b_0$ it follows that u and v are linearly independent on the interval $|x - c| < r$. A similar result applies to three or more series.

11. (abcde) As in Example 3, find a power series for y through terms in x^3 if the differential equation is as follows and $y(0) = c$:

$$y' = xy + 1, \quad y' = e^{xy}, \quad y' = e^{xy} - 1, \quad y' = y \sin x, \quad y' = \ln(1 + y \sin x)$$

12. (abc) Obtain the Taylor series for y through terms in t^3, t^7, or t^{11}:

$$\frac{dy}{dt} = t^2 + y^2, \quad y(0) = 0$$

---------------------------ANSWERS---------------------------

  3. $r = \infty$ except: (f) $r = 1$

  5. $a_n, \quad na_n, \quad n(n-1)a_n, \quad a_{n+1}, \quad n^2 a_n, \quad (n+1)^2 a_{n+1}, \quad n^2 a_{n+1} + a_n.$

11. Partial answer: Coefficient of x^3 is $1/3$, $(2 + c^2/6)$, $c^2/6$, 0, $-c^2/6$

12. The series in Section 14.2, Problem 6 is correct through terms in t^{11}.

We have adroitly defined the infinite in arithmetic by a love-knot, in this manner ∞, but we possess not therefore the clearer notion of it.

                                      —Voltaire

Chapter C

MATRICES

An m-by-n matrix is a set of mn real or complex numbers arranged in a rectangular array of m rows and n columns and enclosed by parentheses, thus:

$$A = \begin{pmatrix} a_{11} & a_{12} & \cdots & a_{1n} \\ a_{21} & a_{22} & \cdots & a_{2n} \\ \vdots & \vdots & \ddots & \vdots \\ a_{m1} & a_{m2} & \cdots & a_{mn} \end{pmatrix}.$$

This matrix can also be described as follows:

$$A = (a_{ij}), \qquad i = 1, 2, \cdots, m; \quad j = 1, 2, \cdots, n.$$

The numbers a_{ij} are the elements or entries of the matrix and the ordered pair (m, n) gives its size, m-by-n. The size is often understood from context, in which case one writes $A = (a_{ij})$ without specifying the range of i and j. When $m = n$ the matrix is square and n is its order. In a square matrix the elements a_{ii} are *diagonal elements* and the remaining a_{ij} are off-diagonal.

If $m = 1$ the matrix consists of the first row only and is a row vector, while if $n = 1$ the matrix consists of the first column only and is a column vector. The elements of a row or column vector are *coordinates* of the vector and the number of elements in a row or column vector is its *dimension*. For legibility the coordinates of a row vector are often separated by commas. However, the commas have no mathematical significance and a row vector such as $(a\ b\ c)$ has the same meaning as $(a,\ b,\ c)$.

Here are four examples:

$$\begin{pmatrix} 1 & 2 & 3 \\ 4 & 5 & 6 \end{pmatrix}, \quad \begin{pmatrix} 3 & 4 \\ 5 & 6 \end{pmatrix}, \quad (a\ b\ c\ d), \quad \begin{pmatrix} p \\ q \end{pmatrix}$$

The first matrix is of size 2-by-3, the second is a square matrix of order 2, the third is a row vector of dimension 4, and the fourth is a column vector of dimension 2.

Two matrices are equal if they have the same size and corresponding elements agree. Thus

$$(a_{ij}) = (b_{ij}) \quad \text{means} \quad a_{ij} = b_{ij}$$

where it is understood that the matrices have the same size and that the equation on the right holds for all relevant i and j. This definition indicates that a matrix is not just a set of numbers, but involves the arrangement of the numbers as well. For example, no two of the following matrices are equal:

$$(1 \quad 2), \quad (2 \quad 1), \quad (1 \quad 2 \quad 0), \quad (0 \quad 1 \quad 2).$$

If the rows and columns of a matrix A are interchanged the resulting matrix is called the transpose of A and is written A^T. When the elements of A^T are replaced by their complex conjugates, the resulting matrix is called the conjugate transpose or adjoint of A and is written A^*. (Hence for real matrices the transpose and the adjoint coincide.) Here are some examples:

$$\begin{pmatrix} a & b \\ c & d \end{pmatrix}^* = \begin{pmatrix} \bar{a} & \bar{c} \\ \bar{b} & \bar{d} \end{pmatrix}, \qquad (a \quad b)^T = \begin{pmatrix} a \\ b \end{pmatrix}, \qquad \begin{pmatrix} p \\ q \end{pmatrix}^T = (p \quad q).$$

The transpose of a row vector is a column vector of the same dimension, and the transpose of a column vector is a row vector of the same dimension. Obviously

$$(A^T)^T = A, \qquad (A^*)^* = A.$$

A matrix A is said to be symmetric if $A = A^T$, self-adjoint if $A = A^*$. Self-adjoint matrices are also called Hermitian, after the nineteenth century French mathematician Charles Hermite. For example, with $i = \sqrt{-1}$ as usual, the first of the following is symmetric and Hermitian, the second is neither symmetric nor Hermitian, the third is symmetric but not Hermitian, and the fourth is Hermitian but not symmetric:

$$\begin{pmatrix} 1 & 3 \\ 3 & 4 \end{pmatrix}, \qquad \begin{pmatrix} 1 & 2 \\ 3 & 4 \end{pmatrix}, \qquad \begin{pmatrix} i & 1+i \\ 1+i & 2 \end{pmatrix}, \qquad \begin{pmatrix} 1 & 1-i \\ 1+i & 2 \end{pmatrix}.$$

In a context involving matrices a real or complex number λ is often called a scalar. The chief operations on matrices are addition, multiplication by scalars, and matrix multiplication, symbolized respectively by

$$A + B, \qquad \lambda A, \qquad AB.$$

These operations are explained next.

(a) Addition and multiplication by scalars

Matrices of the same size are added by adding corresponding elements, and multiplication by a real or complex scalar λ is also done elementwise:

$$(a_{ij}) + (b_{ij}) = (a_{ij} + b_{ij}) \qquad \lambda(a_{ij}) = (\lambda a_{ij}).$$

We define $A\lambda = \lambda A$, $-A = (-1)A$, and $B - A = B + (-A)$. Here are two examples:

$$\begin{pmatrix} 1 & 2 \\ 3 & 4 \end{pmatrix} + \begin{pmatrix} 1 & 4 \\ 2 & 3 \end{pmatrix} = \begin{pmatrix} 2 & 6 \\ 5 & 7 \end{pmatrix}, \qquad 2\begin{pmatrix} 1 & 2 \\ 3 & 4 \end{pmatrix} = \begin{pmatrix} 2 & 4 \\ 6 & 8 \end{pmatrix}.$$

A matrix with all elements zero is denoted by 0, the same notation being used regardless of the size. The main property of 0 is that

$$A + 0 = A, \qquad 0 + A = A, \qquad A - A = 0$$

provided 0 has the same size as the matrix A in the equations. Only a little experience with matrices is needed to realize that *matrices of different sizes cannot be added,* and that is why the size of 0 need not be built into the notation.

For matrices A, B, C of the same size, you can easily convince yourself of the following:

$$A + B = B + A, \quad (A + B) + C = A + (B + C)$$
$$\lambda(A + B) = \lambda A + \lambda B, \quad (\lambda + \mu)A = \lambda A + \mu A, \quad (\lambda\mu)A = \lambda(\mu A)$$
$$(A + B)^T = A^T + B^T, \quad (A + B)^* = A^* + B^*, \quad (\lambda A)^T = \lambda A^T, \quad (\lambda A)^* = \bar{\lambda} A^*.$$

(b) The matrix product

If $R = (r_i)$ is a row vector and $C = (c_j)$ is a column vector of the same dimension, the product RC is a 1-by-1 matrix defined as follows:

$$RC = (r_1 c_1 + r_2 c_2 + \cdots + r_n c_n).$$

For example,

$$(a \quad b) \begin{pmatrix} p \\ q \end{pmatrix} = (ap + bq), \qquad (a \quad b \quad c) \begin{pmatrix} p \\ q \end{pmatrix} \quad \text{is undefined.}$$

Suppose next that A consists of row vectors R_i, and B consists of column vectors C_j. If the common dimension of the R_i is different from that of the C_j the product AB is undefined. But if the dimension of each R_i is the same as the dimension of each C_j the element p_{ij} in $P = AB$ is the product of the ith row of A with the jth column of B. The following example sheds light on this definition of the matrix product:

$$(1) \qquad \begin{pmatrix} a & b \\ c & d \end{pmatrix} \begin{pmatrix} p & 0 & 1 & 1 \\ q & 1 & 0 & 1 \end{pmatrix} = \begin{pmatrix} ap + bq & b & a & a + b \\ cp + dq & d & c & c + d \end{pmatrix}.$$

Taking this example as point of departure, we make some comments.

Each product $R_i C_j$ is a matrix. However, when a 1-by-1 matrix (a) is an element of another matrix, we agree not to distinguish (a) from a. Thus,

$$\begin{pmatrix} (a) & (b) \\ (c) & (d) \end{pmatrix} = \begin{pmatrix} a & b \\ c & d \end{pmatrix}.$$

That is why the elements of the matrix on the right of (1) need not be enclosed in parentheses. See also Problem 14.

If the rows of A have the same dimension as the columns of B we can form the product AB and it is said that A, B, in that order, are compatible. Thus the two matrices on the left of (1) are compatible in the order given, but if the order were reversed, they would not be compatible.

As suggested by (1), quite generally the product AB of two compatible matrices A, B can be obtained by applying A to each column of B. In symbols,

$$(2) \qquad A(C_1, C_2, \cdots, C_n) = (AC_1, AC_2, \cdots, AC_n).$$

Since each AC_j is a matrix, here too we need the above convention regarding omission of parentheses. With that convention, any matrix can be thought to be an ordered set of row vectors R_j or of column vectors C_j.

A counterpart of (2), also suggested by (1), is

$$(3) \qquad \begin{pmatrix} R_1 \\ R_2 \\ \vdots \\ R_m \end{pmatrix} B = \begin{pmatrix} R_1 B \\ R_2 B \\ \vdots \\ R_m B \end{pmatrix}.$$

This says that the kth row in the product AB is obtained when B is multiplied on the left by the kth row of A.

Equations such as (2) and (3) give a good deal of insight into matrix algebra. For example, if the kth row of A is multiplied by a constant, the kth row of AB is multiplied by the same constant. If two rows of A are interchanged, this has the effect of interchanging the same two rows in AB. These results follow from (3) and similar results for columns follow from (2).

(c) Algebraic properties

The algebra of matrices differs from the algebra of real or complex numbers in two very important respects:

(i) In general $AB \neq BA$ even if both products are defined.

(ii) It is possible to have $AB = 0$ even if neither A nor B is 0.

Both statements are illustrated by

$$\begin{pmatrix} 1 & 0 \\ 0 & 0 \end{pmatrix} \begin{pmatrix} 0 & 1 \\ 0 & 0 \end{pmatrix} = \begin{pmatrix} 0 & 1 \\ 0 & 0 \end{pmatrix}, \qquad \begin{pmatrix} 0 & 1 \\ 0 & 0 \end{pmatrix} \begin{pmatrix} 1 & 0 \\ 0 & 0 \end{pmatrix} = \begin{pmatrix} 0 & 0 \\ 0 & 0 \end{pmatrix}.$$

Otherwise the algebra of matrices is much like the algebra of numbers. This is seen by the equations

$$A(B + C) = AB + AC, \qquad (B + C)A = BA + CA$$
$$(\lambda A)B = \lambda(AB), \qquad A(\lambda B) = \lambda(AB), \qquad A(BC) = (AB)C$$

which hold if the sizes of A, B, C are such that the various expressions are meaningful. The first two equations are distributive laws, the next three are associative laws.

For proof, the formula for $R_i C_j$ shows that the product of compatible matrices A, B can be written in the form

$$(a_{ij})(b_{ij}) = (c_{ij}) \quad \text{where} \quad c_{ij} = \sum_{k=1}^{n} a_{ik} b_{kj}.$$

Here n is the common dimension of the row vectors of A and the column vectors of B. If B and C have the same size, and A, B are compatible, the distributive law $A(B+C) = AB + AC$ is obtained when the identity

$$a_{ik}(b_{kj} + c_{kj}) = a_{ik} b_{kj} + a_{ik} c_{kj}$$

is summed on k. If A, B and B, C are compatible, the associative law

$$A(BC) = (AB)C$$

can be deduced from

$$a_{ip}(b_{pq} c_{qj}) = (a_{ip} b_{pq}) c_{qj}$$

as follows: Sum the left-hand member first on q and then on p to get $A(BC)$. Sum the right-hand member first on p and then on q to get $(AB)C$. Since the sum is independent of the order in which the terms are added, the two sides after the summation are equal. Proof of the other equations above is similar and is left to you. It is also left for you to establish the following, which hold whenever AB is meaningful:

$$(AB)^T = B^T A^T, \qquad (AB)^* = B^* A^*.$$

A square matrix in which all the diagonal elements are 1 and all the other elements are 0 is called the *identity matrix* and is denoted by I, the same notation being used regardless of the size. The main property of I is that

$$AI = IA = A$$

whenever A is a square matrix of the same size as I. The proof follows from (2) and (3). Here is an illustration:

$$\begin{pmatrix} a & b & c \\ p & q & r \\ u & v & w \end{pmatrix} \begin{pmatrix} 1 & 0 & 0 \\ 0 & 1 & 0 \\ 0 & 0 & 1 \end{pmatrix} = \begin{pmatrix} a & b & c \\ p & q & r \\ u & v & w \end{pmatrix} = \begin{pmatrix} 1 & 0 & 0 \\ 0 & 1 & 0 \\ 0 & 0 & 1 \end{pmatrix} \begin{pmatrix} a & b & c \\ p & q & r \\ u & v & w \end{pmatrix}.$$

Only a little experience with matrices is needed to realize that *square matrices of different sizes cannot be multiplied* and that is why the order of I need not be built into the notation.

(d) The inverse

Throughout Part (d) we use A, B, M to denote square matrices of the same order. If there exists a matrix M such that $AM = MA = I$ then A is said to be invertible, M is called the inverse of A, and we write $M = A^{-1}$. Thus

$$A^{-1}A = AA^{-1} = I, \qquad A \quad \text{invertible.}$$

If A is invertible and $AM = I$, then $M = A^{-1}$. In other words A can have at most one inverse. To see why, let the equation $AM = I$ be multiplied *on the left* by A^{-1}. The result is

$$A^{-1}(AM) = A^{-1}I, \qquad (A^{-1}A)M = A^{-1}, \qquad IM = A^{-1}, \qquad M = A^{-1}.$$

We have included some unneeded parentheses to emphasize the role of the associative law. Since the matter has now been brought to your attention, unnecessary parentheses are omitted from now on. For example, $(AB)C$ and $A(BC)$ are denoted by ABC when defined. In the same spirit, repeated multiplication of the square matrix A is denoted by A^m, with $A^0 = I$ by definition. For example, $A^3 = AAA$.

One of the most important theorems about inverses is the following:

THEOREM 1. *If $AB = I$ then $BA = I$, so each matrix A, B is the inverse of the other. Hence to test whether a given matrix M is the inverse of A, it suffices to check one of the equations $AM = I$ or $MA = I$ and the other is an automatic consequence.*

Theorem 1 is somewhat deeper than anything we have stated up to now and its proof is postponed to Chapter E. It will be used, however, in both this chapter and the next.

If we try to construct A^{-1} by solving $AM = I$ for M we are led to a set of linear equations, one for each column of M. For example, the equation

$$\begin{pmatrix} a & b \\ c & d \end{pmatrix} \begin{pmatrix} p & r \\ q & s \end{pmatrix} = \begin{pmatrix} 1 & 0 \\ 0 & 1 \end{pmatrix}$$

is equivalent to the two linear systems

$$ap + bq = 1 \qquad ar + bs = 0$$
$$cp + dq = 0 \qquad cr + ds = 1$$

It is left for you to solve each system and thus to get the formula

$$\begin{pmatrix} a & b \\ c & d \end{pmatrix}^{-1} = \frac{1}{ad - bc} \begin{pmatrix} d & -b \\ -c & a \end{pmatrix}.$$

This assumes $ad \neq bc$; if $ad = bc$ the solution does not exist. By Theorem 1 the equation $MA = I$ holds automatically, so $M = A^{-1}$. The result is summarized as follows:

RULE FOR TWO-BY-TWO INVERSES. To get the inverse of an invertible 2-by-2 matrix, interchange the diagonal elements, change the sign of the off-diagonal elements, and divide by the determinant.

replace ( ) by 11

Obviously, the rule cannot be followed if the determinant is 0; but in that case the inverse does not exist.

This rule provides by far the most efficient way to solve systems of two equations in two unknowns. We illustrate in an example.

EXAMPLE 1. To solve the linear system $u + 2v = p$, $3u + 4v = q$ write the coefficient matrix, the inverse, and the solution:

$$\begin{pmatrix} 1 & 2 \\ 3 & 4 \end{pmatrix}, \qquad -\frac{1}{2}\begin{pmatrix} 4 & -2 \\ -3 & 1 \end{pmatrix}, \qquad \begin{pmatrix} u \\ v \end{pmatrix} = \begin{pmatrix} -2 & 1 \\ 3/2 & -1/2 \end{pmatrix}\begin{pmatrix} p \\ q \end{pmatrix}.$$

(e) Calculations with matrices

The following examples are intended chiefly to illustrate the technique of matrix calculation. However, the results have independent interest and will be used whenever convenient. The letters A, B, M denote square matrices of order n.

EXAMPLE 2. Let A be invertible and let Y be any matrix with the same number of rows as A. (For example, Y could be a square matrix of order n or a column vector of dimension n.) Show that the equation $AX = Y$ has a unique solution, and find it.

If the equation $AX = Y$ is multiplied *on the left* by A^{-1} the result is

$$A^{-1}AX = A^{-1}Y, \qquad IX = A^{-1}Y, \qquad X = A^{-1}Y.$$

This shows that the only possible solution is $X = A^{-1}Y$. The following calculation shows that $X = A^{-1}Y$ is in fact a solution:

$$AX = AA^{-1}Y = IY = Y.$$

EXAMPLE 3. Suppose the equation $Ax = y$ has at least one solution x for each n-dimensional column vector y. Show that A is invertible (and hence the conclusion of Example 2 holds).

By (2), the equation $AM = I$ gives an equation of the form $Ax = y$ for the first column of M, where y is the first column of I. This can be solved by hypothesis. In a like manner $AM = I$ gives an equation of the form $Ax = y$ for the second column of M, and so on. This technique was already illustrated for 2-by-2 inverses in Part (d) and will be developed systematically in Chapter D.

EXAMPLE 4. Quotient law. Suppose $Ax = Bx$ for all n-dimensional column vectors x. Show that $A = B$.

If we let x be the first column of I, then the second column of I, and so on, the hypothesis gives $AI = BI$ and hence $A = B$.

EXAMPLE 5. If both A and B are invertible show that AB is also invertible and find its inverse.

Assuming that AB has an inverse M we write $(AB)M = I$ and solve for M as follows:

$$ABM = I, \qquad BM = A^{-1}, \qquad M = B^{-1}A^{-1}.$$

We multiplied *on the left* first by A^{-1} and then by B^{-1}. Thus the only possibility is

$$(AB)^{-1} = B^{-1}A^{-1}.$$

That this is in fact the inverse follows from

$$ABB^{-1}A^{-1} = AIA^{-1} = AA^{-1} = I.$$

EXAMPLE 6. If A is invertible show that A^{-1}, A^T, and A^* are also invertible and

$$(A^{-1})^{-1} = A, \qquad (A^T)^{-1} = (A^{-1})^T, \qquad (A^*)^{-1} = (A^{-1})^*.$$

Let $B = A^{-1}$. Then $AB = I$, $B^T A^T = (AB)^T = I$, $B^* A^* = (AB)^* = I$. In view of Theorem 1, all three statements follow from this.

PROBLEMS

1. Find AB_1, AB_2, AB if:

$$A = \begin{pmatrix} 1 & 2 & 3 \\ 0 & 1 & 2 \\ 0 & 0 & 1 \end{pmatrix} \quad B_1 = \begin{pmatrix} 0 \\ -4 \\ 3 \end{pmatrix} \quad B_2 = \begin{pmatrix} 0 \\ -3 \\ 2 \end{pmatrix} \quad B = \begin{pmatrix} 0 & 0 & 1 \\ -4 & -3 & -2 \\ 3 & 2 & 1 \end{pmatrix}$$

2. Find B^*, $(A - B)^2$, AC, AD, $D^T A$ and in two ways find $D^T AD$, if

$$A = \begin{pmatrix} 1 & 1 \\ 0 & 1 \end{pmatrix} \quad B = \begin{pmatrix} 1 & -1 \\ -1 & 1 \end{pmatrix} \quad C = \begin{pmatrix} 1 & 2 \\ 0 & 0 \end{pmatrix} \quad D = \begin{pmatrix} u \\ v \end{pmatrix}$$

3. Find A^*, AB, BA and in two ways find $C^T AC$ if

$$A = \begin{pmatrix} 0 & 1 & 2 \\ -1 & 0 & 3 \\ -2 & -3 & 0 \end{pmatrix} \quad B = \begin{pmatrix} 9 & -6 & 3 \\ -6 & 4 & -2 \\ 3 & -2 & 1 \end{pmatrix} \quad C = \begin{pmatrix} u \\ v \\ w \end{pmatrix}$$

4. Let

$$A = \begin{pmatrix} 1 & 0 & 1 \\ 1 & -1 & 0 \end{pmatrix} \qquad B = \begin{pmatrix} 1 & 0 \\ 1 & 1 \end{pmatrix} \qquad C = \begin{pmatrix} 1 & 0 & -1 \end{pmatrix}$$

Decide which of the following products are meaningful and compute those that are meaningful in two ways, using the associative law:

$$ABC \qquad ABC^* \qquad BAC^* \qquad C^*BA \qquad BCA^* \qquad CA^*B$$

5. Compute the products

$$\begin{pmatrix} c & d \\ a & b \end{pmatrix} \begin{pmatrix} 1 & 0 & 1 & p \\ 1 & 1 & 0 & q \end{pmatrix} \qquad\qquad \begin{pmatrix} p & 0 & 1 & 1 \\ q & 1 & 0 & 1 \end{pmatrix}^T \begin{pmatrix} a & c \\ b & d \end{pmatrix}$$

6. If $i^2 = -1$ classify as Hermitian, symmetric, or neither:

$$\begin{pmatrix} i & 2 & 3 \\ 2 & 8 & i \\ 3 & i & 9 \end{pmatrix} \qquad \begin{pmatrix} 1 & 1+i & 7 \\ 1-i & 3 & 8 \\ 7 & 8 & 5 \end{pmatrix} \qquad \begin{pmatrix} 1 & 2 & 1 \\ 1 & 2 & 1 \\ 1 & 2 & 1 \end{pmatrix}$$

7. Solve as in Example 1 and check by substitution:

(a) $\begin{cases} u + 2v &= 1 \\ 3u + 5v &= 1 \end{cases}$ (b) $\begin{cases} 2u + 3v &= p \\ 4u + 5v &= q \end{cases}$ (c) $\begin{cases} u\cos\theta + v\sin\theta &= p \\ -u\sin\theta + v\cos\theta &= q \end{cases}$

8P. A square matrix (a_{ij}) in which all the off-diagonal elements are zero is called a diagonal matrix and is written $\mathrm{diag}(a_{ii})$. Prove that

$$\mathrm{diag}(\lambda_i)\,\mathrm{diag}(\mu_i) = \mathrm{diag}(\lambda_i\mu_i)$$

9P. If $\Lambda = \mathrm{diag}(\lambda_i)$ deduce from Problem 8 that $\Lambda^m = \mathrm{diag}(\lambda_i^m)$ for nonnegative integers m.

10. Let $A = E\Lambda E^{-1}$ where E and Λ are square matrices of the same size with E invertible. Verify the equation $A^m = E\Lambda^m E^{-1}$ for $m = 3$, and show that it holds for all positive integers m.

11. (abcd) If A, B, C, X are square matrices of the same size, and if the needed inverses exist, solve for X:

$$AXB = C \qquad A(X - B) = C \qquad BX^*A^* = (CB^*)^* \qquad XA = XB + C$$

12. A real matrix A is said to be orthogonal if $A^T A = I$, and a real or complex matrix A is said to be unitary if $A^* A = I$. Show that the following are orthogonal. Are they unitary?

$$\begin{pmatrix} 0 & 0 & 1 \\ 1 & 0 & 0 \\ 0 & 1 & 0 \end{pmatrix} \qquad \frac{1}{\sqrt{6}} \begin{pmatrix} 1 & -2 & 1 \\ \sqrt{2} & \sqrt{2} & \sqrt{2} \\ \sqrt{3} & 0 & -\sqrt{3} \end{pmatrix} \qquad \begin{pmatrix} 1 & 0 & 0 \\ 0 & \cos\theta & \sin\theta \\ 0 & -\sin\theta & \cos\theta \end{pmatrix}$$

13. In this problem A, B, E are square matrices of order n and u, v are column vectors of dimension n.

(a) Show that AB is orthogonal if A and B are each orthogonal, and AB is unitary if A and B are each unitary.

(b) If E is unitary show that $E^*AE = B \iff A = EBE^*$.

(c) If A is Hermitian show that $(u^*Av)^* = v^*Au$.

14. Block multiplication. The rule for multiplication of matrices can be extended to allow the elements a_{ij} themselves to be matrices. For example, if A_i and B_i are matrices such that the right-hand side is meaningful,

$$\begin{pmatrix} A_1 & A_2 \\ A_3 & A_4 \end{pmatrix} \begin{pmatrix} B_1 & B_2 \\ B_3 & B_4 \end{pmatrix} = \begin{pmatrix} A_1B_1 + A_2B_3 & A_1B_2 + A_2B_4 \\ A_3B_1 + A_4B_3 & A_3B_2 + A_4B_4 \end{pmatrix}$$

(a) Let $A_1 = (3)$, $A_2 = (-1 \quad 1)$, $B_1 = (2)$, $B_2 = (2 \quad 0)$, and

$$A_3 = \begin{pmatrix} p \\ q \end{pmatrix}, \qquad A_4 = \begin{pmatrix} 1 & 2 \\ 2 & 3 \end{pmatrix}, \qquad B_3 = \begin{pmatrix} u \\ v \end{pmatrix} \qquad B_4 = \begin{pmatrix} -1 & 0 \\ 3 & 1 \end{pmatrix}$$

Obtain 3-by-3 matrices by inserting these values for A_i and B_i on the left-hand side of the above equation and dropping parentheses. Then show that the product of these 3-by-3 matrices agrees with the right-hand side.

(b) Make a similar check when A_i and B_i are 2-by-2 matrices of your own choosing; the left-hand side is now a product of 4-by-4 matrices. Your example should lead to easy calculations, yet should not have so many zero elements that it becomes trivial.

————————————ANSWERS————————————

1. First column in A^T, second column in A^T, and A^T itself.
2. B, $2I$, C, $-$, $-$, $(u^2 + uv + v^2)$ 3. $-A$, 0, 0, 0
4. Partial answer: Two are meaningful, so you have to compute eight matrix products. The computation is self-checking.
5. (a) Interchange the two rows, and the first and last column, in the matrix on the right of (1). (b) Transpose of (1).   6. SHN
11. (a) $A^{-1}CB^{-1}$ (b) $B + A^{-1}C$ (c) $A^{-1}C$ (d) $C(A - B)^{-1}$
12. Yes, because $A^T = A^*$ for real matrices A.

The essence of mathematics is its freedom.

—Georg Cantor

LINEAR EQUATIONS

Throughout this chapter $A = (a_{ij})$ is a square matrix of order n, and x and y are column vectors of dimension n. The linear equations

$$\sum_{j=1}^{n} a_{ij}x_j = y_i, \qquad i = 1, 2, \cdots, n$$

can be written in the compact form $Ax = y$. This follows from the definition of matrix multiplication. For example, the system

$$a_{11}x_1 + a_{12}x_2 + a_{13}x_3 = y_1$$
$$a_{21}x_1 + a_{22}x_2 + a_{23}x_3 = y_2$$
$$a_{31}x_1 + a_{32}x_2 + a_{33}x_3 = y_3$$

is equivalent to

$$\begin{pmatrix} a_{11} & a_{12} & a_{13} \\ a_{21} & a_{22} & a_{23} \\ a_{31} & a_{32} & a_{33} \end{pmatrix} \begin{pmatrix} x_1 \\ x_2 \\ x_3 \end{pmatrix} = \begin{pmatrix} y_1 \\ y_2 \\ y_3 \end{pmatrix}.$$

Linear systems can be solved by a procedure that is called the method of pivot elements, the method of detached coefficients, or Gauss-Jordan elimination. (The latter is named after Gauss and the German geodesist Wilhelm Jordan.) Instead of using letters to represent the unknowns, one specifies the system by giving the augmented matrix (A, y). This is the coefficient matrix A with an extra column y at the right. As an illustration, the augmented matrix of the 3-by-3 system above is

$$\begin{pmatrix} a_{11} & a_{12} & a_{13} & y_1 \\ a_{21} & a_{22} & a_{23} & y_2 \\ a_{31} & a_{32} & a_{33} & y_3 \end{pmatrix}.$$

EXAMPLE 1. Solve the system on the left below, with the augmented matrix at the right:

$$u + 3v = 5 \qquad \begin{pmatrix} 1 & 3 & 5 \\ 2 & 4 & 6 \end{pmatrix}.$$
$$2u + 4v = 6$$

idiocy

We subtract twice the first equation from the second, which means we subtract twice the first row of the augmented matrix from the second row. The element 1 in the augmented matrix plays a special role in this calculation and is called a pivot; a formal definition is given later. The result of this first step is the left-hand matrix below:

$$\begin{pmatrix} 1 & 3 & 5 \\ 0 & -2 & -4 \end{pmatrix} \qquad \begin{pmatrix} 1 & 3 & 5 \\ 0 & 1 & 2 \end{pmatrix} \qquad \begin{pmatrix} 1 & 0 & -1 \\ 0 & 1 & 2 \end{pmatrix}$$

The next matrix is obtained when the second equation is divided by -2. The last matrix is obtained when -3 times the second equation is added to the first equation. The linear system corresponding to the last matrix yields $u = -1$, $v = 2$, which gives the solution of the original system.

(a) Elementary row operations

Here are the main algebraic operations used in calculations such as those above:

(i) Multiply an equation by a nonzero constant.

(ii) Add a multiple of one equation to another equation.

(iii) Interchange two equations.

These correspond to operations on the augmented matrix that are called *elementary row operations* and are defined as follows:

(i) Multiply a row by a nonzero constant.

(ii) Add a multiple of one row to another row.

(iii) Interchange two rows.

The objective is to get an equivalent system that can be solved by inspection, as in Example 1.

In solving a linear system by Gauss-Jordan reduction, we always simplify from left to right, and we require that subsequent simplifications shall not disrupt columns already simplified. This suggests the following definition:

DEFINITION. A pivot is any element of the coefficient matrix that is the first nonzero element of its row.

The value of a pivot is usually reduced to 1 by division, but this requirement is not built into the definition.

A column that has a pivot and at least one other nonzero element is unreduced. All other columns are reduced. We can now state the following:

RULE OF REDUCTION FOR LINEAR SYSTEMS. Find the leftmost unreduced column, reduce it by use of any pivot in it, and repeat until every column is reduced.

EXAMPLE 2. Solve by elementary row operations:

$$x_1 - x_2 + x_3 = 3$$
$$x_1 + 2x_2 + 4x_3 = 9$$
$$2x_1 + 3x_2 + 4x_3 = 7.$$

We begin by simplifying the augmented matrix, as follows:

$$\begin{pmatrix} 1 & -1 & 1 & 3 \\ 1 & 2 & 4 & 9 \\ 2 & 3 & 4 & 7 \end{pmatrix} \begin{matrix} \\ R_2 - R_1 \\ R_3 - 2R_1 \end{matrix} \begin{pmatrix} 1 & -1 & 1 & 3 \\ 0 & 3 & 3 & 6 \\ 0 & 5 & 2 & 1 \end{pmatrix} R_2/3 \begin{pmatrix} 1 & -1 & 1 & 3 \\ 0 & 1 & 1 & 2 \\ 0 & 5 & 2 & 1 \end{pmatrix}.$$

The element 1 was used as a pivot to eliminate all other elements from the first column. The symbol $R_2 - R_1$ between the first two matrices indicates that the first row was subtracted from the second. It is placed opposite the second row because the second row was changed by the operation. Interpretation of $R_3 - 2R_1$ and $R_2/3$ is similar.

To simplify the second column we add the second row to the first and we subtract 5 times the second row from the third. The result is the first matrix on the left below:

$$\begin{pmatrix} 1 & 0 & 2 & 5 \\ 0 & 1 & 1 & 2 \\ 0 & 0 & -3 & -9 \end{pmatrix} -R_3/3 \begin{pmatrix} 1 & 0 & 2 & 5 \\ 0 & 1 & 1 & 2 \\ 0 & 0 & 1 & 3 \end{pmatrix} \begin{matrix} R_1 - 2R_3 \\ R_2 - R_3 \\ \\ \end{matrix} \begin{pmatrix} 1 & 0 & 0 & -1 \\ 0 & 1 & 0 & -1 \\ 0 & 0 & 1 & 3 \end{pmatrix}.$$

Continuing the process we get the matrix on the right, which corresponds to the system $x_1 = -1$, $x_2 = -1$, $x_3 = 3$. Merely writing this system gives the solution to the original equations.

In the next example we have chosen the numbers so that no row has to be divided by a constant. Such division complicates the arithmetic but does not illustrate any new concepts.

EXAMPLE 3. Here is a possible choice of pivots:

$$\begin{pmatrix} -1 & 0 & -4 & 2 & 7 \\ -2 & 0 & -8 & 3 & 9 \\ 1 & 1 & 6 & 0 & 4 \\ 1 & 2 & 8 & 1 & 10 \end{pmatrix} \begin{matrix} R_1 + R_3 \\ R_2 + 2R_3 \\ \\ R_4 - R_3 \end{matrix} \begin{pmatrix} 0 & 1 & 2 & 2 & 11 \\ 0 & 2 & 4 & 3 & 17 \\ 1 & 1 & 6 & 0 & 4 \\ 0 & 1 & 2 & 1 & 6 \end{pmatrix} \begin{matrix} R_1 - R_4 \\ R_2 - 2R_4 \\ R_3 - R_4 \\ \\ \end{matrix}$$

Subtracting the fourth row from the others, as indicated, we get

$$\begin{pmatrix} 0 & 0 & 0 & 1 & 5 \\ 0 & 0 & 0 & 1 & 5 \\ 1 & 0 & 4 & -1 & -2 \\ 0 & 1 & 2 & 1 & 6 \end{pmatrix} \begin{matrix} R_1 - R_2 \\ \\ R_3 + R_2 \\ R_4 - R_2 \end{matrix} \begin{pmatrix} 0 & 0 & 0 & 0 & 0 \\ 0 & 0 & 0 & 1 & 5 \\ 1 & 0 & 4 & 0 & 3 \\ 0 & 1 & 2 & 0 & 1 \end{pmatrix}.$$

The final system is $x_4 = 5$, $x_1 + 4x_3 = 3$, $x_2 + 2x_3 = 1$. Hence we can set $x_3 = c$ where c is any constant and express the solution as follows:

$$x_1 = 3 - 4c, \qquad x_2 = 1 - 2c, \qquad x_3 = c, \qquad x_4 = 5.$$

There are infinitely many solutions, and the above formulas give them all. If we had set $x_1 = c$ or $x_2 = c$ the resulting formulas would look different, but would represent the same set of solutions.

The rule of reduction does not require the third elementary row operation, interchange of two rows. However, for many purposes it is desirable to arrange the reduced matrix in row-echelon form. This is a form in which the leftmost pivot is in the top row only, the next leftmost pivot is in the second row only, and so on. For example, the row-echelon form of the final matrix obtained in Example 3 is

$$\begin{matrix} R_3 \\ R_4 \\ R_2 \\ R_1 \end{matrix} \begin{pmatrix} 1 & 0 & 4 & 0 & 3 \\ 0 & 1 & 2 & 0 & 1 \\ 0 & 0 & 0 & 1 & 5 \\ 0 & 0 & 0 & 0 & 0 \end{pmatrix}.$$

Here the R_3 opposite the first row means that the top row is the third row of the original matrix, and similarly in other cases. Going from a reduced matrix to a matrix in row-echelon form requires only one more step, no matter how large the matrix may be.

As defined above, a row-echelon form is not unique. A matrix in row-echelon form is in *reduced row-echelon form* if each pivot element is equal to 1 and is the only nonzero element in its column. Informally, this means that the process of simplification has been carried as far as possible.

It can be shown that the reduced row-echelon form is unique. In particular, if each column has a pivotal element, the reduced row-echelon form of the coefficient matrix is I, the reduced augmented matrix has the form (I, z), and the solution of the corresponding linear system is $x = z$. For example the augmented matrix

$$\begin{pmatrix} 1 & 0 & 0 & z_1 \\ 0 & 1 & 0 & z_2 \\ 0 & 0 & 1 & z_3 \end{pmatrix}$$

corresponds to the system $x_1 = z_1$, $x_2 = z_2$, $x_3 = z_3$, which gives $x = z$.

(b) Computation of the inverse

The following two features of the Gauss-Jordan method are helpful when linear systems are used to compute the inverse of a given matrix A:

(i) As soon as the coefficient matrix is reduced to I, you can read the solution by looking at the right-hand column of the augmented matrix.

(ii) The procedure by which the reduction (i) is effected is independent of the right-hand column of the augmented matrix.

Because of these features the n linear systems needed to get the columns of M from $AM = I$ can be solved all at once, as in the following example.

EXAMPLE 4. We shall compute the inverse of the 3-by-3 matrix that forms the left half of the first matrix below. The initial steps are:

$$\begin{pmatrix} 1 & 0 & -1 & 1 & 0 & 0 \\ 2 & 1 & 1 & 0 & 1 & 0 \\ 1 & 1 & 3 & 0 & 0 & 1 \end{pmatrix} \begin{matrix} \\ R_2 - 2R_1 \\ R_3 - R_1 \end{matrix} \begin{pmatrix} 1 & 0 & -1 & 1 & 0 & 0 \\ 0 & 1 & 3 & -2 & 1 & 0 \\ 0 & 1 & 4 & -1 & 0 & 1 \end{pmatrix}$$

The next step is to subtract the second row from the third. Thus we get the matrix on the left below, and the process is completed to give the matrix on the right:

$$\begin{pmatrix} 1 & 0 & -1 & 1 & 0 & 0 \\ 0 & 1 & 3 & -2 & 1 & 0 \\ 0 & 0 & 1 & 1 & -1 & 1 \end{pmatrix} \begin{matrix} R_1 + R_3 \\ R_2 - 3R_3 \\ \end{matrix} \begin{pmatrix} 1 & 0 & 0 & 2 & -1 & 1 \\ 0 & 1 & 0 & -5 & 4 & -3 \\ 0 & 0 & 1 & 1 & -1 & 1 \end{pmatrix}.$$

The right half of the last matrix is the desired inverse. As a check we compute

$$\begin{pmatrix} 1 & 0 & -1 \\ 2 & 1 & 1 \\ 1 & 1 & 3 \end{pmatrix} \begin{pmatrix} 2 & -1 & 1 \\ -5 & 4 & -3 \\ 1 & -1 & 1 \end{pmatrix} = \begin{pmatrix} 1 & 0 & 0 \\ 0 & 1 & 0 \\ 0 & 0 & 1 \end{pmatrix}.$$

The general principle illustrated here is that if you start with the augmented matrix $(A\ I)$ and reduce the coefficient matrix A to I by row operations, in the final agumented matrix $(I\ M)$ the right-hand half M gives A^{-1}. Though surprising, this result does not require a new theory. It is just an application of the procedure developed in Part (a) to several systems at once.

In the rest of this chapter we discuss the notion of a vector space. This important concept allows a geometric interpretation of some aspects of linear systems and conversely, the algebra of linear systems gives insight into the geometry of vector spaces. The relevance to differential equations will become clear as the discussion progresses.

(c) Definition of a vector space

For the moment, let $\mathbf{x}$, $\mathbf{y}$, $\mathbf{z}$ represent real column vectors of dimension n and let s, t denote real numbers. If the set of all such column vectors is denoted by $\mathbf{V}$, we have the *closure condition*

(C) $$\mathbf{x} \in \mathbf{V}, \quad \mathbf{y} \in \mathbf{V} \Rightarrow \mathbf{x} + \mathbf{y} \in \mathbf{V}, \quad s\mathbf{x} \in \mathbf{V}$$

the *addition laws*

(A) $\mathbf{x}+\mathbf{y} = \mathbf{y}+\mathbf{x}, \quad \mathbf{x}+(\mathbf{y}+\mathbf{z}) = (\mathbf{x}+\mathbf{y})+\mathbf{z}, \quad \mathbf{x}+\mathbf{0} = \mathbf{x}, \quad \mathbf{x}+(-\mathbf{x}) = \mathbf{0},$

and the *multiplication laws*

(M) $s(t\mathbf{x}) = (st)\mathbf{x}, \quad (s+t)\mathbf{x} = s\mathbf{x} + t\mathbf{x}, \quad s(\mathbf{x}+\mathbf{y}) = s\mathbf{x} + s\mathbf{y}, \quad 1\mathbf{x} = \mathbf{x}.$

These properties arise in so many contexts that a special name, *vector space*, has been given to any system that satisfies them. Here is the formal definition:

VECTOR SPACE. A vector space $\mathbf{V}$ consists of a set of elements, including an element $\mathbf{0}$, such that the laws (CAM) hold for all $\mathbf{x}$, $\mathbf{y}$, $\mathbf{z} \in \mathbf{V}$ and all real numbers s, t. The elements of $\mathbf{V}$ are called vectors and the real numbers are called scalars.

More specifically, this defines a vector space *over the real field*, because the scalars are required to be real. For a vector space over the complex field the scalars are complex. These two cases are not the only ones that are important in linear algebra, but they suffice for the study of differential equations.

Vector spaces have many properties familiar from elementary algebra. For example, it is left for you to show that $0\mathbf{x} = \mathbf{0}$ and $(-1)\mathbf{x} = -\mathbf{x}$ for all $\mathbf{x} \in \mathbf{V}$. The following discussion is written for the case of real scalars but needs only minor changes when they are complex.

EXAMPLE 5. The set $\mathbf{V}$ of real-valued functions f defined on $(-\infty, \infty)$ is a vector space. To see why, recall that the sum $f + g$, the product sf, and equality $f = g$ are defined respectively by

$$(f+g)(t) = f(t) + g(t), \quad (sf)(t) = sf(t), \quad f(t) = g(t).$$

The needed algebraic properites of $f \in \mathbf{V}$ follow from corresponding properties of the real numbers. The set of differentiable functions on $(-\infty, \infty)$ forms a vector space that is contained in the vector space defined above.

It often happens that a subset of vectors in $\mathbf{V}$ forms a vector space in its own right under the same operations as those in $\mathbf{V}$. Such a subset is called a *subspace*. Since a subspace is a vector space, it must be closed under addition, subtraction, and multiplication by scalars. Here is a converse:

THEOREM 1. *A nonempty subset $\mathbf{S}$ of a vector space $\mathbf{V}$ is a subspace of $\mathbf{V}$ if $\mathbf{S}$ is closed under addition and under multiplication by scalars.*

For proof, pick $\mathbf{x} \in \mathbf{S}$, as we can do because $\mathbf{S}$ is not empty. Closure under multiplication by scalars gives $\mathbf{0} = 0\mathbf{x} \in \mathbf{V}$ and $-\mathbf{x} = (-1)\mathbf{x} \in \mathbf{V}$. Since the other algebraic properties we need are inherited from those in $\mathbf{V}$, this completes the proof.

EXAMPLE 6. The set of real solutions of $y'' + y = 0$ consists of the functions

$$(1) \qquad\qquad y = c_1 \cos t + c_2 \sin t$$

where c_1 and c_2 can take any real values. Since this set is closed under addition and under multiplication by scalars, it forms a subspace of the vector space considered in Example 5. By Theorem 1, a similar result holds for solutions of $Ty = 0$ where T is any linear differential operator.

The expression for y in (1) is a linear combination of $\cos t$ and $\sin t$, the functions $\cos t$ and $\sin t$ are linearly independent, and they form a basis for the set of solutions of $y'' + y = 0$ because the representation (1) is unique. This terminology was introduced in Sections 4.1, 6.1, and 11.2 respectively. It is now generalized.

(d) Span, linear dependence, basis

Throughout the following discussion, $\mathbf{B}$ denotes a sequence

$$\mathbf{b}_1, \quad \mathbf{b}_2, \quad \cdots, \quad \mathbf{b}_n$$

of elements of a given vector space $\mathbf{V}$. A *linear combination* of the $\mathbf{b}_j$ is an expression

$$(2) \qquad\qquad \mathbf{v} = v_1 \mathbf{b}_1 + v_2 \mathbf{b}_2 + \cdots + v_n \mathbf{b}_n$$

where the v_j are scalars. By Theorem 1, the set of all such linear combinations is a subspace of $\mathbf{V}$. This subspace is called the *span* of $\mathbf{B}$ and is written span $\mathbf{B}$. If every vector $\mathbf{v} \in \mathbf{S}$ can be represented *uniquely* in the form (2), the sequence $\mathbf{B}$ is called a *basis* for $\mathbf{S}$. Here is an example.

EXAMPLE 7. The polynomials

$$1, \quad t, \quad t^2, \quad \cdots, \quad t^{n-1}$$

form a basis for the vector space $\mathbf{V}$ of all solutions of $y^{(n)} = 0$. This is true because every solution can be written in the form

$$y(t) = c_1 + c_2 t + c_3 t^2 + \cdots c_n t^{n-1}$$

and the representation is unique by Taylor's formula.

To investigate uniqueness of the representation (2) suppose we have, besides (2), a second representation

$$(3) \qquad \mathbf{v} = w_1\mathbf{b}_1 + w_2\mathbf{b}_2 + \cdots + w_n\mathbf{b}_n.$$

With $c_j = v_j - w_j$ subtraction of (2) and (3) gives

$$(4) \qquad c_1\mathbf{b}_1 + c_2\mathbf{b}_2 + \cdots + c_n\mathbf{b}_n = 0.$$

If this implies $c_j = 0$ for each j then $v_j = w_j$ and uniqueness follows. Conversely, if the equations have a solution with at least one $c_j \neq 0$ then the representation of the zero vector is not unique.

By definition, the sequence of vectors $\mathbf{B}$ is *linearly independent* if (4) implies that each $c_j = 0$. Otherwise $\mathbf{B}$ is *linearly dependent*. The foregoing discussion can now be summarized as follows:

THEOREM 2. *A sequence of vectors $\mathbf{B}$ forms a basis for $\mathbf{V}$ if and only if $\mathbf{B}$ spans $\mathbf{V}$ and is linearly independent.*

EXAMPLE 8. The sets of real and complex n-dimensional column vectors are denoted by $\mathbf{R}^n$ or $\mathbf{C}^n$, respectively. Let us show that the vectors

$$\mathbf{b}_1 = \begin{pmatrix} 1 \\ 2 \end{pmatrix}, \qquad \mathbf{b}_2 = \begin{pmatrix} 1 \\ 1 \end{pmatrix}$$

form a basis for $\mathbf{R}^2$. This means that any vector $\mathbf{x} \in \mathbf{R}^2$ can be written uniquely in the form

$$\mathbf{x} = c_1\mathbf{b}_1 + c_2\mathbf{b}_2.$$

The above equation is equivalent to

$$\begin{pmatrix} x_1 \\ x_2 \end{pmatrix} = c_1 \begin{pmatrix} 1 \\ 2 \end{pmatrix} + c_2 \begin{pmatrix} 1 \\ 1 \end{pmatrix}$$

which in turn is equivalent to the system

$$x_1 = c_1 + c_2, \qquad x_2 = 2c_1 + c_2.$$

Subtracting these equations we get $c_1 = x_2 - x_1$ and the first equation then gives c_2. Since the c_j are unique, the vectors $\mathbf{b}_j$ form a basis.

EXAMPLE 9. Show that the three columns of the matrix below are linearly independent unless $k = 0$, in which case they are linearly dependent:

(5)
$$\begin{pmatrix} 1 & 2 & 1 \\ 0 & 1 & k \\ 1 & 0 & 1 \end{pmatrix}.$$

The equation

$$c_1 \begin{pmatrix} 1 \\ 0 \\ 1 \end{pmatrix} + c_2 \begin{pmatrix} 2 \\ 1 \\ 0 \end{pmatrix} + c_3 \begin{pmatrix} 1 \\ k \\ 1 \end{pmatrix} = \begin{pmatrix} 0 \\ 0 \\ 0 \end{pmatrix}$$

is equivalent to a linear homogeneous system with the matrix (5) as coefficient matrix. The reduction

$$\begin{pmatrix} 1 & 2 & 1 \\ 0 & 1 & k \\ 1 & 0 & 1 \end{pmatrix} \overset{R_1 - R_3}{\longrightarrow} \begin{pmatrix} 0 & 2 & 0 \\ 0 & 1 & k \\ 1 & 0 & 1 \end{pmatrix}$$

yields a matrix for which the corresponding homogeneous equations are

$$2c_2 = 0, \quad c_2 + kc_3 = 0, \quad c_1 + c_3 = 0.$$
$$c_2 + c_3 = 0$$

If $k \neq 0$ the only solution is $c_j = 0$, hence the vectors are independent. But if $k = 0$ there is a nontrivial solution $(1, 0, -1)$ and the vectors are linearly dependent.

(e) The concept of dimension

In general, a space has dimension n when points in it can be specified by n coordinates. For example, the real line is one-dimensional and points on it can be described by a single coordinate (x). The two-dimensional plane requires two coordinates (x, y), three-dimensional space requires three, and so on. To be sure, the geometric visualization fails when the dimension exceeds three, but visualization is of less importance than the analytic apparatus for coping with geometric problems.

The following theorem marks an important step towards a definition of *dimension* in every vector space $\mathbf{V}$:

THEOREM 3. *Suppose a vector space* $\mathbf{V}$ *is spanned by the n vectors* $\{\mathbf{b}_j\}$. *Then every sequence of $n + 1$ vectors in* $\mathbf{V}$ *is linearly dependent.*

For proof, let $\{\mathbf{a}_j\}$ be a sequence of $n + 1$ vectors in $\mathbf{V}$. Since the $\mathbf{b}_j$ span $\mathbf{V}$ we can write

$$\mathbf{a}_j = \sum_{i=1}^{n} a_{ij} \mathbf{b}_i$$

where the a_{ij} are real numbers. For any real constants c_j we have

$$\sum_{j=1}^{n+1} c_j \mathbf{a}_j = \sum_{j=1}^{n+1} c_j \sum_{i=1}^{n} a_{ij} \mathbf{b}_i = \sum_{i=1}^{n} \sum_{j=1}^{n+1} c_j a_{ij} \mathbf{b}_i.$$

The coefficient of each $\mathbf{b}_i$ is 0 if

$$\sum_{j=1}^{n+1} a_{ij} c_j = 0, \qquad i = 1, 2, \cdots, n.$$

This is a linear homogeneous system of n equations in $n + 1$ unknowns c_j. Since there are more unknowns than equations, it follows from Chapter E, Theorem 3 that there is a nontrivial solution $\{c_j\}$. This shows that the vectors $\mathbf{a}_j$ are linearly dependent and completes the proof.

THEOREM 4. *If a vector space* $\mathbf{V}$ *has a finite basis, every basis has the same number of elements.*

By definition, a *finite basis* is a basis with finitely many elements. For proof, suppose there is a basis $\{\mathbf{b}_j\}$ with n elements. By Theorem 3 every set of more than n elements is linearly dependent, hence cannot be a basis. Thus the number of elements in every basis is *at most* n. But if there is a basis with fewer than n elements, Theorem 3 shows that the n vectors $\mathbf{b}_j$ would be linearly dependent, contradicting the fact that they form a basis. Hence the number of elements in any basis is *at least* n. This completes the proof.

DIMENSION. The dimension of a vector space $\mathbf{V}$ *is denoted by* $\dim \mathbf{V}$. *If* $\mathbf{V}$ *has only the single element* $\mathbf{0}$, *then* $\dim \mathbf{V} = 0$. *If* $\mathbf{V}$ *has a finite basis,* $\dim \mathbf{V}$ *is the number of elements in any basis. In all other cases* $\dim \mathbf{V} = \infty$.

The main statement is the second one, which depends on Theorem 4. We have to consider the vector space $\{\mathbf{0}\}$ separately because this space contains no linearly independent vectors.

THEOREM 5. *In a vector space of dimension* n, *every basis has* n *linearly independent vectors and every set of* n *linearly independent vectors forms a basis.*

The first statement has already been proved. For the second, let $\{\mathbf{v}_j\}$ be a set of n linearly independent vectors in $\mathbf{V}$ and let $\mathbf{v}$ be any vector in $\mathbf{V}$. By Theorem 3 the $n + 1$ vectors $\mathbf{v}_j$, $\mathbf{v}$ are linearly dependent. Thus there are constants c and c_j, not all 0, such that

$$c\mathbf{v} + c_1\mathbf{v}_1 + c_2\mathbf{v}_2 + \cdots + c_n\mathbf{v}_n = \mathbf{0}.$$

We have $c \neq 0$ since the $\mathbf{v}_j$ are linearly independent. Hence we can solve for $\mathbf{v}$ and deduce that $\mathbf{v}$ is in the space spanned by the $\mathbf{v}_j$. Since the $\mathbf{v}_j$ are linearly independent and span $\mathbf{V}$, they form a basis. This completes the proof.

PROBLEMS

1. Suppose $Ax = y$ has two distinct solutions $x = u$ and $x = v$. Show that $Ax = 0$ and $Ax = y$ both have infinitely many solutions.

2. (abc) The following are augmented matrices (A, y) for a linear system $Ax = y$ of three equations in three unknowns. Solve by row operations. You need not reduce to echelon form:

$$\begin{pmatrix} 1 & 2 & 3 & 0 \\ 2 & 3 & 1 & 4 \\ 3 & 4 & 2 & 5 \end{pmatrix} \quad \begin{pmatrix} 4 & 7 & -1 & 3 \\ 3 & 2 & 3 & 6 \\ 2 & 4 & 6 & 8 \end{pmatrix} \quad \begin{pmatrix} 3 & 2 & 8 & 4 \\ 4 & 8 & 0 & 0 \\ 2 & 4 & 7 & 7 \end{pmatrix}$$

3. (abc) If possible, proceed as in Problem 3:

$$\begin{pmatrix} 1 & -1 & 2 & 1 \\ 3 & 4 & 2 & 13 \\ 5 & 1 & 4 & 11 \end{pmatrix} \quad \begin{pmatrix} 1 & 2 & 3 & 1 \\ 3 & 7 & 10 & 4 \\ 2 & 4 & 6 & 5 \end{pmatrix} \quad \begin{pmatrix} 1 & 2 & 3 & 10 \\ 2 & 4 & 5 & 19 \\ 1 & 3 & 6 & 15 \end{pmatrix}$$

4. (abcde) For a homogeneous system $Ax = 0$ the rightmost column of the augmented matrix remains 0 during the reduction and need not be written down. Here are coefficient matrices for homogeneous systems with $n = 3$. After reduction, obtain a nontrivial solution with $x_3 = 1$ if possible. If not, try $x_2 = 1$. If that fails too, try $x_1 = 1$:

$$\begin{pmatrix} 1 & 0 & 0 \\ 0 & 0 & 0 \\ 0 & 1 & 4 \end{pmatrix} \begin{pmatrix} 0 & 0 & 1 \\ 1 & 2 & 0 \\ 0 & 0 & 0 \end{pmatrix} \begin{pmatrix} 0 & 0 & 0 \\ 1 & 0 & 1 \\ 0 & 1 & 1 \end{pmatrix} \begin{pmatrix} 0 & 1 & 2 \\ 0 & 1 & 1 \\ 0 & 1 & 3 \end{pmatrix} \begin{pmatrix} 1 & -1 & -6 \\ 0 & 0 & 0 \\ 1 & 2 & 3 \end{pmatrix}$$

5P. To get the inverse of the 3-by-3 matrix at the left, start as follows:

$$\begin{pmatrix} 0 & 3 & 3 & 1 & 0 & 0 \\ 0 & 1 & 2 & 0 & 1 & 0 \\ 3 & 6 & 9 & 0 & 0 & 1 \end{pmatrix} R_3/3, R_{13} \begin{pmatrix} 1 & 2 & 3 & 0 & 0 & 1/3 \\ 0 & 1 & 2 & 0 & 1 & 0 \\ 0 & 3 & 3 & 1 & 0 & 0 \end{pmatrix}$$

The notation $R_3/3$, R_{13} means that the third row is divided by 3 and that the first and third rows are then interchanged. The next step would be indicated by writing $R_1 - 2R_2$ after the first row and $R_3{'} - 3R_2$ after the third. Complete the calculation and check by computing AA^{-1} or $A^{-1}A$. Since you are asked to check, no answer is provided.

6. (abcde) As in Problem 6, find the inverse and check:

$$\begin{pmatrix} 1 & 2 & 3 \\ 0 & 1 & 4 \\ 0 & 0 & 1 \end{pmatrix} \quad \begin{pmatrix} 1 & 0 & 0 \\ 3 & 1 & 0 \\ 7 & 4 & 1 \end{pmatrix} \quad \begin{pmatrix} 1 & 0 & 1 \\ 2 & 3 & 1 \\ 1 & 1 & 1 \end{pmatrix} \quad \begin{pmatrix} 2 & 2 & 1 \\ 3 & 2 & 1 \\ 4 & 3 & 2 \end{pmatrix} \quad \begin{pmatrix} 2 & 1 & 1 \\ 3 & 3 & 2 \\ 4 & 5 & 5 \end{pmatrix}$$

7. (abcd) Compute the inverse by reducing (A, I) and check:

$$\begin{pmatrix} 1 & 2 & 3 & 4 \\ 0 & -1 & 2 & 3 \\ 0 & 0 & 1 & 5 \\ 0 & 0 & 0 & -1 \end{pmatrix} \quad \begin{pmatrix} 1 & 0 & 0 & 0 \\ 3 & 1 & 0 & 0 \\ 4 & 3 & 1 & 0 \\ 5 & 1 & 1 & 1 \end{pmatrix} \quad \begin{pmatrix} 1 & 1 & 0 & 0 \\ 1 & 0 & 1 & 0 \\ 1 & 1 & 0 & 1 \\ 1 & 0 & 0 & 0 \end{pmatrix} \quad \begin{pmatrix} 1 & 1 & 0 & 0 \\ 1 & 0 & 1 & 0 \\ 1 & 1 & 0 & 1 \\ 1 & 1 & 1 & 1 \end{pmatrix}$$

8. (abcde) Show that the columns of the matrices in Problem 4 are linearly dependent.

9. (abcde) Show that the columns of the matrices in Problem 6 form a basis for $\mathbf{R}^3$ and (fghi) that the columns of the matrices in Problem 7 form a basis for $\mathbf{R}^4$.

10. If A_j are the columns of A, show that the equation

$$x_1 A_1 + x_2 A_2 + \cdots + x_n A_n = y$$

is equivalent to $Ax = y$. Hence the system $Ax = y$ has a solution if, and only if, y is in the vector space spanned by the columns of A. This gives a further illustration of the connection between vector spaces and linear systems.

————————————————ANSWERS————————————————

2. (a) 1, 1, −1   (b) 1, 0, 1   (c) −2, 1, 1
3. (a) 1, 2, 1   (b) no solution   (c) 3, 2, 1
4. $x^T = (0 \ -4 \ 1)$ $(-2 \ 1 \ 0)$ $(-1 \ -1 \ 1)$ $(1 \ 0 \ 0)$ $(3 \ -3 \ 1)$

Rigor is no foe of simplicity.

—Otto Blumenthal

Chapter E

DETERMINANTS

Let $A = (a_{ij})$ be a square matrix of order at least 2 and let the parentheses ( ) in the displayed form of A be replaced by vertical bars | |. The result is a numerical-valued function of A, which is written $\det A$ or $|A|$ and is called the determinant. Here are two examples:

(1)
$$\begin{vmatrix} a_1 & a_2 \\ b_1 & b_2 \end{vmatrix} = a_1 b_2 - a_2 b_1$$

$a_1 b_2 c_3 + a_3 b_1 c_2 + a_2 b_3 c_1 - a_1 b_3 c_2$

$a_1 (b_2 c_3 - b_3 c_2) = a_1 \begin{vmatrix} b_2 & b_3 \\ c_2 & c_3 \end{vmatrix}$

(2)
$$\begin{vmatrix} a_1 & a_2 & a_3 \\ b_1 & b_2 & b_3 \\ c_1 & c_2 & c_3 \end{vmatrix} = a_1 \begin{vmatrix} b_2 & b_3 \\ c_2 & c_3 \end{vmatrix} - a_2 \begin{vmatrix} b_1 & b_3 \\ c_1 & c_3 \end{vmatrix} + a_3 \begin{vmatrix} b_1 & b_2 \\ c_1 & c_2 \end{vmatrix}.$$

Note that the determinants on the right of (2) can be evaluated by (1). Note also that the vertical bars | | replace the parentheses ordinarily associated with a matrix; we do not use | | in addition to ( ).

Although the value of a determinant is a number, not a matrix, terminology is extended in a natural way from A to $\det A$. For example, the determinant on the left of (2) is of order 3, the first row of the determinant (1) is $(a_1\ a_2)$, the elements of a determinant are the elements of the corresponding matrix, and so on. A determinant of order 1 has only a single element and is defined by $\det(a) = a$. The notation $|a|$ is not used in this case because it could be confused with the absolute value.

By analogy to (1) and (2), a determinant of order $n \geq 2$ consists of a square array

(3)
$$\det A = |A| = \begin{vmatrix} a_{11} & a_{12} & \cdots & a_{1n} \\ a_{21} & a_{22} & \cdots & a_{2n} \\ \vdots & \vdots & \ddots & \vdots \\ a_{n1} & a_{n2} & \cdots & a_{nn} \end{vmatrix}$$

to which a numerical value is attached as follows: Let M_{1i} be the determinant obtained when the first row and the ith column of the original determinant are deleted. Then

(4) $\quad |A| = a_{11} M_{11} - a_{12} M_{12} + a_{13} M_{13} - a_{14} M_{14} + \cdots + (-1)^{n+1} a_{1n} A_{1n}.$

This definition is inductive; a determinant of order n is defined in terms of those of order $n-1$. You can easily convince yourself that the definition gives (1) and (2) when $n = 2$ and $n = 3$, respectively. It should be emphasized that *only a square matrix can have a determinant* and, with that in mind, we use $A = (a_{ij})$ throughout this chapter to denote a square matrix of order n. We also use x and y to denote column vectors of dimension n.

The following example shows how determinants are reduced to determinants of lower order by successive expansion on the first row.

EXAMPLE 1. No matter what numbers are represented by the stars,

$$\begin{vmatrix} a & 0 & 0 & 0 \\ * & b & 0 & 0 \\ * & * & c & 0 \\ * & * & * & d \end{vmatrix} = a \begin{vmatrix} b & 0 & 0 \\ * & c & 0 \\ * & * & d \end{vmatrix} = ab \begin{vmatrix} c & 0 \\ * & d \end{vmatrix} = abcd.$$

(a) Laplace expansion

The determinant M_{ij} obtained when the ith row and jth column of (3) are deleted is the *minor* of the element a_{ij} and the signed minor

$$A_{ij} = (-1)^{i+j} M_{ij}$$

is the *cofactor* of a_{ij}. As an illustration, the 2-by-2 determinants on the right of (2) are the minors of the elements a_1, a_2, a_3 respectively, and when provided with alternating signs $+$, $-$, $+$ they become the corresponding cofactors. Hence (2) can be written

$$|A| = a_1 A_{11} + a_2 A_{12} + a_3 A_{13}.$$

In a like manner, the definition (4) is

(5) $$\det A = a_{11} A_{11} + a_{12} A_{12} + \cdots + a_{1n} A_{1n}.$$

Equation (5) is a *Laplace expansion* on the first row, named after the French mathematician Pierre Simon de Laplace. More generally we can make a Laplace expansion on the ith row:

$$|A| = a_{i1} A_{i1} + a_{i2} A_{i2} + \cdots + a_{in} A_{in},$$

or also on the jth column:

$$|A| = a_{1j} A_{1j} + a_{2j} A_{2j} + \cdots + a_{nj} A_{nj}.$$

The fact that all such expansions give the same value is the content of the following theorem, which provides a very convenient approach to the theory of determinants:

EXPANSION THEOREM. A determinant can be evaluated by Laplace expansion on any row or on any column.

The main ideas needed for the proof are outlined in DETA, Chapter E. The following examples show how the theorem is used.

EXAMPLE 2. Evaluate the determinant D below:

$$D = \begin{vmatrix} 1 & 0 & -1 & 2 \\ 6 & 0 & 4 & 3 \\ 4 & 7 & 0 & 2 \\ 2 & 0 & 2 & 3 \end{vmatrix} \qquad S = \begin{vmatrix} + & - & + & - \\ - & + & - & + \\ + & - & + & - \\ - & + & - & + \end{vmatrix}$$

The sign matrix S at the right interprets the $(-1)^{i+j}$ rule for going from minors to cofactors. Expanding D on the second column, and using signs as given by the second column of S, we get

$$D = 0(-M_{12}) + 0(+M_{22}) + 7(-M_{32}) + 0(+M_{42}).$$

The only surviving minor M_{32} is obtained by crossing out the row and column containing the element 7 in D. This gives the first equality below. The second is obtained when M_{32} is expanded on its first row:

$$D = (-7)\begin{vmatrix} 1 & -1 & 2 \\ 6 & 4 & 3 \\ 2 & 2 & 3 \end{vmatrix} = (-7)(6 + 12 + 8) = -182.$$

EXAMPLE 3. No matter what numbers are represented by the stars,

$$\begin{vmatrix} 1 & 2 & 0 & * \\ * & * & 4 & * \\ 2 & 1 & 0 & * \\ 0 & 0 & 0 & 3 \end{vmatrix} = -4 \begin{vmatrix} 1 & 2 & * \\ 2 & 1 & * \\ 0 & 0 & 3 \end{vmatrix} = -12 \begin{vmatrix} 1 & 2 \\ 2 & 1 \end{vmatrix} = 36.$$

Here the 4-by-4 determinant was expanded on the third column and the 3-by-3 determinant was then expanded on the third row.

(b) Elementary properties

The following properties of determinants are established in DETA, Chapter E:

(i) If each element of a row or column is 0, the determinant is 0.

(ii) If each element of a row or column is multiplied by λ the determinant is multiplied by λ.

(iii) If each element of a row or column is a sum of two terms the determinant is a sum of two corresponding determinants; for example,

$$(6) \qquad \begin{vmatrix} a_1 & a_2 & a_3 \\ b_1 + d_1 & b_2 + d_2 & b_3 + d_3 \\ c_1 & c_2 & c_3 \end{vmatrix} = \begin{vmatrix} a_1 & a_2 & a_3 \\ b_1 & b_2 & b_3 \\ c_1 & c_2 & c_3 \end{vmatrix} + \begin{vmatrix} a_1 & a_2 & a_3 \\ d_1 & d_2 & d_3 \\ c_1 & c_2 & c_3 \end{vmatrix}.$$

(iv) If two rows or two columns are interchanged the determinant changes sign.

(v) If rows and columns are interchanged the determinant is unaltered; in other words, $\det A = \det A^T$.

(vi) If two rows are equal, or two columns are equal, the determinant is 0.

(vii) If one row is a multiple of another, or one column is a multiple of another, the determinant is 0.

(viii) If a multiple of one row is added to another row, or a multiple of one column is added to another column, the determinant is unchanged.

(ix) If A and B are square matrices of the same size, $|AB| = |A||B|$.

These properties show that a determinant can be simplified by elementary operations on the matrix, and we can operate on columns as well as on rows. Here is an example.

EXAMPLE 4. Explain this reduction:

$$\begin{vmatrix} 1 & 2 & 3 & 4 \\ 3 & 6 & 6 & 9 \\ 4 & 6 & 8 & 10 \\ 3 & 4 & 5 & 6 \end{vmatrix} = 6 \begin{vmatrix} 1 & 2 & 3 & 4 \\ 1 & 2 & 2 & 3 \\ 2 & 3 & 4 & 5 \\ 3 & 4 & 5 & 6 \end{vmatrix} = 6 \begin{vmatrix} 0 & 0 & 1 & 1 \\ 1 & 2 & 2 & 3 \\ 0 & -1 & 0 & -1 \\ 0 & -2 & -1 & -3 \end{vmatrix}.$$

In the first step 3 was factored out of the second row and 2 out of the third row. In the next step the element 1 was used as a pivot to reduce the first column by row operations; this does not change the value of the determinant. We now expand the determinant on elements of the first column. It is left for you to complete the calculation and show that the determinant is 0.

(c) Determinants and linear systems

The following theorem is of fundamental importance both in linear algebra and in the theory of differential equations:

THEOREM 1. If $|A| = 0$ the equation $Ax = 0$ has a solution other than the trivial solution $x = 0$. Conversely, existence of a nontrivial solution implies $|A| = 0$.

When $n = 1$ the equation is $0x_1 = 0$, which obviously has a nontrivial solution. Taking $n \geq 2$, we assume that the first statement in Theorem 1 is known for matrices of order $n-1$ and we use mathematical induction. There are two cases.

Case 1. If the first column of A is 0 we can choose $x_1 = 1$ and all the other unknowns to be 0. This gives a nontrivial solution.

Case 2. If at least one element $a_{i1} \neq 0$ we assume, by interchanging rows, that $a_{11} \neq 0$. Dividing by a_{11} makes $a_{11} = 1$. Let $a_{11} = 1$ be used as a pivotal element to reduce the first column of the augmented matrix. The right-hand column is still 0, and the coefficient matrix now looks like this:

$$C = \begin{pmatrix} 1 & a_{12} & \cdots & a_{1n} \\ 0 & b_{22} & \cdots & b_{2n} \\ \vdots & \vdots & \ddots & \vdots \\ 0 & b_{n2} & \cdots & b_{nn} \end{pmatrix}.$$

The hypothesis $|A| = 0$ yields $|C| = 0$ and, expanding on the first column, we see that $|C| = |B|$ where $B = (b_{ij})$. Hence $|B| = 0$. By the induction hypothesis we can choose $x_2, x_3, \cdots, x_n$, not all 0, satisfying the system associated with B. If x_1 is now chosen so as to satisfy the first equation, we get a nontrivial solution for the system associated with C and hence for the original system. This completes the proof of the first statement. The second statement is taken up after the proof of Theorem 2.

A counterpart to Theorem 1 is:

THEOREM 2. *If $|A| \neq 0$ the equation $Ax = y$ has a unique solution x for each y. Conversely, if $Ax = y$ has a unique solution for at least one y, then $|A| \neq 0$.*

Let $|A| \neq 0$. When $n = 1$ the equation is $ax_1 = y_1$ with $a \neq 0$. This has the unique solution $x_1 = y_1/a$. Taking $n \geq 2$, we assume that the first statement in Theorem 2 is known for matrices of order $n - 1$ and we use mathematical induction. If the first column of A is 0 then $|A| = 0$, which is contrary to the hypothesis. Hence Case 1 above is ruled out automatically. The rest of the argument is similar to that in Case 2 and is left to you.

So far, we have proved (a) that there is a nontrivial solution u of $Au = 0$ if $|A| = 0$, and (b) that there is a unique solution x of $Ax = y$ if $|A| \neq 0$. These proofs were similar but independent.

Result (b) shows that $x = 0$ is the sole solution of $Ax = 0$ when $|A| \neq 0$, and this gives the second statement in Theorem 1. When $|A| = 0$, Result (a) shows that the equation $Ax = y$ has infinitely many solutions if it has any; namely, it has solutions $x + cu$ for any constant c. This gives the second statement in Theorem 2 and completes the proof of both theorems.

THEOREM 3. *A homogeneous linear system with more unknowns than equations always has a nontrivial solution.*

This is true because we can add extra equations with zero coefficients to convert the system to the form $Ax = 0$ where A is square. Since at least one row of A is 0 the determinant is 0 and the result follows from Theorem 1.

When $|A| \neq 0$, Part (b) of the following gives a formula for the unique solution in Theorem 2:

CRAMER'S RULE. Let D_i be the determinant obtained when the ith column of $|A|$ is replaced by y.

(a) *If $Ax = y$ then $|A|x_i = D_i$ for $i = 1, 2, \cdots, n$.*
(b) *If $|A| \neq 0$, the vector x in (a) satisfies $Ax = y$.*

All essential features of the proof are contained in the case $n = 3$. In view of this, we consider the simultaneous equations

(7)
$$
\begin{aligned}
a_1 x_1 + a_2 x_2 + a_3 x_3 &= y_1 \\
b_1 x_1 + b_2 x_2 + b_3 x_3 &= y_2 \; . \\
c_1 x_1 + c_2 x_2 + c_3 x_3 &= y_3
\end{aligned}
$$

By Rules (ii) and (viii) of Part (a)

$$
x_1 \begin{vmatrix} a_1 & a_2 & a_3 \\ b_1 & b_2 & b_3 \\ c_1 & c_2 & c_3 \end{vmatrix} = \begin{vmatrix} x_1 a_1 & a_2 & a_3 \\ x_1 b_1 & b_2 & b_3 \\ x_1 c_1 & c_2 & c_3 \end{vmatrix} = \begin{vmatrix} x_1 a_1 + x_2 a_2 + x_3 a_3 & a_2 & a_3 \\ x_1 b_1 + x_2 b_2 + x_3 b_3 & b_2 & b_3 \\ x_1 c_1 + x_2 c_2 + x_3 c_3 & c_2 & c_3 \end{vmatrix} .
$$

Hence, if (7) holds,

$$
x_1 \begin{vmatrix} a_1 & a_2 & a_3 \\ b_1 & b_2 & b_3 \\ c_1 & c_2 & c_3 \end{vmatrix} = \begin{vmatrix} y_1 & a_2 & a_3 \\ y_2 & b_2 & b_3 \\ y_3 & c_2 & c_3 \end{vmatrix} .
$$

In the notation of Cramer's rule, the above result and the corresponding results for x_2 and x_3 are

$$
Dx_1 = D_1, \qquad Dx_2 = D_2, \qquad Dx_3 = D_3.
$$

This gives Part (a) and Theorem 2 then gives Part (b).

EXAMPLE 5. In the system on the left below, the value of x_2 is given by the formula on the right:

$$
\begin{aligned}
x_1 + 2x_3 &= 4 \\
x_1 + x_2 &= 5 \\
2x_2 + 4x_3 &= 6
\end{aligned}
\qquad
\begin{vmatrix} 1 & 0 & 2 \\ 1 & 1 & 0 \\ 0 & 2 & 4 \end{vmatrix} x_2 = \begin{vmatrix} 1 & 4 & 2 \\ 1 & 5 & 0 \\ 0 & 6 & 4 \end{vmatrix} .
$$

The determinants are easily evaluated when the first row is subtracted from the second. The result is $8x_2 = 16$ or $x_2 = 2$.

(d) The inverse revisited

Let us recall that when the elements of a row or column are combined with the corresponding cofactors, as in the expansion theorem, the result is the determinant. For example

(8)
$$\begin{vmatrix} a_{11} & a_{12} & a_{13} \\ a_{21} & a_{22} & a_{23} \\ a_{31} & a_{32} & a_{33} \end{vmatrix} = a_{12}A_{12} + a_{22}A_{22} + a_{32}A_{32}.$$

It is of considerable interest that if the elements of one row or column are combined with the cofactors of another row or column the result is 0. The meaning of this statement and the reason for its truth are both explained by the calculation

(9)
$$\begin{vmatrix} a_{11} & a_{13} & a_{13} \\ a_{21} & a_{23} & a_{23} \\ a_{31} & a_{33} & a_{33} \end{vmatrix} = a_{13}A_{12} + a_{23}A_{22} + a_{33}A_{32}.$$

The sum on the right of (9) is the Laplace expansion of the determinant on the second column and is 0 because the determinant has two equal columns. This same sum would be obtained by combining elements of the third column of the determinant (8) with cofactors of the second column.

A similar result, with similar proof, is valid in general. In general if we combine elements of the ith column with cofactors of the jth column, where $i \neq j$, the result is an expansion of a determinant with two equal columns and is therefore 0. The same applies to rows.

These facts are conveniently summarized by use of the Kronecker delta, which is defined as follows:

$$\delta_{ij} = 1 \quad \text{if } i = j, \qquad \delta_{ij} = 0 \quad \text{if } i \neq j.$$

In terms of this symbol we have

(10)
$$\sum_{k=1}^{n} a_{ik}A_{jk} = \delta_{ij}|A| = \sum_{k=1}^{n} a_{ki}A_{kj}.$$

When $i = j$ the result of the summation is $|A|$ by the expansion theorem, and when $i \neq j$ the result is 0 by the remarks above.

The left-hand sum in (10) would agree with the formula for a matrix product if the summand were $a_{ik}A_{kj}$ instead of $a_{ik}A_{jk}$. Similarly for the sum on the right. Since $(A_{jk})^{T} = (A_{kj})$ we introduce the *cofactor transpose*, which is denoted by adj A and is defined as follows:

$$\text{adj } A = (A_{ij})^{T}.$$

The abbreviation adj A is in deference to earlier terminology, in which the cofactor transpose was called the adjoint. When the sums (10) are interpreted by matrix products the result is

(11) $$A(\text{adj } A) = (\text{adj } A)A = |A|I.$$

For example,

$$\begin{pmatrix} a_{11} & a_{12} & a_{13} \\ a_{21} & a_{22} & a_{23} \\ a_{31} & a_{32} & a_{33} \end{pmatrix} \begin{pmatrix} A_{11} & A_{21} & A_{31} \\ A_{12} & A_{22} & A_{32} \\ A_{13} & A_{23} & A_{33} \end{pmatrix} = \begin{pmatrix} |A| & 0 & 0 \\ 0 & |A| & 0 \\ 0 & 0 & |A| \end{pmatrix}.$$

Theorem 1 shows that if $|A| = 0$ then A has no inverse; otherwise $Ax = 0$ would have the unique solution $x = A^{-1}0 = 0$. Hence, if A is invertible, we can divide (11) by $|A|$ to get the following generalization of the rule for 2-by-2 inverses given in Chapter C:

THEOREM 4. *To obtain the inverse of an invertible matrix A, write the matrix of cofactors (A_{ij}), take its transpose, and divide by the determinant.*

It is important that the matrix $A^{-1} = |A|^{-1}\text{adj } A$ given by Theorem 3 satisfies *both* equations

$$A^{-1}A = I, \qquad AA^{-1} = I.$$

This follows from (11).

EXAMPLE 6. For the matrix A below, the matrix of minors is M and the matrix of cofactors is C:

$$A = \begin{pmatrix} 1 & 2 & 3 \\ 0 & 1 & 1 \\ 2 & 0 & 4 \end{pmatrix}, \quad M = \begin{pmatrix} 4 & -2 & -2 \\ 8 & -2 & -4 \\ -1 & 1 & 1 \end{pmatrix}, \quad C = \begin{pmatrix} 4 & 2 & -2 \\ -8 & -2 & 4 \\ -1 & -1 & 1 \end{pmatrix}.$$

When the cofactor transpose adj $A = C^T$ is multiplied by A we get

$$(\text{adj } A)A = \begin{pmatrix} 4 & -8 & -1 \\ 2 & -2 & -1 \\ -2 & 4 & 1 \end{pmatrix} \begin{pmatrix} 1 & 2 & 3 \\ 0 & 1 & 1 \\ 2 & 0 & 4 \end{pmatrix} = \begin{pmatrix} 2 & 0 & 0 \\ 0 & 2 & 0 \\ 0 & 0 & 2 \end{pmatrix}.$$

The equation shows that $|A| = 2$. Dividing C^T by 2 yields the inverse.

EXAMPLE 7.   Let A and B be square matrices such that $BA = I$. Multiplying the equation $Ax = 0$ by B we get $x = 0$. Theorem 2 shows that $|A| \neq 0$ and Theorem 4 now shows that A has a two-sided inverse. Multiplying $BA = I$ on the right by A^{-1} we get $B = A^{-1}$, hence $AB = I$.

(e) The expanded form

A *permutation* of the sequence $1, 2, \cdots, n$ is a rearrangement of its terms. A permutation can be regarded as a function p whose domain and range both consist of the set $\{1, 2, \cdots, n\}$. For example, the permutation taking $1, 2, 3, 4$ into $4, 2, 3, 1$ is the function p defined as follows:

$$(12) \qquad p(1) = 4, \qquad p(2) = 2, \qquad p(3) = 3, \qquad p(4) = 1.$$

In general, $p(1)$ can be any of the n numbers $1, 2, \cdots, n$. After $p(1)$ is chosen there remain $n - 1$ choices for $p(2)$, and so on. Hence there are $n!$ different permutations p.

In a sequence such as $4, 2, 3, 1$ an inversion is a pair of numbers that are not in their natural order; that is, the first number of the pair is larger than the second. This particular sequence has five inversions,

$$(4, 2), \ (4, 3) \ (4, 1) \ (2, 1) \ (3, 1).$$

The sign of a permutation p is plus if the number of inversions in it is even, minus if it is odd. For example the permutation (12) has sign $p = -$ since there are 5 inversions. We state without proof that the expanded form of an nth-order determinant is

$$(13) \qquad\qquad |A| = \sum \text{sign } p \; a_{1\,p(1)} a_{2\,p(2)} \cdots a_{n\,p(n)}$$

where the sum is over all $n!$ permutations p.

PROBLEMS

1P. By Equation (2) the determinant of $A = (a_{ij})$ is

$$a_{11}a_{22}a_{33} + a_{12}a_{23}a_{31} + a_{13}a_{21}a_{32} - a_{11}a_{23}a_{32} - a_{13}a_{22}a_{31} - a_{12}a_{21}a_{33}$$

when $n = 3$. Open the book to a random page and note the last digit of the page number. Then expand the determinant (2) according to the following scheme:

$$(0, 1) = 2R \qquad (2, 3) = 3R \qquad (4, 5) = 1C \qquad (6, 7) = 2C \qquad (8, 9) = 3C$$

For example, if the last digit of the page number is 6 or 7, expand the determinant on its second column.

2. Evaluate by using rules (ii) and (viii) to get a row with three zeros:

(a) $\begin{vmatrix} 1 & 2 & 3 & 4 \\ 3 & 6 & 6 & 9 \\ 4 & 6 & 8 & 10 \\ 3 & 4 & 6 & 5 \end{vmatrix}$ (b) $\begin{vmatrix} 1 & 1 & 1 & 1 \\ 2 & 3 & 4 & 1 \\ 3 & 4 & 1 & 2 \\ 4 & 1 & 2 & 3 \end{vmatrix}$ (c) $\begin{vmatrix} 3 & 5 & 7 & 9 \\ 7 & 9 & 3 & 5 \\ 5 & 9 & 3 & 7 \\ 9 & 7 & 5 & 3 \end{vmatrix}$

3. (abc) Evaluate the determinants above by using rules (ii) and (viii) to get a column with three zeros.

4. Write equations corresponding to the following augmented matrices and solve by Cramer's rule:

(a) $\begin{pmatrix} 3 & 2 & 1 & 6 \\ 1 & 1 & 1 & 2 \\ 1 & 2 & 4 & 1 \end{pmatrix}$ (b) $\begin{pmatrix} 2 & 2 & 1 & 8 \\ 2 & 8 & 1 & 8 \\ 2 & 8 & 4 & 5 \end{pmatrix}$ (c) $\begin{pmatrix} 3 & 1 & 4 & 1 \\ 5 & 9 & 2 & 6 \\ 5 & 3 & 5 & 8 \end{pmatrix}$

5P. (abcde) Find the inverse as in Example 6. Since the method incorporates a check, no answer is provided:

$\begin{pmatrix} 1 & 0 & 1 \\ 1 & 1 & 0 \\ 0 & 1 & 1 \end{pmatrix}$ $\begin{pmatrix} 1 & 0 & -1 \\ 2 & 1 & 0 \\ 0 & 1 & 1 \end{pmatrix}$ $\begin{pmatrix} 1 & 2 & 3 \\ 3 & 2 & 1 \\ 2 & 1 & 3 \end{pmatrix}$ $\begin{pmatrix} 1 & 1 & 1 \\ 1 & 2 & 1 \\ 1 & 1 & 3 \end{pmatrix}$ $\begin{pmatrix} 1 & 2 & 3 \\ 4 & 5 & 6 \\ 7 & 8 & 9 \end{pmatrix}$

6. If the elements of a determinant $|A|$ are differentiable functions, show that its derivative $|A|'$ can be obtained by differentiating each row separately and adding the results, or also by differentiating each column separately and adding the results.

Hint: Differentiate the defining relation (5) and use mathematical induction. The fact that you can differentiate by columns follows from $|A^T| = |A|$.

──────────────────ANSWERS──────────────────

2. 12, 16, 0

4. (a) $(1, 2, -1)$ (b) $(9/2, 0, -1)$ (c) $(191, -67, -122)/18$

───

> As the sun eclipses the stars by his brilliancy, so the man of knowledge will eclipse the fame of others in assemblies of the people if he proposes algebraic problems, and still more if he solves them.
>
> —Brahmagupta

INDEX

Typography

Each chapter starts on verso and has an even number of pages. The book was set in ten-point Computer Modern by Dan Port and the author using TeX. It was magnified 20% for trial teaching and editing, then photographically reduced 9% to give a final size of eleven points at a resolution of 1285 dots per inch. Both TeX and the typeface are creations of Donald E. Knuth.

An informal reminder

If introduced without explanation, c, C, c_j, C_j denote constants

Complete solution: Family containing all solutions

General solution: Enough solutions to satisfy initial conditions

Implicit solution: Solution defined by $F(x, y) = c$

Solution by quadrature: Solution expressed by integrals

Linear operator: $T(u + v) = Tu + Tv, \quad T(cu) = cTu$

Linear equation: $Ty = f$ where T is a linear operator

Linear homogeneous equation: $Ty = 0$ with T as above

Change of independent variable: If $y' = p$ then $y'' = p \, dp/dy$

Exponential shift theorem: $P(D)e^{kt}u = e^{kt}P(D + k)u$

Euler-de Moivre formula: $e^{ix} = \cos x + i \sin x$

Complex numbers 112   Taylor series 425

Bounded: $|f(p)| \le M$ for all relevant p

Lipschitzian: $|f(x, y) - f(x, z)| \le K|y - z|$ for all relevant x, y, z

Existence: The problem has at least one solution

Uniqueness: The problem has at most one solution

Fundamental matrix: Solution of $U' = AU$ with $\det U \ne 0$

Eigenvalue: Scalar λ such that $Ac = \lambda c$ has a solution $c \ne 0$

Eigenvector: Nonzero vector c as above

Admissible: Finitely many discontinuities and bounded

Class E: Admissible on $[0, a]$ for each a and of exponential growth

12 inches = 1 foot   3 feet = 1 yard   5280 feet = 1 mile

1 meter is about 1.1 yards   1 kilometer is about 0.6 miles

One kilogram at the earth's surface weighs about 2.2 pounds